DNA Systematics

Volume I:
Evolution

Editor

S. K. Dutta, Ph.D.

Professor
Departments of Botany,
Genetics and Human Genetics,
and Oncology
Howard University
Washington, D.C.

CRC Press, Inc.
Boca Raton, Florida

Library of Congress Cataloging-in-Publication Data
Main entry under title:

DNA systematics.

   Includes bibliographies and indexes.
   Contents: v. 1. Evolution—v. 2. Plants.
   1. Deoxyribonucleic acid—Collected works.
2. Recombinant DNA—Collected works. 3. Chemotaxonomy
—Collected works. I. Dutta, S. K. (Sisir K.)
QP624.D19   1986   574.87′3282   85-21285
ISBN 0-8493-5820-5 (v. 1)
ISBN 0-8493-5821-3 (v. 2)

Direct all inquiries to CRC Press, Inc., 2000 Corporate Blvd., N.W., Boca Raton, Florida, 33431.

© 1986 by CRC Press, Inc.
International Standard Book Number 0-8493-5820-5 (Volume I)
International Standard Book Number 0-8493-5821-3 (Volume II)

Library of Congress Card Number 85-21285
Printed in the United States

FOREWORD

In the early 1960s in the now defunct Biophysics Section of the Carnegie Institution of Washington, Department of Terrestrial Magnetism, Ellis Bolton, Brian McCarthy and I were amazed that we could readily compare the reassociation of mammalian DNA-RNA and DNA-DNA; this, especially, since current dicta said that, because of great complexity, meaningful reassociation did not take place. Our examinations of unfractionated, sheared eukaryotic nucleic acids were mostly made possible by the presence of repeated families of sequences. These families were soon recognized and analyzed by Roy Britten, David Kohne and Michael Waring. Ultimately, Britten and Eric Davidson produced a theory in which the repeated DNA sequences were used as part of control mechanisms governing transcription functions; testing of this theory is still ongoing. At the same period of time, and before, another Carnegie Institution of Washington Staff Member, Barbara McClintock, was performing genetic experiments with maize which indicated "jumping genes". These experiments, finally recognized by a Nobel Prize, have stimulated a resurgence of interest in DNA control elements which are treated herein along with other forms of transcription translation controls such as the tRNAs and rRNAs. In this book series, an international mix of authors has been assembled to address current progress in control and genome comparison. The next 20 years should provide a very great increase in knowledge of these systems. What will the next related volume in this series contain?

Bill Hoyer

INTRODUCTION

In recent years, numerous studies have been performed that use various characteristics of DNA to estimate the diversity or relatedness between both closely and distantly related species. Some of these studies have been concerned with experimental microevolution dealing with the accumulation of relatively small changes within a species, and some with macroevolution involving taxonomic categories and measurements taken over long periods of time or on a geological scale. Most of this explosion of knowledge has been due to the utilization of the powerful methods of recombinant-DNA technology, and a new interdisciplinary science has evolved spontaneously which may be called "DNA Systematics". Included in this science are the characterization of DNA in nuclear and cytoplasmic genomes; DNA:DNA reassociation kinetics of repeated and nonrepeated DNA sequences; thermal stability measurements of heteroduplexes; restriction enzyme patterns analysis of specific nuclear and non-nuclear DNA segments, rDNAs, and mitochondrial and chloroplast DNAs; the rate of evolution of cell organelle genomes vs. nuclear genomes; the implication of gene duplication and gene fusion in evolution and the evolutionary history of specific genes like rRNA genes and hemoglobin genes; evolutionary trends in regulation; and species specificity and DNA sequencing of processing sites of introns and different RNA maturation sites.

Historically, DNA systematics studies were initiated more than 20 years ago by Ellis Bolton, Roy Britten, Bill Hoyer, David Kohne, Brian McCarthy, and others at the Carnegie Institution of Washington, Washington, D.C., and a few other scientists of England, France, and of U.S.S.R. whose work has been reviewed by Belozersky and Antonov during 1972 and 1980 at Moscow University Press, U.S.S.R. Their techniques were mostly based on DNA:DNA hybridization, which is now claimed as the "most favorable of all" methods for revealing family trees, as discussed by Lewin.* Based on these techniques Charles Sibley and Jon Ahlquist (see footnote) have proposed a "DNA clock" to construct phylogenetic trees. Antonov and his associates from the U.S.S.R proposed similar DNA clocks earlier. These molecular clock(s) are becoming very popular. So much so, that they are "now in danger of becoming the dogma that the fossils once were".* Unfortunately, there is no available comprehensive treatise of this vast amount of new knowledge particularly on microevolution based on studies in DNA systematics using new tools of recombinant-DNA technology. In the present work we attempt the first comprehensive review of new information on DNA systematics.

The enormous amount of accumulated information has been reviewed by authors who are active in their respective areas and then organized into three volumes. Volume I is devoted to general topics of DNA systematics with respect to general evolution. Hobish reviews the present state-of-the-art use of computers for storage and retrieval of DNA research data. The role of movable elements in evolution and species formation is reviewed by Georgiev and his associates. It is well established now that mobile DNA sequences provide variability for natural selection and for evolutionary jumps. It makes genomes flexible, and mobile sequences are widespread in living creatures. Studies have been performed on the evolutionary significance of various control mechanisms which regulate speciation and evolution. Studies dealing with the regulation of ribosomal RNA processing sites, regulation of transcription, and analysis of various small RNAs along with their phylogenetic significance are reviewed by Crouch and Bachellerie; Huang; and Beljanski and Le Goff, respectively. The examination of mitochondrial genomes from mammals, *Drosophila,* and fungi has produced models of mt-DNA variation and offers a comparative treatment of evolution with nuclear genomes; these investigations are reviewed by Birley and Croft. Two of the most important gene sets were selected for discussion in this volume. One is the histone gene set, most genes of which do not have introns, and the other is the hemoglobin gene set,

---

* Lewin, R., DNA reveals surprises in human family tree: the application of DNA-DNA hybridization, *Science,* 226, 1179, 1984.

which does have introns. Enormous amounts of information on these gene sets on the evolution of *Xenopus*, avians, rodents, and higher primates including humans are reviewed by Marzluff; and Winter, respectively.

The second volume is devoted primarily to the DNA systematics of plants, although, where necessary, reference species other than plants have been included to present a complete story. This volume starts with the classical approach to plant systematics using knowledge obtained from the contents of plant nuclear DNAs; this material is reviewed by Ohri and Khoshoo. Plant species, particularly higher green plants, show polyploidy and have 70 to 75% repeated DNA sequences. Studies made on these repeated and single copy DNA sequences of monocot and dicot plants are reviewed by Mitra and Bhatia; and Antonov, respectively. Appels and Honeycutt; and Troitsky and Bobrova have reviewed extensive studies done on ribosomal RNA genes of plants along with information obtained from other species and have given extensive analysis of phylogenetic significance of these studies. The DNA systematics of some fungal species are reviewed by Ojha and Dutta. The chloroplast genomes of green plants have provided excellent information on plant systematics; this information is reviewed by Palmer, whose group has done extensive work with chloroplast genomes of various plants. A critical glossary of different terminologies used in plant DNA systematics is given by A. K. Sharma.

The third volume, now in preparation in collaboration with Dr. William P. Winter as Co-Editor, is devoted primarily to the DNA systematics and evolution of *Homo sapiens* and related higher primate species. This volume will treat some of the newer insights into relationships within the higher primates and the origin of modern man from anthropoid ancestors. Also included will be discussions of the relationships between the races of man as determined by DNA analysis. In addition, several genes which are of vital concern to humans like neuron-specific genes, lipoprotein genes, HLA genes and others will be discussed. Dr. Ronald L. Nagel of Albert Einstein College of Medicine, New York; Dr. C. G. Sibley of Yale University; Drs. M. G. George, R. M. Millis, Mukesh Verma and S. K. Dutta, and William P. Winter of Howard University; Drs. T. B. Rajaveshisth, A. J. Lusis and others of the University of California at Los Angeles; Dr. I. A. Levedevan of Vavilov Institute of Human Genetics, Moscow, U.S.S.R.; Dr. R. L. Honeycutt of Harvard University; and Dr. R. D. Schmickel of the University of Pennsylvania will be writing chapters for this third volume.

These three volumes are expected to be valuable references, not only to students of evolution but also to others interested in efficient germ plasm resource maintenance and utilization, and fields which are vital for planning plant and animal breeding programs. Knowledge of DNA markers correlating the geographic distribution of genes responsible for heritable diseases such as human sickle cell anemia should be of profound importance to physicians and epidemiologists.

In addition to contributing authors, who have also helped in reviewing several chapters, several other authors have helped in organizing and improving various chapters. I would like to acknowledge particularly Fransciso Ayala of the University of California, Davis; Igor Dawid, H. Westphal and A. Schecter of the National Institutes of Health, Bethesda, Md; Professor A. K. Sharma of Calcutta University, Calcutta, India; R. L. Peterson, George Mathew and D. R. Maglott of Howard University, Washington, D.C.; H. James Price, Texas A&M University, College Station, Texas; H. R. Chen, National Biomedical Research Foundation, Georgetown University Medical Center, Washington, D.C.; Bill Hoyer of Georgetown University; E. S. Weinberg of the University of Pennsylvania, Philadelphia; and G. N. Wilson, Pediatric Genetics, The University of Michigan, Ann Arbor.

S. K. Dutta
Editor

THE EDITOR

Sisir K. Dutta, Ph.D., is Professor of Molecular Genetics in the Department of Botany; and Adjunct Professor in the Departments of Genetics and Human Genetics; and in the Department of Oncology at Howard University, Washington, D.C.

Dr. Dutta obtained his B.S. degree from Dacca University, Bangladesh in 1949, and thereafter, served for 6 years as Research Assistant in Genetics and Plant Breeding for the government of West Bengal, India. He received his M.S. and Ph.D. degrees in genetics from Kansas State University, Manhattan, Kansas in 1958 and 1960, respectively. He was a research associate and/or visiting scientist at the University of Chicago, Columbia and Rockefeller Universities in New York, the Pasteur Institute in Paris, the National Institutes of Health in Bethesda, Maryland, and Rice University in Houston, Texas. He was Chief Research Officer-*cum*-Director of the National Pineapple Research Institute of Malaysia from 1960 to 1964, Chairman of the Division of Natural Sciences, and Chairman of the Biology Department of the Christian University Affiliated College at Hawkins, at Texas Southern University from 1964 to 1967. In 1967 he assumed his present duties at Howard University.

He has been the organizer, chairman, and speaker of several national and international symposia held in the U.S., U.S.S.R., Europe, and Asia. He has been a member of the editorial board of the *East Pakistan Agricultural Journal*, a reviewer and panelist of several government and private agencies. He has been inducted as a personality in America's Hall of Fame for his contribution in molecular genetics, has appeared in *Who's Who in the World*, *Who's Who in America*, and *Who's Who in Frontier Sciences and Technology*. He is a member of several national and international professional societies, author or coauthor of more than 100 papers including monographs, and book chapters and editor of four books. He has been a recipient of several research awards for the U.S. National Science Foundation, National Institutes of Health, Department of Energy, Environmental Protection Agency, Research Corporation, Anna Fuller Fund, and several other agencies including the United Nations Development Projects.

His current research interest is in the areas of regulation of ribosomal RNA transcription and processing, molecular genetics of neuron-specific genes, and molecular evolution.

CONTRIBUTORS

Volume I

Jean-Pierre Bachellerie
Charge de Recherche
Centre Recherche Biochemie and
  Genetique Cellulaire
Centre National de la Recherche
  Scientifique
Toulouse, France

Mirko Beljanski
Master in Scientific Research
Department of Pharmacodynamie
Faculté des Sciences Biologiques et
  Pharmaceutiques
Chatenay-Malabry, France

A. J. Birley
Lecturer
Department of Genetics
University of Birmingham
Birmingham, England

J. H. Croft
Lecturer
Department of Genetics
University of Birmingham
Birmingham, England

Robert J. Crouch
Research Chemist
Laboratory of Molecular Genetics
National Institutes of Health
Bethesda, Maryland

Georgii P. Georgiev
Professor, Head
Department of Nucleic Acid Biosynthesis
Institute of Molecular Biology
U.S.S.R. Academy of Sciences
Moscow, U.S.S.R.

Tatiana I. Gerasimova
Chief
Group of Mobile Elements
Institute of General Genetics
U.S.S.R. Academy Sciences
Moscow, U.S.S.R.

Mitchell K. Hobish
Assistant Research Scientist
Laboratory of Chemical Evolution
Department of Chemistry
University of Maryland
College Park, Maryland

Pien Chien Huang
Professor of Biochemistry
Johns Hopkins University School of
  Hygiene and Public Health
Baltimore, Maryland

Yurii V. Ilyin
Doctor of Biological Science
Head, Department of Genome Mobility
Institute of Molecular Biology
U.S.S.R. Academy of Sciences
Moscow, U.S.S.R.

Liliane Le Goff
Docteur de Sciences
Department of Pharmacodynamie
Faculté des Sciences Biologiques et
  Pharmaceutique
Chatenay-Malabry, France

William F. Marzluff
Professor of Chemistry
Florida State University
Tallahassee, Florida

Alexei P. Ryskov
Senior Researcher
Department of Nucleic Acids
  Biosynthesis
U.S.S.R. Academy of Science
Moscow, U.S.S.R.

William P. Winter
Senior Biochemist
Center for Sickle Cell Disease
Associate Professor of Genetics and
  Human Genetics and Medicine
Howard University
Washington, D.C.

Volume II

Andrew S. Antonov
Professor
A. N. Belozersky Laboratory of
  Molecular Biology
Department of Evolutionary Biochemistry
Moscow State University
Moscow, U.S.S.R.

R. Appels
Principal Research Scientist
Division of Plant Industry
CSIRO
Canberra, ACT, Australia

Chittranjan R. Bhatia
Head
Nuclear Agriculture Division
Bhabha Atomic Research Centre
Trombay, Bombay, India

V. K. Bobrova
Department of Evolutionary Biochemistry
A. N. Belozersky Laboratory of
Molecular Biology and Bioorganic
  Chemistry
Moscow State University
Moscow, U.S.S.R.

Sisir K. Dutta
Professor in Molecular Genetics
Department of Botany
Howard University
Washington, D.C.

Rodney L. Honeycutt
Assistant Professor of Biology
Department of Organismic and
  Evolutionary Biology
Harvard University
Cambridge, Massachusetts

T. N. Khoshoo
Distinguished Scientist
CSIRO
New Delhi, India

Ranjit K. Mitra
Scientific Officer
Nuclear Agriculture Division
Bhabha Atomic Research Centre
Trombay, Bombay, India

Deepak Ohri
Scientist
Cytogenetics Laboratory
National Botanical Research Institute
Lucknow, India

Mukti Ojha
Maitre d'Enseignement et de Recherche
Biologie Vegetale
Universite de Geneve
Geneva, Switzerland

Jeffrey D. Palmer
Arthur F. Thurnau Assistant Professor of
  Molecular Genetics
Division of Biological Sciences
University of Michigan
Ann Arbor, Michigan

Arun Kumar Sharma
Indian National Academy of Science
Professor and Programme Coordinator
Centre of Advanced Study
Department of Botany
University of Calcutta
Calcutta, India

A. V. Troitsky
Department of Evolutionary Biochemistry
A. N. Belozersky Laboratory of
  Molecular Biology and Bioorganic
  Chemistry
Moscow State University
Moscow, U.S.S.R.

TABLE OF CONTENTS

Volume I

Volume II

Chapter 1

THE ROLE OF THE COMPUTER IN ESTIMATES OF DNA NUCLEOTIDE SEQUENCE DIVERGENCE

Mitchell K. Hobish

TABLE OF CONTENTS

I. INTRODUCTION

The computer has taken on several roles with respect to the analysis of protein and nucleic acid sequences. Indeed, without the computer, we may never have seen the current boom in sequence information. The roles played by the computer cover not only the initial acquisition and subsequent storage and retrieval of sequence data, but also the analysis of those data to determine such features as the location of control regions, structural homology, and estimates of divergence of sequences for the inference of phylogenetic relationships. In this chapter, descriptions of the state-of-the-art in these various areas will be presented, along with some comments and suggestions for optimizing the interaction between the experimenter and the information, as facilitated by the medium of the computer.

Modern scientific instrumentation has revolutionized not only the way in which research is carried out, but even the very research being undertaken. Phenomena whose existence had heretofore only been theorized may now be analyzed with instrumentation capable of measurements over ranges and sensitivities that have improved by several orders of magnitude in just a few years. Concomitantly, there have come significant strides in the ability of computers to capture, store, process, and analyze the data obtained with such instrumentation. Recent advances in the design and manufacture of microprocessors have led to the ubiquity of microcomputers, which have capacities and capabilities that were available only on minicomputers and mainframes only a few years ago. This form of inexpensive, distributed processing augers well for experimental design and execution. As a result, in some cases even an unskilled novice may do in minutes what had previously taken a highly skilled technician several days.[1] The acquisition of data, its analysis, and even its final preparation for presentation may now be almost entirely automated. Analysis of data may be done almost on a real-time basis, with concomitant re-design of experimental protocols as necessary. One result of this process is that the interaction between the investigator and the research underway has been facilitated.

In the specific case of the application of computers and automation to research on nucleic acids, most of the impact has been felt at the data manipulation end of the scale, with comparatively little effect on the acquisition of those data. In the following pages, the role of the computer in such acquisition, as well as the storage and retrieval of nucleic acid sequence data will be examined, with special emphasis on how use of these machines may be used to analyze those data to infer sequence divergence, and hence, phylogeny. This chapter is not designed to present everything you always wanted to know about the inference of phylogenetic trees, but rather a general description of how the computer has been used in that area of research. This topic(s) is presented in other chapters in this volume.

II. COMPUTERS IN THE ACQUISITION OF NUCLEOTIDE SEQUENCE DATA

The development and refinement of rapid techniques for DNA sequencing[2,3] have contributed directly to the present boom in nucleic acid sequence research. Also contributory has been the availability of low-cost computational facilities to the laboratories where the sequencing is being done. This local capability has given rise to systems that aid in the acquisition of sequence data and its subsequent preliminary analysis.

As a data acquisition tool, the computer acts as an extension of the bench techniques which generate raw sequence data as autoradiographs. Several programs have been written[4-7] which enable the operator to directly read DNA sequences from such gel autoradiographs using a graphics tablet (a digitizer) as a transducer. In this way the operational information is converted to a form directly accessible by a computer, with subsequent storage for later manipulation. Errors in reading the gels or in typographical transcription of the sequences are obviated. Such programs offer the operator several options beyond data transduction, including, but not limited to, portrayal of the sequence and verification thereof (by replicate scanning of the gel), and the ability to write the sequence data to a mass storage device for later manipulation or transmission to another program or computer. In addition, by carefully defining the way in which the gels are read, irregularities in spacing of the bands are included in the data set. These irregularities arise due to compression of the gel as a result of the interaction between regions of secondary structure induced in the DNA fragments and the gel matrix.[5] This information itself may be of use in the investigation of such structures.[8,9] A necessary feature in accurate transcription of gel data is real-time quality control on the part of the operator. The program should provide the operator some means of ensuring that the base entered into the computer is correct from the standpoint of reading the gel. For example, Lautenberger[4] uses a simple audio feedback system, while Gingeras et al.[5] employ recent advances in the area of computer-generated speech to actually tell the operator what is happening.

Once sequences have been determined, whether aided by computer or not, true manipulation of those data may begin. These operations may range from simple calculation of base composition, to translation of the sequence in several reading frames, and pattern matching in an attempt to identify regions of interest such as palindromes,[10] control regions,[11] and restriction sites,[12-14] in which information may be used in further restriction analysis.[15] Such utility is subject to several constraints, as described by Gallant,[16] who demonstrates how sequence reconstruction by proper alignment and overlap of cleavage products is "computationally intractable" for large numbers of input data. The problem may be handled, however, by defining several boundary conditions on the input data set, which leads to a limited number of candidate sequences. This limited set may now be used for further sequence refinement using cleaving agents with increasing specificity. Except for this proviso, these programs may be used on a real-time basis to determine what sequencing strategy may next be followed. This real-time basis defines an apparent need for local computing power; desktop or microcomputers are ideally suited for this purpose, except for some constraints on the length of sequences with which a user deals. In contrast to the data acquisition level of use, the type of computer used (or, more exactly, the clock speed of the computer), the amount of memory available, and the implementation of a specific algorithm will provide the limiting factors. For this reason, only relatively short sequences may be dealt with, unless the programs are implemented on a mini- or mainframe system. Since the original sequence data are obtained in relatively short pieces, this is, at least at this stage, no hardship. Similarly, from the standpoint of rapid pattern recognition, interpreted BASIC is not well suited here. Since the sequences may often be treated as strings of characters, a language without adequate string handling capacity would also limit efficiency. On the other hand, a language built around operations involving strings, arrays, or tables could reduce processing time. A language well suited for such array analysis is APL. Unfortunately, this language is not often implemented on microcomputers. Furthermore, the language suffers from a case of terminal unreadability with increasing temporal distance from the creation of a program.

Indeed, it is the string-of-characters representation of sequence data which provides a complicating factor in the presentation of the raw data to the investigator. Representing the sequence data on a CRT or hard-copy device as a string of bases or base symbols may limit the utility of the data, since there is no obvious delineation of regions of interest. There has been at least one attempt to overcome several limitations inherent in string representation of nucleic acid sequence data,[17] but this seems to have its greatest utility in examining the overall structure of large sequences. Similar approaches to this problem have been taken by Staden,[18] who presents the output of a program (ANALYSEQ) showing alignments between protein coding regions, splice junctions, ribosome-binding sites, and poly-A sites.

The increasing appearance of inexpensive medium and high resolution color graphics terminals provides an interesting possibility for representation of pattern analysis of nucleic acid sequences. For example, regions of interest, as mentioned above, could all be delineated in different colors. This technique has been taken virtually to state-of-the-art for microcomputers with a program described by Watanabe et al.[19] In this paper, the authors describe a microcomputer program which produces a three-dimensional, color display of protein and nucleic acid structure. Rotation and enlargement of defined regions may also be done. Eventually it may be possible to compare nucleic acid sequence structures by superposition and cancellation of homologous regions, leaving only the difference structures for comparison.

The graphic representation of sequences and/or structures notwithstanding, several programs which aid in pattern recognition have been written. The more complicated of these[20-23] have such extensive sequence handling options that they must be implemented on a mainframe system. For example, the packages mentioned above have been used on DEC System 10 and 20 systems, and are being licensed to the commercial community by Intelligenetics. These programs, and others like them,[24,25] are generally run in an interactive mode and allow rapid pattern analysis of sequences whose lengths prohibit analysis on memory-limited microcomputers. The search for rapid and efficient analysis by character pattern recognition continues, and has recently been addressed by several groups.[26-28] However, it should be noted that while pattern recognition is possible by computer analysis, it is very often the human aspect of the system which can recognize patterns that may be just barely above noise. Recent advances in artificial intelligence and its application to "expert systems" will certainly have an impact in this area.

III. NUCLEIC ACID SEQUENCE DATABASES

Historically, even as the amount of sequence data was growing, the research community began to recognize a need for a centralized, standardized sequence data base. Despite reports of sequences as early as 1978, the first of these, the Nucleotide Sequence Data Library of the European Molecular Biology Laboratory was not established until April 1982. A similar facility for U.S. researchers had been considered as early as 1979, and by the end of that year Dayhoff's group (NBRF, the National Biomedical Research Foundation)[29] in Washington, D.C., and Goad's at Los Alamos,[25,30] had submitted proposals for development to the National Institutes of Health (NIH). Many issues had to be dealt with, including security of "sequences in progress", and the need to determine the most effective means of implementing the data network. Even the Department of Defense became involved, since the ARPANET was to be used as the common information nexus. An early attempt to centralize research resulted in the GENET package.[31] The package was implemented on the SUMEX-AIM (Standford University Medical Experimental computer-Artificial Intelligence in Medicine) system and demonstrated the utility of artificial intelligence methods to a "knowledge-based" information system. One advantage of this method is the ability to apply sophisticated pattern recognition algorithms to large numbers of sequences, thereby increasing the utility of the computer in nucleic acid research by test cloning (simulating) sequencing strategies in advance of an actual experiment.[32] Unfortunately, the desire for access to the information available on SUMEX through GENET eventually resulted in complete shutdown of that account. Individual users were left to their own devices, with the result that there has been significant redundancy of effort in the generation of sequence analysis software.

Eventually, Bolt Beranek and Newman, Inc. (BBN), a private corporation located in Cambridge, Mass., with expertise in computer communications, won a contract to develop a national sequence database,[33] now called GenBank™, the Genetic Sequence Data Bank. GenBank, a trademark of the NIH, is a U.S. government-sponsored internationally available repository of all reported nucleic acid sequences greater than 50 nucleotides in length, cataloged, and annotated for sites of biological interest and checked for accuracy. GenBank™ was created by the National Institue of General Medical Sciences of NIH in 1982. Co-

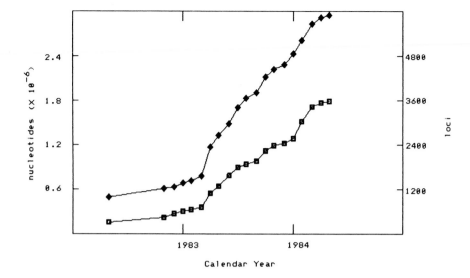

FIGURE 1.   Growth of GenBank[℗], 1982 to the present. Data were obtained from Bolt Beranek and Newman, Inc. Diamonds (♦) represent the number of nucleotides; squares (■) represent the number of loci.

sponsors include the National Cancer Institute, the National Institute of Allergy and Infectious Disease, the Division of Research Resources of NIH, the National Institute of Arthritis, Diabetes and Digestive and Kidney Diseases, as well as the National Science Foundation, the Department of Energy, and the Department of Defense.

Despite this investment, recently the NIH awarded $5.6 million (1985) to IntelliGenetics of Palo Alto, Calif., for the purpose of establishing a national computer resource for molecular biology.[34] The database, to be called Bionet, will provide a centralized resource of nucleic acid and protein sequences, and provide sophisticated software for manipulation of the data. In addition, it is hoped that the network would provide an electronic community of investigators, which could facilitate communication between research groups about topics of common interest. The IntelliGenetics effort is an outgrowth of the GENET network, but conceived of and executed with regard for the problems encountered by its predecessor. For example, the number of researchers accessing Bionet will start at 10, and may be increased to 15, as compared with GENETs 2 lines. Furthermore, Bionet will be accessible via local numbers from all states. The users will comprise about 300 research groups and perhaps 1000 individual subscribers, all linked in a loose confederation of investigators in a huge "laboratory without walls".[35] Thus, in addition to providing the sequence information itself, the network will provide several of the functions of electronic conference call, allowing routine and essentially instantaneous communication between researchers across the country.

Despite the lack of a unique centralized resource there has been tremendous growth in the number of sequences which have been reported and stored in the several databanks currently available. For example, at its inception in 1978, the NBRF database had some 11 sequences, comprised of 24,658 nucleotides. By January 1984, this number had risen to 1234 sequences, of 1,822,759 nucleotides. Impressive as this is, it is eclipsed by the data for GenBank: as of April 1984, there were 3576 loci recorded, consisting of 2,945,001 nucleotides. The growth rate is evident from Figure 1.

As an example of the type of information available in these databanks, we shall examine a few features of GenBank. The information in GenBank is provided in the form of 11 tables, of 31 columns, and as many rows as there are entries in the database. The 31 columns, contain such information as the locus (named), a short description of the locus, the actual sequence, the published reference, annotated structural regions, and more. A typical entry,

```
GETGENBANKENTRY <GO>
[ACCESSING GENBANK ENTRIES ...DONE]


WHICH LOCUS DO YOU WISH TO SEE?RABHBA<GO>
[RETRIEVING  RABHBA ...DONE]


Annotated listing for sequence RABHBA   3/11/84

DEFINITION:
RABBIT ALPHA-GLOBIN MRNA.


          10        20        30        40        50
          :         :         :         :         :
    1 ACACTTCTGG TCCAGTCCGA CTGAGAAGGA ACCACCATGG TGCTGTCTCC
   51 CGCTGACAAG ACCAACATCA AGACTGCCTG GGAAAAGATC GGCAGCCACG
  101 GTGGCGAGTA TGGCGCCGAG GCCGTGGAGA GGATGTTCTT GGGCTTCCCC
  151 ACCACCAAGA CCTACTTCCC CCACTTCGAC TTCACCCACG GCTCTGAGCA
  201 GATCAAAGCC CACGGCAAGA AGGTGTCCGA AGCCCTGACC AAGGCCGTGG


          260       270       280       290       300
          :         :         :         :         :
  251 GCCACCTGGA CGACCTGCCC GGCGCCCTGT CTACTCTCAG CGACCTGCAC
  301 GCGCACAAGC TGCGGGTGGA CCCGGTGAAT TTCAAGCTCC TGTCCCACTG
  351 CCTGCTGGTG ACCCTGGCCA ACCACCACCC CAGTGAATTC ACCCCTGCGG
  401 TGCATGCCTC CCTGGACAAG TTCCTGGCCA ACGTGAGCAC CGTGCTGACC
  451 TCCAAATATC GTTAAGCTGG AGCCTGGGAG CCGGCCTGCC CTCCGCCCCC


          510       520       530       540       550
          :         :         :         :         :
  501 CCCATCCCCG CAGCCCACCC CTGGTCTTTG AATAAAGTCT GAGTGAGTGG
  551 CA

** Composition**

  113  A
  198  C
  144  G
   97  T
Length = 552


TYPE:
MRNA
```

FIGURE 2.   Sample entry in GenBank©™. (Courtesy of Bolt Beranek and Newman, Inc.)

in this case, for rabbit hemoglobin α mRNA, is reproduced in Figure 2. The row and column structure enable rapid access and display of information. For example, the command:

DISPLAY PHAGESEQUENCES ROWS 1 TO 10 COLUMNS 1 TO 5

would result in the information found in Figure 3.

The entire database is found in the table GENBANK. The remaining ten tables contain grouped sequences: mammalian sequences, other vertebrate sequences, invertebrate sequences, plant sequences, organelle sequences, bacterial sequences, structural RNA sequences, viral sequences, bacteriophage sequences, and synthetic sequences. As of April 1984, the 3576 loci presented arise from 4471 individual reports.

Recently, BBN has provided an alternative means of manipulating the databank through the medium of "user-friendly" menus. The user is guided step-by-step through the process, and helpful hints and reminders about response format are provided throughout the menu system. At all stages, help about the menu under scrutiny is available, so there appear to be few places where even a neophyte could get into serious trouble. This could be of great utility when it comes to minimizing on-line charges, which, as of April 1984, amounted to

```
DATE:
UPDATED
11/01/83

ACCESSION:
J00658

Keywords:
complementary DNA; globin; alpha-globin.

SOURCE:
RABBIT.

Organism;
Oryctolagus cuniculus
Eukaryota; Metazoa; Chordata; Vertebrata; Tetrapoda; Mammalis;
Eutheria; Lagomorpha.

REFERENCE:
1   (BASES 45 TO 357)
HEINDELL, H.C., LIU,A., PADDOCK, G.V., STUDNICKA, G.M. AND SALSER, W.A.
THE PRIMARY SEQUENCE OF RABBIT ALPHA-GLOBIN MRNA
CELL 15, 43-54 (1978)
2   (BASES 1 TO 44)
BARALLE,F.E.
STRUCTURE-FUNCTION RELATIONSHIP OF 5' NON-CODING SEQUENCE OF RABBIT
ALPHA- AND BETA-GLOBIN MRNA
NATURE 267, 279-281 (1977)
3( BASES 358 TO 552)
PROUDFOOT, N.J., GILLAM,S., SMITH,M. AND LONGLEY, J.I.
CELL 11, 807-818 (1977)


COMMENT:
COMPARED WITH SUMEX TAPE. [ 3] COMPARES RABBIT
ALPHA-GLOBIN CDNA TO
HUMAN ALPHA-GLOBIN CDNA.

FEATURES:
FEATURES         FROM  TO/SPAN     DESCRIPTION
    PEPT          37     465       ALPHA-GLOBIN

SITES:
   SITES
    - >PEPT        37     1        ALPHA-GLOBIN GENE START
    REFNUMBR       40     1        NUMBERED 1 IN [1]; ZERO NOT USED
    PEPT <-       465     1        ALPHA-GLOBIN GENE END
    MRNA <-       552     1        POLY A ADDITION SITE
    REFNUMBR      552     1        NUMBERED 1 IN [2],[3];   3' TO 5'

ORIGIN:
36 BP 5-PRIME TO 5-PRIME
ECORII SITE


WOULD YOU LIKE TO SEE ANOTHER LOCUS?NO<GO>
```

FIGURE 2.   Continued.

$18/hr during ''business'' hours, and $9/hr thereafter; a $50/quarter minimum is required. The database may be accessed through the NIH PROPHET system, or directly through the GenBank system. Conversely, a complete local library may be obtained through BBN through their ability to provide the sequence information in standard ASCII-format magnetic tape (9-track). The tape provides the same databases available via interactive mode, as well as some documentation and several descriptive directories of the database. This enables the investigator to use custom software to manipulate the database. Some standardization has now been suggested by the provision of a consistent set of software tools for the analysis of such off-line data.[36] The system describes benefits from a modular approach, which should aid in its transportability between systems.

It has been noted that creation of a centralized library would not obviate the requirement for local processing ability.[37] In this area the microcomputer, used first as a ''smart terminal''

```
PHAGESEQUENCES:  GENETIC SEQUENCE DATA BANK
GENBANK(TM)  RELEASE 18.0  (12 MAR 1984)
    PHAGE SEQUENCES  84R X 31C
```

0 LOCUS	1 DEFINITION
1. ALPHA3ORI	BACTERIOPHAGE ALPHA3 ORIGIN OF DNA REPLICATION.
2. B01TR3	BACTERIOPHAGE B01 3'-TERMINAL REGION RNA.
3. BZ13TR3	BACTERIOPHAGE BZ13 3'-TERMINAL REGION RNA.
4. D108A	BACTERIOPHAGE D108 GENE A 5' END.
5. F1	BACTERIOPHAGE F1; COMPLETE GENOME.
6. F1C	BACTERIOPHAGE F1 COMPLETE GENOME.
7. F1VVIIVIII	BACTERIOPHAGE F1 GENES V, VII AND VIII.
8. FD	BACTERIOPHAGE FD, STRAIN 478, COMPLETE GENOME.
9. FITR3	BACTERIOPHAGE FI 3'-TERMINAL REGION RNA.
10. FR1TR3	BACTERIOPHAGE FR1 3'-TERMINAL REGION RNA.

0 LOCUS	2 SEQUENCE	3 TYPE	4 DATE	5 ACCESSION
1. ALPHA3ORI	CCGGTCGTGTGATT- GGTACCATCGCTAC- GACTCAGGTTATTC-		PRE-ENTRY 01/06/83	J02444
2. B01TR3	TTCCCTCAGGAGTG- TTGGCCAGCGAGCT- CTCCTCGGTAGCTG-		PRE-ENTRY 03/01/83	J02445
3. BZ13TR3	TGAAACCCTATCTT- CCGCCAGGTTTAGG- TGCAAACCTAACTC-		PRE-ENTRY 03/01/83	J02446
4. D108A	ATTTGGCCGTCTCG- ATACTAGGTGCGCT- ATGAAAGAATGGTA-		PRE-ENTRY 09/01/83	J02447
5. F1	AACGCTACTACTAT- TAGTAGAATTGATG- CCACCTTTTCAGCT-		PRE-ENTRY 03/01/83	J02448
6. F1C	AACGCTACTACTAT- TAGTAGAATTGATG- CCACCTTTTCAGCT-	SS-DNA CIRCULAR	PRE-ENTRY 06/01/83	J02449
7. F1VVIIVIII	AATCGCATAAGGTA- ATTCACAATGATTA- AAGTTGAAATTAAA-	SS-DNA	PRE-ENTRY 10/03/83	J02450
8. FD	AACGCTACTACCAT- TAGTAGAATTGATG- CCACCTTTTCAGCT-	CIRCULAR	UPDATED 11/01/83	J02451
9. FITR3	TTGAGCCCCGAGAG- AGAGAAAGAAAGAA- AACTCCCTCTTTGA-		PRE-ENTRY 03/01/83	J02452
10. FR1TR3	TTCCCTCAGGAGTG- TTGGCCAGCGAGCT- CTCCTCGGTAGCTG-		PRE-ENTRY 03/01/83	J02453

FIGURE 3.   Sample of output available from interactive, command-driven level of BBNs GenBank™. (Courtesy of Bolt Beranek and Newman, Inc.)

and later as an information processor, may be used to great advantage, since sequences may be down-loaded from the national database for local storage and subsequent manipulation and analysis. Floppy disk drives are now reliably storing upwards of one to two megabytes of information, and hard disks can store more than ten times as much. One piece of technological wizardry that may have some utility in this regard is the optical disk, used now for entertainment through storage of music and/or movies. Several firms[38] are investigating the use of this medium for computer storage, opening up the possibility of storing gigabytes of information locally.

IV. ASSESSMENT OF NUCLEOTIDE SEQUENCE HOMOLOGY

The combination of extensive (and still growing) sequence databases and sophisticated, easy-to-use software for sequence manipulation allows complete analysis of those sequences in relation to each other. This assessment of nucleotide sequence homology in turn allows examination of possible mechanisms of genetics and evolution.

Since the basic tenets of Darwinian evolution became generally accepted as a paradigm by which to investigate biological relationships, several schemes have been used to aid in classifying biological units. An excellent in-depth discussion of this may be found in Sneath and Sokal.[39] It is quite clear from that presentation that (at least) pairwise comparison of quantifiable biological factors is the most objective measure of interrelatedness between taxonomic units, overcoming several problems associated with organizing schemes based on more qualitative criteria (e.g., phenetics). Application of such numerical methods to protein sequence data was early seen as a potential means of assessing phylogenetic relationships.[40] The simplest level upon which such analysis can operate is at comparison of two sequences at a time. As pointed out by Peacock[41] and others, when performing such analyses, one must distinguish between homology and analogy. Furthermore, parallel evolution must be taken into account.

Several procedures have been developed for the computation of sequence homology. Perhaps the first of these, that of Fitch,[42] set early guidelines for subsequent work. Fitch's algorithm compared fragments up to 30 amino acids long from 2 sequences to be compared. For each pairing, a mutation value was assigned, equal to the minimum number of nucleotide substitutions required to convert one observed amino acid to the other. Fitch later refined these techniques to allow assessment of sequences of unequal length,[43] and to distinguish between homology and analogy.[44]

The procedures of Needleman and Wunsch[45] defined the matrix method for homology analysis. Their procedure tests for statistically significant similarity between a pair of sequences while allowing for gaps in all possible places. Sequences are numbered starting at N-terminus (for proteins), and arrayed as an $n \times m$ matrix, where n and m are the number of monomers comprising the respective sequences. Each cell is then scored on the similarity between the two entries in that cell. By a relatively simple transformation, a pathway which provides the maximum similarity between the sequences may be found, with a perfect match producing a simple diagonal. Gaps which may occur in this pairwise test result in deviation from the diagonal. Any realistic alignment requires a penalty score for introduction of gaps, although assignment of such a penalty factor is somewhat arbitrary. A sample plot of such an alignment matrix is shown in Figure 4. The significance of a calculated match is assessed by Monte Carlo methods using randomly constructed sequences which have the same overall composition as the protein under test. Such tests for significance added to the computing time, already proportional to $n \times m$ (see above). However, the sensitivity of this algorithm is increased greatly as a result, and is well worth the computational penalty. Modifications to this basic procedure have been explored,[46] and applied to sequences of both nucleic acids as well as proteins. The reader should keep in mind that the original sequence homology

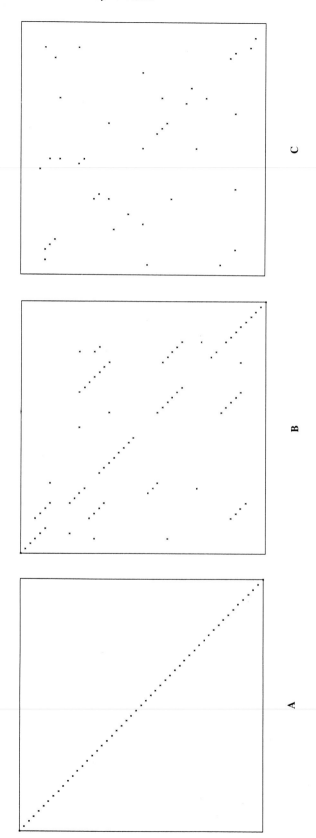

FIGURE 4. Simulated output from sequence homology matrix program. (A) Comparison of two sequences which are perfectly homologous, resulting in a perfect diagonal. (B) Comparison of two sequences which are closely homologous, but with gaps in one or both of the sequences. (C) Comparison of two sequences which are not closely homologous. Note the appearance of a slight diagonal, indicating some homology despite random appearance of matrix.

programs were written for the analysis of amino acid sequences, and that modifications must be made in order to apply the same algorithms to analysis of nucleic acid sequences.

Other methods though not as frequently used in subsequent programs, were those of Dayhoff[47] and Sankoff.[48] (For an excellent review of the details of these and other homology algorithms, see Peacock.[41])

Current approaches to assessment of sequence homology rely heavily on the routines described above. The effort now seems to be on refining these procedures, incorporating sophisticated algorithms for the comparison, and matching, scoring, filtering, and representation of sequence homology data. For example, Staden[49] has developed a program that depends on a preprocessing step which entails pattern recognition by the investigator in order to define a region of interest for further examination by computational means. Two scoring methods are used, both dependent on excluding insertions or deletions. For sequences of lengths m and n, memory requirements are of the order of $2(m + n)$ for the perfect scoring algorithm, and $2m + 3n$ for the proportional. As a result, the claim is made that these algorithms are suitable for implementation on small computers. The program has been written in FORTRAN 77 as implemented on VAX 11/780, and has some graphics terminal-specific aspects, but it appears that transportability to other systems is possible.

All sequence homology determinations require corrections for multiple substitutions. This may be defined as noise. Several nonphyletic corrections for such noise have been implemented, including, but not limited to minimal mutation distance, Poisson correction, random evolutionary hits, accepted point mutation correction, and augmentation correction.[50] Noise filtration is the major benefit of a program described by Pustell and Kafatos.[51] Their program filters noise through the use of a weighted exponential, applied across up to 100 contiguous bases. In this respect it is similar to the "best fit" algorithm of Steinmetz et al.[52] Despite this increase in signal to noise ratio, the program is claimed to run quickly and in little memory. They claim analysis of a 500×500 matrix using filtering over 40 bases in 9 min, from start-up to printed output, using a BBN C-70. Another added advantage of their algorithm is that compression of the data matrix is possible, allowing the user to first examine large sequences, then later zoom in on regions of particular interest. The basic program has been undergoing constant development, and may be run under any of FORTRAN 66, FORTRAN 77, FORTRAN IV, FORTRAN IV +, and others.[53]

Fast processing speed is the benefit of an algorithm used by Wilbur and Lipman,[54] but at the expense of the resolution provided by the original Needleman and Wunsch approach. The authors claim that a match of a 350-residue test protein sequence against the Protein Data Bank of the National Biomedical Research Foundation takes less than 3 min, and that a 500-base test sequence against the Los Alamos Nucleic Acid Data Base takes less than 2 min. Their program has been written in PASCAL as implemented on a DEC KL-10 mainframe. An even more impressive entry in the speed contest is a program by Taylor[55] which allows alignment in a single pass, with a number of steps equal to the product of the lengths of the two sequences under examination. The execution time was found to be approximately 2.5 min for sequences of length 500, and only 39.5 min for sequences of length 2000.

Several papers describing the use of programs which claim greater efficiency and/or optimal alignments have recently been published. Fitch and Smith[56] have developed an algorithm that takes into account a general treatment of gaps. Other algorithms assume linear, continuous gaps and apply a weighting factor appropriately. However, a linear gap assumption may not be the optimum method of handling insertions or deletions in a given sequence. The claim is made that homologous alignment between chicken α- and β-hemoglobin sequences may now be discerned, whereas it may be missed by other methods. Nussinov[57] describes an algorithm which allows alignment of two sequences of any length, incorporating gap constraints in the manner of Needleman and Wunsch. In this work, the program was applied to the ΦX-174 sequence. An interesting approach to optimization of alignment has been taken by Fickett,[58] who states determination of optimal alignment in 10% of the time previously required. This is accomplished by reordering the alignment matrix so that alignment may be assessed when only a small fraction of the entire matrix has been filled.

Any conclusions drawn from the application of a given algorithm to a given data set may need to be refined using, for example, the Needleman and Wunsch algorithm or others. Indeed, some statistical test of the proposed alignment should always be applied. Kanehisa[59] deals well with this topic, and Lipman et al.[60] address themselves to the always important question of the relevance of a proposed alignment to biology. The major disadvantage with their method is that computation time may become a practical problem. The authors state, however, that the tests they propose may not have to be implemented if several relatively minor boundary constraints (e.g., preservation of nearest neighbor frequencies) are applied to the proposed alignment.

The application of biological concepts to the heretofore mechanical alignment problem has resulted in the proposal of a new rule for detection of homology in nucleic acid sequences.[61] The premise of the proposal is that translation phase must be taken into account, since homologous sequences, presumably derived from a common ancestor, should show the same phase dependence; similarities that are in-phase should be more frequent than those considered from out-of-phase alignments, or alignments between nonhomologous or random sequences. Application of this rule to several sequences appears to bear out this contention.

V. INFERENCE OF PHYLOGENETIC RELATIONSHIPS

The distance measurements which result from homology assessment provide an excellent data set for the application of standard numerical taxonomic methods, as described by Sneath and Sokal.[39] Many of these methods were developed for use with phenetic classifications,[62,63] and therefore had to be modified appropriately to deal with quantitative data such as amino acid and nucleotide sequences. Assessment of sequence homology allows us to continue increasing the complexity of the analysis of sequence data. Once we know how two sequences may be related, we are in a position to ask what is the nature of the evolutionary relationships between the organisms represented by the nucleotide sequences. In practice, we must limit ourselves to relatively small data sets, since the computation time increases dramatically with the addition of more data. However, evolutionary relationships may be inferred over large phylogenetic distance, although confidence in such predictions decreases as the phylogentic distance increases.

Relatedness between species has long been analyzed through the construction of phylogenetic trees, based on morphological or other biological parameters. The possibility of more in-depth analysis through the use of protein sequence data quickly became apparent when these data became available. The basic principles behind such molecular analysis apply to the use of nucleotide sequences in the development of phylogenetic trees, although the details of the algorithms employed must be modified to make use of such data.

Each point on a tree corresponds to a specific time, species, and/or macromolecular sequence. Through common usage, the earliest time on the tree is at the bottom, with increasing proximity to the present as we move vertically up the tree. In the analysis of any given tree, the features to be aware of are the points of branching (or, the topology of the tree), and perhaps secondarily from the standpoint of phylogenetics, the lengths of the individual branches. The branching order may be derived from the alignment of comparable sequences from the species under test (ancestral sequence methods) or by using the matrix values obtained from such an alignment (matrix methods). In constructing a tree, all the data must be taken into account. It is this all-inclusive requirement that makes the use of a computer de rigeur, especially in light of the evolutionary noise level which is observed, usually found in the form of parallel evolution. Most assays of topology are based on "logical" changes which may have occurred through time in the sequences, such that amino acid changes, or nucleotide mutations, must make sense. This "sense" is generally measured in terms of conformity with various rules and defined internally to a given program, using a metric which depends on known mutation or physicochemical relatedness between amino acids.

It is difficult, if not impossible, to arrive at a specific, unique tree for any set of sequences. If the exact number of changes which have occurred since species divergence is known, including those sites where successive changes have occurred, it should be possible to generate such a unique tree. This does not prove practical, however, since only one change at a given site may be observed. Multiple mutations at the same site would lead to erroneous conclusions. This is also a noise-generating problem, which must be dealt with.

The topic of the development of phylogenetic trees has been treated in fine fashion by several relatively recent publications,[41,64] and will not be discussed in any further detail here. Rather, some of the fundamental approaches to the inference of phylogenetic trees will be described where relevant to our discussion of the use of the computer in such activity.

In general, the process of constructing a phylogenetic tree consists of taking sequences and considering where they may be added to a proposed tree to result in the maximization or minimization of some criterion. The algorithms move through the tree, rearranging where necessary to generate an improvement. There are three major approaches to the prediction of phylogenetic trees: methods using parsimony, methods based on compatibility, and maximum likelihood approaches. It should be noted that parsimony is strictly an heuristic principle, based on no mathematical or statistical foundation, and that results obtained using this approach do not guarantee the generation of the unique phylogenetic tree. On the other hand, compatibility methods are based on the well-known phenomenon that different data sets, or the order in which the individual sequences in a given data set are used for the analysis, can lead to different phylogenies. Indeed, compatibility, or preferably, incompatibility, has been mathematically proven by Estabrook et al.[65] and others. Although proven for cladistic characters, its application to sequences is clear, as long as parallelism is not too great.[41]

To this end then, Fitch and Margoliash[62] use a distance method designed to choose the phylogenetic tree that minimized the difference between the tree distances (phyletic distances) and the observed sequence differences. Matrix methods, such as that of Farris[66] use a table of sequence differences among all possible pairs of sequences. These require only the number of sequence differences and not the actual changes. Maximum parsimony methods construct trees with the minimum total length. Dayhoff's approach builds a phylogeny by simultaneously determining a branch order and an ancestral sequence at each branch point. This determines a minimum length tree, but does not consider the genetic code. In all cases, there is a need for indication of possible ambiguities in branching order due to nearly equivalent alternative trees ("confidence limits"). Also required is some measure of similarity between molecular and nonmolecular phylogenetic trees, as well as between the trees produced from different sorts of macromolecules for a given set of species.

A method which may rectify this confidence limit problem has been advanced by the work of Felsenstein,[67] which seems to deal well with some of the constraints described above. For example, both compatiblity and parsimony methods assume that there have been few changes at a given site, and that changes per se are improbable. However, most data sets incorporate moderate to large changes; hence parsimony methods are of limited utility on these data sets. According to Felsenstein, such methods tend to produce inconsistent estimates of the tree, converging to the wrong tree as more sequences are used for the same set of species. The maximum likelihood approach attempts to make explicit and efficient use of all the information in a sequence by formulating a probabilistic model of evolution, and applying known statistical methods.[66] Early efforts in this direction suffered from inefficient computing power since the program must search for the maximum likelihood tree through effecting small changes, while simultaneously determining its likelihood. Such a search is made computationally feasible if the rates of base substitution are allowed to differ among lineages. Under these conditions, the iterative method he proposes will guarantee a continued increase in the likelihood.[68] It is noted by Felsenstein that the likelihood of a tree is not identical with the probability that the proposed tree is the "correct" (unique) tree. However, a feature inherent in maximum likelihood approaches is that results so obtained have some estimate of the error involved in the analysis. There are problems associated with the algorithm as described by Felsenstein, in that deletions and insertions are not incorporated into this model, and the program, written in PASCAL, is very slow. The model was tested on several eukaryotic 55 RNAs, and found to provide information that was unattainable by other methods. Felsenstein is the provider of a package, PHYLIP, which incorporates these and several other algorithms for inference of phylogenies, including that of Camin and Sokal,[69] polymorphism parsimony programs for discrete state data, and several programs for DNA sequence data, specifically. The package has been written in a standard subset of PASCAL, and should be compatible with most computers (e.g., IBM 370, CDC 6000, DEC 10, and others including microcomputers) with little or no modification.

Since the earlier papers were published, advances along the lines of parsimony and compatibility have largely taken the form of refinements and modifications, rather than wholly new algorithms. For example, Sankoff et al.[70] have devised a total weighted mutational distance criterion for the evaluation of two data sets, one of phenylalanine tRNA sequences, and one of 5S RNA sequences. The key to their procedure is the imposition of restrictions on several optimal trees. These restrictions arise from patterns in the data, and from certain biological knowledge (i.e., spinach and bean chloroplast data should be grouped together, and separate from that of *Euglena*). This restriction limits the number of trees to be searched to a finite subset of all possible trees, and makes the problem computationally tractable. Their program has been written in FORTRAN, and implemented on a CYBER 173 computer. Since it does not appear to involve any system-specific calls, it should be transportable to other systems, as well.

The state-of-the-art in construction of evolutionary trees from sequence data seems to rely on the inclusion of procedures that are at odds with earlier assumptions; for example, the assumption of constant rates of evolution.[64] To this end, there have been several papers which compare several algorithms through the use of simulations. Blanken et al.[71] present three methods for construction of evolutionary trees, based on corrections for nonconstant evolutionary rates. In particular, the authors compare their methods with those of Dayhoff[72] and Fitch and Margoliash.[62] Both these papers use procedures that choose a correct topology over all possible topologies, and hence, require long computation times. Linear programming had previously been used to circumvent this necessity.[73] Blanken et al.[71] subtract from each element of the difference matrix the distance of the two sequences being compared from a common ancestor of all the organisms under examination. Then, using cluster analysis, the correct topology will be generated. They address themselves to the lack of any real information on such common ancestors, but provide justification for placement of this fixed reference at any point on an evolutionary tree. By the test used in the paper, the authors find that the new method is at least as good as, if not better than, those of Dayhoff[72] and Fitch and Margoliash.[62] These comparisons were accomplished using extensive computer simulations with a WANG 2200 computer. No information has been given concerning the language used.

Tateno et al.[74] use molecular data to test the accuracy of the UPGMA (unweighted pair group method),[39] the Fitch and Margoliash[62] (F/M), and a modification of the Farris[63] method, as proposed by them. Implicit in their analyses was the assumption that sequence changes were proportional to evolutionary time, an assumption that has, as yet, not been proven to a certainty.[50] They used computer simulations to follow the evolutionary change of 300 nucleotides over as many as 32 operational taxonomic units, according to the several models. Nucleotide substitution was assumed to occur according to a Poisson distribution. They found that the Farris and modified Farris methods were better than UPGMA and F/M when the coefficient of variation of the branch lengths was large. However, their modification gave better results when calculating estimates of the number of nucleotide substitutions for each branch. The UPGMA approach was found to be the best when the coefficient of variation of the branch lengths was small. No mention was made of the language used for the program or of the hardware used.

VI. CONCLUDING REMARKS

It would appear from the above discussion that the tools are well in hand for analysis of phylogenetic relationships. The computer has shown itself to be invaluable in the acquisition, storage, manipulation, and analysis of nucleic acid sequence data, to the extent of giving fine distinctions between taxonomic groups. As mentioned, certain algorithms are best used under specific circumstances. In most cases, the limiting factors of such manipulations seem to be constraints due to software. Perhaps the best means of optimizing present use of computers in nucleic acid research would be by decreasing the use of specific system calls for input and output in specific programs, and by using standard implementations of the languages used. There already exists a free exchange of programs between groups. This should be encouraged. Increases in overall processing speed may best be approached from the standpoint of examining the hardware used. Indeed, analysis of the very foundation of present-day processing, the serial von Neumann approach, indicates that larger and more complex computers may be required to analyze really large data sets. The use of array processors and other parallel processing schemes[75] may help here, as would excursions outside the von Neumann realm.

ADDENDUM

I am pleased to note that, while this chapter was in press, a paper entitled, "Nucleic Acid and Protein Sequence Databases", was published by G. G. Kneale and M. J. Bishop, of the University of Cambridge, U.K., in *Computer Applications in the Biosciences*. I am encouraged by this confirmation of some of the points raised herein.

ACKNOWLEDGMENTS

The author would like to thank Professor Cyril Ponnamperuma for his enthusiastic support during the generation of this work. I would also like to thank Drs. Robert Donnelly, Chris Overton, and Allen Place for helpful discussions. In addition, I would like to thank Dr. Frances Lewiter of Bolt Beranek and Newman and Kathryn Sidman of the National Biomedical Research Foundation, for providing information on the growth rates of nucleotide sequence databases.

REFERENCES

1. **Stone, T. W. and Potter, K. N.,** A DNA analysis program designed for computer novices working in an industrial-research environment, *Nucl. Acids Res.,* 12, 367, 1984.
2. **Sanger, F., Nicklen, S., and Coulson, A. R.,** Chain sequencing with chain-terminating inhibitors, *Proc. Natl. Acad. Sci. U.S.A.,* 74, 5463, 1977.
3. **Maxam, A. M. and Gilbert, W.,** Sequencing end-labeled DNA with base-specific chemical cleavages, *Meth. Enzymol.,* 65, 499, 1980.
4. **Lautenberger, J. A.,** A program for reading DNA sequence gels using a small computer equipped with a graphics tablet, *Nucl. Acids Res.,* 10, 27, 1982.
5. **Gingeras, T. R., Rice, P., and Roberts, R. J.,** A semi-automated method for the reading of nucleic acid sequencing gels, *Nucl. Acids Res.,* 10, 103, 1982.
6. **Staden, R.,** A computer program to enter DNA gel reading data into a computer, *Nucl. Acids Res.,* 12, 499, 1984.
7. **Komaromy, M. and Govan, H.,** An inexpensive semi-automated sequence reader for the Apple II computer, *Nucl. Acids Res.,* 12, 675, 1984.
8. **Shapiro, B. A., Lipkin, L. E., and Maizel, J.,** An interactive technique for the display of nucleic acid secondary structure, *Nucl. Acids Res.,* 10, 7041, 1982.
9. **Lapalme, G., Cedergren, R. J., and Sankoff, D.,** An algorithm for the display of nucleic acid secondary structure, *Nucl. Acids Res.,* 10, 8351, 1982.
10. **Day, G. R. and Blake, R. D.,** Statistical significance of symmetrical and repetitive segments in DNA, *Nucl. Acids Res.,* 10, 8323, 1982.
11. **Harr, R., Haggstrom, M., and Gustafsson, P.,** Search algorithm for pattern match analysis of nucleic acid sequences, *Nucl. Acids Res.,* 11, 2943, 1983.
12. **Tolstoshev, C. M. and Blakesley, R. W.,** RSITE: a computer program to predict the recognition sequence of a restriction enzyme, *Nucl. Acids Res.,* 10, 1, 1982.
13. **Lilley, D. M. J.,** A simple computer program for calculating, modifying and drawing circular restriction maps, *Nucl. Acids Res.,* 10, 19, 1982.
14. **Pearson, W. R.,** Automatic construction of restriction site maps, *Nucl. Acids Res.,* 10, 217, 1982.
15. **Staden, R.,** Automation of the computer handling of gel reading data produced by the shotgun method of DNA sequencing, *Nucl. Acids Res.,* 10, 4731, 1982.
16. **Gallant, J. K.,** The complexity of the overlap method for sequencing biopolymers, *J. Theor. Biol.,* 101, 1, 1983.
17. **Hamori, E. and Ruskin, J.,** H curves, a novel method of representation of nucleotide series especially suited for long DNA sequences, *J. Biol. Chem.,* 258, 1318, 1983.
18. **Staden, R.,** Graphic methods to determine the function of nucleic acid sequences, *Nucl. Acids Res.,* 12, 521, 1984.
19. **Watanabe, K., Yasukawa, K., and Iso, K.,** Graphic display of nucleic acid structure by a microcomputer, *Nucl. Acids Res.,* 12, 801, 1984.
20. **Brutlag, D. L., Clayton, J., Friedland, P., and Kedes, L. H.,** SEQ: a nucleotide sequence analysis and recombination system, *Nucl. Acids Res.,* 10, 279, 1982.
21. **Bach, R., Friedland, P., Brutlag, D. L., and Kedes, L.,** MAXAMIZE. A DNA sequencing advisor, *Nucl. Acids Res.,* 10, 295, 1982.
22. **Clayton, J. and Kedes, L.,** GEL, a DNA sequencing project management system, *Nucl. Acids Res.,* 10, 305, 1982.
23. **Friedland, P., Kedes, L., Brutlag, D., Iwasaki, Y., and Bach, R.,** GENESIS, a knowledge-based genetic simulation system for representation of genetic data and experiment planning, *Nucl. Acids Res.,* 10, 323, 1982.
24. **Queen, C. L. and Korn, L. J.,** Computer analysis of nucleic acids and proteins, *Meth. Enzymol.,* 65, 595, 1980.
25. **Kanehisa, M. I.,** Los Alamos sequence analysis package for nucleic acids and proteins, *Nucl. Acids Res.,* 10, 183, 1982.
26. **Abarbanel, R. M., Wieneke, P. R., Mansfield, E., Jaffe, D. A., and Brutlag, D. L.,** Rapid searches for complex patterns in biological molecules, *Nucl. Acids Res.,* 12, 263, 1984.
27. **Bucher, P. and Bryan, B.,** Signal search analysis: a new method to localize and characterize functionally important DNA sequences, *Nucl. Acids Res.,* 12, 287, 1984.
28. **Staden, R.,** Computer methods to locate signals in nucleic acid sequences, *Nucl. Acids Res.,* 12, 505, 1984.
29. **Orcutt, B. C., George, D. G., and Dayhoff, M. O.,** Protein and nucleic acid sequence database systems, *Ann. Rev. Biophys. Bioeng.,* 12, 419, 1983.
30. **Kanehisa, M., Fickett, J. W., and Goad, W. B.,** A relational database system for the maintenance and verification of the Los Alamos sequence library, *Nucl. Acids Res.,* 12, 149, 1984.

31. **Friedland, P., Kedes, L., Brutlag, D., Iwasaki, Y., and Bach, R.,** GENESIS, a knowledge-based genetic simulation system for representation of genetic data and experiment planning, *Nucl. Acids Res.,* 10, 323, 1982.
32. **Bach, R., Iwasaki, Y., and Friedland, P.,** Intelligent computational assistance for experimental design, *Nucl. Acids Res.,* 12, 11, 1984.
33. **Lewin, R.,** Long-awaited decision on DNA database, *Science,* 217, 817, 1982.
34. **Lewin, R.,** National networks for molecular biologists, *Science,* 223, 1379, 1984.
35. **Watt, P.,** Biologists map genes on-line, *InfoWorld,* 43, May 7, 1984.
36. **Calverie, J.-M.,** A common philosophy and FORTRAN 77 software package for implementing and searching sequence databases, *Nucl. Acids Res.,* 12, 397, 1984.
37. **Schneider, T. D., Stormo, G. D., Haemer, J. S., and Gold, L.,** A design for computer nucleic-acid-sequence storage, retrieval, and manipulation, *Nucl. Acids Res.,* 10, 3013, 1982.
38. **Whieldon, D.,** New lift for mass storage, *Comput. Decisions,* 15, 172, 1983.
39. **Sneath, P. H. A. and Sokal, R. R.,** *Numerical Taxonomy,* W. H. Freeman, San Francisco, 1973.
40. **Zuckerkandl, E. and Pauling, L.,** Molecules as documents of evolutionary history, *J. Theor. Biol.,* 8, 357, 1965.
41. **Peacock, D.,** Data handling for phylogenetic trees, in *Biochemical Evolution,* Gutfreund, H., Ed., Cambridge University Press, London, 1981, chap. 3.
42. **Fitch, W. M.,** An improved method for detecting evolutionary homology, *J. Mol. Biol.,* 16, 9, 1966.
43. **Fitch, W. M.,** Further improvements in the method of testing for evolutionary homology among proteins, *J. Mol. Biol.,* 49, 1, 1970.
44. **Fitch, W. M.,** Distinguishing homologous from analogous proteins, *Syst. Zool.,* 19, 99, 1970.
45. **Needleman, S. B. and Wunsch, C. D.,** A general method applicable to the search for similarities in the amino acid sequence of two proteins, *J. Mol. Biol.,* 48, 443, 1970.
46. **Boswell, D. R. and McLachlan, A. D.,** Sequence comparison by exponentially-damped alignment, *Nucl. Acids Res.,* 12, 457, 1984.
47. **Dayhoff, M. O.,** *Atlas of Protein Sequence and Structure,* National Biomedical Research Foundation, Silver Springs, Md., 1969.
48. **Sankoff, D.,** Matching sequences under deletion/insertion contraints, *Proc. Natl. Acad. Sci. U.S.A.,* 69, 4, 1972.
49. **Staden, R.,** An interactive graphics program for comparing and aligning nucleic acid and amino acid sequences, *Nucl. Acids Res.,* 10, 2951, 1982.
50. **Wilson, A. C., Carlson, S. S., and White, T. J.,** Biochemical evolution, *Ann. Rev. Biochem.,* 46, 573, 1977 (see discussion and references therein).
51. **Pustell, J. and Kafatos, F.,** A high speed, high capacity homology matrix: zooming through SV40 and polyoma, *Nucl. Acids Res.,* 10, 4675, 1982.
52. **Steinmetz, M., Frelinger, J. G., Fisher, D., Hunkapillar, T., Pereira, D., Weissman, S. M., Vehara, H., Natenson, S., and Hood, L.,** *Cell,* 24, 125, 1981.
53. **Pustell, J. and Kafatos, F. C.,** A convenient and adaptable package of computer programs for DNA and protein sequence management, analysis, and homology determination, *Nucl. Acids Res.,* 12, 643, 1984.
54. **Wilbur, W. J. and Lipman, D. J.,** Rapid similarity searches of nucleic acid and protein data banks, *Proc. Natl. Acad. Sci. U.S.A.,* 80, 726, 1983.
55. **Taylor, P.,** A fast homology program for aligning biological sequences, *Nucl. Acids Res.,* 12, 447, 1984.
56. **Fitch, W. M. and Smith, T. F.,** Optimal sequence alignments, *Proc. Natl. Acad. Sci. U.S.A.,* 80, 1382, 1983.
57. **Nussinov, R.,** An efficient code searching for sequence homology and DNA duplication, *J. Theor. Biol.,* 100, 319, 1983.
58. **Fickett, J. W.,** Fast optimal alignment, *Nucl. Acids Res.,* 12, 175, 1984.
59. **Kanehisa, M.,** Use of statistical criteria for screening potential homologies in nucleic acid sequences, *Nucl. Acids Res.,* 12, 203, 1984.
60. **Lipman, D. J., Wilbur, W. J., Smith, T. F., and Waterman, M. S.,** On the statistical significance of nucleic acid similarities, *Nucl. Acids Res.,* 12, 215, 1984.
61. **Paetkau, V.,** A new rule for analyzing homologous coding sequences in DNA, *Nucl. Acids Res.,* 12, 159, 1984.
62. **Fitch, W. M. and Margoliash, E.,** Construction of phylogenetic trees, *Science,* 155, 279, 1967.
63. **Farris, J. S.,** Methods for computing Wagner trees, *Syst. Zool.,* 19, 83, 1970.
64. **Felsenstein, J.,** Numerical methods for inferring evolutionary trees, *Q. Rev. Biol.,* 57, 379, 1982.
65. **Estabrook, G. F., Johnson, C. S., Jr., and McMorris, F. R.,** A mathematical foundation for the analysis of cladistic character compatibility, *Math. Biosci.,* 29, 181, 1976.
66. **Farris, J.,** Estimating phylogenetic trees from distance matrices, *Am. Nat.,* 106, 645, 1972.
67. **Felsenstein, J.,** Evolutionary trees from DNA sequences: a maximum likelihood approach, *J. Mol. Evol.,* 17, 368, 1981.

68. **Felsenstein, J.,** Evolutionary trees from gene frequencies and quantitative characters: finding maximum likelihood estimates, *Evolution,* 35, 1229, 1981.
69. **Camin, J. H. and Sokal, R. R.,** A method for deducing branching sequences in phylogeny, *Evolution,* 19, 311, 1965.
70. **Sankoff, D., Cedergren, R. J., and McKay, M.,** A strategy for sequence phylogeny research, *Nucl. Acids Res.,* 10, 421, 1982.
71. **Blanken, R. L., Klotz, L. C., and Hinnebusch, A. G.,** Computer comparison of new and existing criteria for constructing evolutionary trees from sequence data, *J. Mol. Evol.,* 19, 1, 1982.
72. **Dayhoff, M. O.,** Survey of new data and computer methods of analysis, in *Atlas of Protein Sequence and Structure,* Dayhoff, M. O., Ed., National Biomedical Research Foundation, Washington, D.C., 1978.
73. **Ratner, V. A., Zharkikh, A. A., and Rodin, S. N.,** in *Mathematical Models of Evolution and Selection,* Ratner, V. A., Ed., Novosibirsk, U.S.S.R., 1977.
74. **Tateno, Y., Nei, M., and Tajima, F.,** Accuracy of estimated phylogenetic trees from molecular data. I. Distantly related species, *J. Mol. Evol.,* 19, 387, 1982.
75. **Collins, J. F. and Coulson, A. F. W.,** Applications of parallel processing algorithms for DNA sequence analysis, *Nucl. Acids Res.,* 12, 181, 1984.

Chapter 2

MOBILE DNA SEQUENCES AND THEIR POSSIBLE ROLE IN EVOLUTION

Georgii P. Georgiev, Yurii V. Ilyin, Alexei P. Ryskov, and Tatiana I. Gerasimova

TABLE OF CONTENTS

I. INTRODUCTION

A. History

The first data on the existence of unstable mutations were obtained as early as the 1920s by Demerec[1-3] who worked on *Drosophila virilis*. However, their significance remained obscure.

In the 1940s, McClintock, after a brilliant analysis of similar unstable mutations in maize, noted the existence in genomes of the elements capable of controlling the functioning of other genes and of transposing from one site of the chromosome to another.[4,5] Thus, mobile genetic elements were discovered, and recognized functionally.

Molecular studies of mobile elements were started by Starlinger and Saedler[6] who discovered transposable elements in bacterial cells. Much data have been accumulated in this field.

Molecular studies on eukaryotic cells developed with some delay. Green[7-9] and Golubovsky[10-12] obtained the genetic data suggesting the existence of movable elements in *D. melanogaster*. Only after the introduction of genetic engineering were the movable elements isolated and studied at the molecular level.

In the 1970s, Finnegan et al.[13] and Ilyin et al.[14] cloned genomic DNA sequences of *D. melanogaster* which were actively transcribed in tissue culture cells. These sequences were found to be in multiple copies in the genome and scattered throughout chromosomes. We were the first to demonstrate that their localization in chromosomes was remarkably variable.[14,15] Such observations immediately suggested that these elements were movable in the eukaryotic genome. This result was subsequently confirmed by many other authors.[16-22] Because of the mobility of the cloned sequences, we designated them as "mobile dispersed genetic elements" or mdg. The American investigators usually use the term "copia-like elements" since the first elements cloned by Hogness' group were referred to as copia.[13,21]

Another important observation was the discovery of long terminal repeats at the ends of mdg elements.[13,23] Their sequencing gave strong support to the idea that mdg elements were transposons.[24-26] At the same time, the data on the organization of retroviral proviruses that were integrated into the host genome appeared. It then became clear that the nature of mdg elements and retroviral proviruses was similar.[24-27] In particular, their long terminal repeats (or LTRs) were organized in essentially the same manner.

The general feature of transposable elements was the almost obligatory presence of short direct repeats flanking the transposons, which originated from the duplication of a target sequence at the site of transposon insertion.[28] It was thus found that a number of different repetitive sequences, for example, ubiquitous repetitive sequences discovered in the mammalian genome — mouse B1 and B2 sequences[29-31] and human Alu sequences,[32,33] were framed with such short repeats and therefore belonged to transposons. These repeats and some other sequences (for instance, pseudogenes) seemed to be inserted through reverse transcription of RNAs transcribed from corresponding genomic elements.[34]

Recently, important progress has been achieved owing to the discovery of the P-factor in *D. melanogaster*.[35] This element is a transposable sequence capable of being moved by itself; in addition, it encodes the factor (transposase) which can mobilize not only the P-element, but also some other tansposons. It is quite possible that similar elements will be discovered in the near future in other systems as well.

B. Active and Passive Transposons

We propose to classify transposable elements of eukaryotes in two major groups: the active and the passive. Active transposons are those which carry the information for the transposition machinery. A mechanism of transposition seems to proceed via direct excision of a DNA sequence which may then be inserted into a novel position. The P-element is very probably a typical representative of movable elements providing the cell with such machinery. Another means of transposition is through a reverse transcription of RNAs. The reverse transcriptase may be encoded by some retroviral proviruses nondefective in the pol gene as well as by some of mdg elements.

Passive transposons are those for which insertion uses the enzyme machinery provided by the active transposons or by the cell itself. Sometimes the active transposons (such as mdg) may become passive and use the transposase induced by the P-element. On the other hand, most of the passive transposons (such as short repetitive DNA sequences or pseudogenes) seem to use reverse transcription for the insertion.

Here, we (1) list the main types of movable elements of animal cells using the above-mentioned classification, (2) describe different possible effects of their transposition on the properties of the cell or organisms, (3) discuss the factors that control the rate of transposition, and finally (4) analyze the role which such events may play in the evolution.

II. ACTIVE TRANSPOSONS OF ANIMAL CELLS

A. Mdg Elements and Retroviral Proviruses

1. Mdg Elements of Drosophila melanogaster

a. Properties of Mdg Elements

There are about 20 families of mdg elements in the genome of *D. melanogaster*, each represented by a number of copies (from few to 150 per haploid genome). Among the most extensively studied are mdg 1,[36-39] mdg 3,[20,40] copia,[16,17,41-45] 412 or mdg 2,[16,17,45,46] mdg 4 or gypsy,[47-49] and some others.[26,50,51] The size of mdg varies from 5 to 10 kb. At least in the case of *D. melanogaster* different copies of the same family are very similar. The members of each family are scattered throughout all chromosomes of *D. melanogaster*. The localization of mdg elements is variable. In different strains of *D. melanogaster* most of the sites are usually different. Prominent differences in localization of mdg elements could even be detected among the individuals belonging to the same strain. Their positions are stable only in inbred stocks. The varying location is one of the fundamental properties of mdg elements.[13-24,52]

b. Long Terminal Repeats

All mdg elements studied contain long direct repeats at their ends referred to as "long terminal repeats'' (LTRs). The size of LTRs varies from 250 to 600 bp. A number of LTRs have been sequenced. These are LTRs of mdg 3,[40] copia,[41,42] mdg 1,[38,39] 412 or mdg 2,[46] B104,[50] 297, 17.6[26,51] and mdg 4 or gypsy.[48,49] All of them possess a very similar organization: at their beginning and end they have slightly mismatched inverted repeats. Mdg elements are flanked by short direct repeats 4 to 5 bp long. In the cells where mdg is absent from this particular site such a 4 to 5 bp sequence is not repeated. Thus, its duplication takes place in the course of the mdg insertion. The ends of some bacterial transposons are organized in a similar manner.[28]

Another interesting feature of LTRs is that in both strands they often contain sequences which may be used as signals for initiation of transcription (TATA boxes) and the AATAAA blocks which are known to be the signals for RNA polyadenylation.[26] In general the organization of LTRs in mdg elements of *D. melanogaster* and in retroviral proviruses of vertebrates is quite the same.

c. Transcription Patterns

Mdg elements are rather efficiently transcribed, especially in cultured cells. The major transcript is a full-length copy of the mdg element starting and terminating within LTR.[20,26,37,44,50] Besides the major transcript, some additional smaller transcripts may be observed which seem to be the products of splicing of the original. A feature of mdg elements is that in addition to transcripts from the major strand, transcripts from another strand are also formed, though the efficiency of transcription from another strand is about 20 times lower than that of the major strand. As a result, a long, double-stranded RNA could be obtained by annealing *D. melanogaster* RNA and this dsRNA which, if melted, efficiently hybridizes to the DNA of mdg elements.[53,54] The partially symmetric transcription of mdg elements seems to be the result of the presence of initiation signals in both the strands of LTR sequences.

2. Mdg Elements in Other Species, Retroviral Proviruses
a. Ty1 Element in Yeast

Soon after the discovery of mdg in *D. melanogaster,* a similar element designated as Ty1 was found in yeast.[19] Ty1 is represented by approximately 35 copies in the haploid yeast genome. It is 5.6 kb long and framed by LTRs of approximately 330 bp called δ. In addition to the δ sequences associated with Ty elements, the yeast genome contains at least 100 unique δ sequences.[19,55] In the case of *D. melanogaster,* solo LTRs were found only for gypsy.[47] Ty1 δ were sequenced.[55-57] δ Sequences are reminiscent of a typical LTR except that the inverted repeats at the ends of δ do not exist. The transcription is started and ended within the δ sequence.[55]

b. Retroviral Proviruses

In vertebrates a class of sequences representing the proviruses of retroviruses integrated into genomes has been known for a long time. As the data on the structural organization of mdg elements and retroviral proviruses were being accumulated,[23,27,58] it became clear that two types of sequences were extremely similar and could be considered as elements of the same type. In particular, both have the same size and are flanked by LTR elements of similar size and organization.[26,27] Furthermore, retrovirus-like particles containing RNA hybridizing to mdg copia have recently been isolated from the nuclear fraction of *D. melanogaster* cultured cells.

Retroviral proviruses are usually represented by several families in the genome. In the mouse genome, the most abundant ($\sim 10^3$ per haploid genome) are the so-called genes for intracisternal A-particles (IAP-genes).[60,61] They are somewhat more heterogeneous than *D. melanogaster* mdg elements. At an early stage of development these genes are actively expressed, giving rise to IAP. Some other endogenous proviruses (represented by several dozen members) are also present in the mouse genome.[27,62,63] The cells may also be infected by exogenous retroviruses. In this case, new copies of provirus are integrated. Usually their transcription is higher due to the presence of a stronger enhancer.

In contrast to mdg elements of *D. melanogaster*, the information content of retroviruses is well known. They encode at least three types of proteins: external envelope (gene env), the RNP complex (gag), and reverse transcriptase (pol).[27,64]

Thus, endogenous retroviral proviruses may serve as suppliers of the cell with reverse transcriptase which can be detected in endogenous retrovirus-like particles. Some proviruses may be defective in one or even all functions.

We attempted to isolate mdg-like sequences from mouse genome using hybridization of cloned mouse DNA with long double-stranded RNA isolated from mouse cells (dsRNA of the A-type).[54,65] A number of clones were obtained, and two sequences occurring most frequently were referred to as A1 and A2. A1 was found to be identical to IAP genes. The A2 sequence was identified with a major mouse long repeat, represented by approximately 2×10^4 copies (6% of the whole DNA) in the genome. Its relation to mdg elements has not been interpreted so far. At the termini, the regions of homology were detected, but it is not yet clear whether they represent real LTR elements. Other vertebrates, including man, also contain many families of retroviral proviruses and long transposon-like repetitive elements.[27,66-67]

It is of interest that different repetitive DNA elements, in some respect, interact with one another, giving new composite transposon-like elements. For example, neither the mouse BAM5 element[69] nor the mouse R element[70] is surrounded by direct repeats, whereas their linked entity (Bam 5 + R) is surrounded by a 15 bp direct repeat.[71]

c. Mechanisms of Transposition

The mechanism of provirus insertion is more or less clear in the case of exogenous infection. RNA injected into the cell is reverse transcribed and the linear DNA copy with two LTR elements is formed in the cytoplasm. Afterward, it moves to the nucleus and is circularized. As a result, the extrachromosomal circular DNA with two or one LTR element is formed in the infected cells.[27,76,77] This may be considered as an intermediate in the insertion process.

In noninfected cells, the amount of DNA circles is very low. Nevertheless, they are detectable at least in some cells. For example, in Ehrlich ascites carcinoma, we could detect the closed supercoiled DNA ~7 kb long which hybridized to the A1 sequence and seemed to represent IAP gene circles.[78] Circular DNA complementary to mdg elements copia, mdg 1, mdg 3, and mdg 4 were also found in lysates of *D. melanogaster* cultured cells.[79-82]

The major portion of these elements contains one LTR, but in the case of copia, molecules with two LTRs were also found. One can suggest that endogenous retroviral proviruses and mdg elements insert at a new chromosomal site through the reverse transcription of their RNA products. This may be one of the more important methods of transposon mobilization.

It should be pointed out that insertion based upon reverse transcription leads to gene amplification as the original copy is not lost. In *D. melanogaster* cultured cells, an mdg amplification takes place.[16,20,36,52] In filogenesis of mice amplification of IAP genes could also be detected.[83,84]

d. Species Specificity

Various species are very different both quantitatively and qualitatively with respect to their mdg elements. For example, the species *D. simulans* and *D. mauritiana* evolutionarily close to *D. melanogaster* contain some mdg elements of *D. melanogaster*, but their copy number is much lower.[22,85,86] On the other hand, *D. virilis* more distant from *D. melanogaster* does not contain these mdg elements. Instead, *D. virilis* possesses an mdg of its own (named

Dv) which is absent form *D. melanogaster*.[87] Mdg Dv is rather heterogeneous in size, possibly representing a more ancient mdg family.

In European strains of mice the IAP genes are very numerous (800 to 1000 copies). The Asiatic mice contain fewer copies of this mdg-like element (only 25 to 50 copies).[84] Different endogenous retroviruses are detected in chicken, mice, and monkeys. Many of them reveal considerable sequence homology. Some homology can even be detected between mdg elements of *D. melanogaster* 297 and 17.6 and the avian leukosis-sarcoma retrovirus (AL-SV).[26]

One may speculate that different retroviruses (mdg elements) have common ancestors, but they evolve rather rapidly. From time to time new variants appear which may infect and expand within a particular species. Such successive infections can give rise to a specific set of mdg elements for each species.

An important feature of mdg is that within each family the elements seem to be very similar, at least in the case of *D. melanogaster*.[16,17,20,36] Such homogeneity may be explained in terms of gene conversion or other correction mechanisms. The mechanism of gene conversion remains obscure, especially in the case of distant genes. It is possible that recombination with gene copies synthesized by reverse transcriptase is involved. If such a mechanism does exist, the selection of a copy (copies) to be used as template would depend on its relative rate of transcription and reverse transcription. The poorly transcribed (and reverse transcribed) genes would diverge much faster than those with a high level of transcription, as the rate of their correction would be decreased.

B. P-Element

1. Properties of the P-Element

The P-element has been cloned from unstable mutants obtained in the course of P-M hybrid dysgenesis. After crosses of P-strain males with M-strain females a number of unstable, possibly transposon-induced mutations appear. Rubin et al.[35] found that several *w* mutations depended on the presence of insertions in the locus white which was cloned previously.[88] Using the cloned sequences as a probe, the authors isolated an element 2.9 kb long which was present in the genome of flies of P-strains, but not of M-strains. This element was called a "P-element". It is present in approximately 50 copies in P-strains and is scattered throughout all chromosomes. Its location varies within different P-strains.[89]

Besides full-lenth copies, a number of shorter sequences are present in the genome.[88,89] A full-length 2.9 kb copy of the P-element contains 4 open reading frames in the internal part of the sequence. On the termini of P-element, short 31 bp inverted repeats are located. These mismatched inverted repeats remained in defective copies of the P-elements. Thus, the size decrease in defective P-elements depends on a loss of internal sequences.[90]

In P-strains themselves, P-elements are rather stable, possibly due to the presence of some cytoplasmic immune factors. However, after a cross with females of the M-cytotype the P-element transposes at a high rate, creating a number of unstable mutations. P-elements frequently transpose to the loci singed and white though they also may be observed at other sites by *in situ* hybridization. The mutations induced by P-elements are usually unstable and may reverse to wild type. In these cases, a precise excision of the P-element was observed.[89]

2. Mechanism of P-Element Transposition

The existence of precise excision of the P-element suggests the mechanism for transposition other than reverse transcription. Most probably in the course of transposition excision and reinsertion of the excised P-element take place.[35] Another possibility is insertion of a replicated P-element.

The cloned P-element efficiently inserts into the genome after injection into an oocyte.[91,92] A full-length copy of the P-element should necessarily be present in injected material. On the other hand, even defective P-elements 0.3 to 0.5 kb long are capable of moving into the genome in the presence of the complete P-factor. Any sequence flanked by the termini of the P-element may be used in this system for insertion.

One can conclude that the P-element carries a gene(s) for enzymes inducing excision-insertion, i.e., "transposase", and the termini which may serve as a recognition site for transposase. Possibly these termini correspond to mismatched inverted repeats of the P-element. The requirements for P-element excision are the same. Excision readily occurs with 0.3 to 0.5 kb insertions if the termini are conserved.[91,92]

Possibly there are elements in the genome other than P-elements which can also encode "transposase". Being expressed, such elements might induce the excision-insertion of sequences with appropriate organization of the termini.

III. PASSIVE TRANSPOSABLE ELEMENTS

A. "Transposase"-Dependent Elements

1. Mdg Elements in P-M Dysgenesis

Several different sequences can be mobilized in P-M hybrid dysgenesis due to activation of transposase encoded by the P-element. Among them are the mdg elements, which in this case can be considered as passive transposans. Several insertion mutations in the white locus appearing in the course of P-M hybrid dysgenesis were induced by insertion of the mdg element copia.[89]

A large family of unstable mutations in the cut locus were found to depend on insertion of mdg 4 at the corresponding region of the X-chromosome (7B), again induced by P-M dysgenesis. In this case, the original mutation ct^{MR2} and many of its derivatives remained unstable for a long time (ct^{MR2} for ~1.5 years or ~50 generations). After this, the rate of $ct \rightarrow ct^+$ reversions and mutation changes was reduced. However, the instability was restored upon a cross with ct^{MR2} females with MRh12/Cy males (which are carriers of a number of P-elements). All stable $ct \rightarrow ct^+$ reversions were accompanied by the excision of mdg 4 from the cut locus, while at a different ct mutation and unstable reversions it remained there, probably changing its position.[93,94] Thus, mdg elements may become a substrate for the mobilization induced by hybrid dysgenesis, i.e., by the P-element. The mobilization of several other mdg elements was also observed in the above-mentioned experiments.[95]

2. Fold-Back Sequences

An interesting type of sequence was described by Potter et al.[96-98] It represents long inverted repeats. Each branch of a palindrome was 250 to 1400 bp in length. The element may either consist solely of palindrome itself, or include a piece of nonpalindromic DNA of a varying length inserted between the homologous sequences. Sometimes this inserted DNA is homologous to the branches of a palindrome.

These elements, called fold-back (FB) elements, are represented by more than 30 copies per haploid genome of *D. melanogaster*. They are scattered throughout the chromosomes, and like other movable elements have quite variable locations. On the flanks of FB-elements, short direct repeats, originating from duplication of the target sequence, were detected which provided for insertion of the elements.[97,98]

The nucleotide sequence of simple FB elements shows that they do not encode proteins. Thus, they do not produce a machinery for transposition and should be considered as passive transposons. The presence of inverted repeats at their ends suggests the involvement of an excision-insertion transposition mechanism using "transposase(s)". Indeed, it was found that in P-M dysgenesis the transposition of FB-elements, possibly induced by the P-factor product, was strikingly increased.[95]

The presence of numerous short repeats 20 to 30 bp long in palindrome branches seems to increase the transposition rate of FB-elements. These repeats may become the substrate for cellular transposases, whose level should be very low.[98] FB sequences incorporated into genes may create unusual mutations in homozygous strains which are unstable even in the absence of the P-element, i.e., of exogenous transposases. For example, highly unstable w^{DZL} and w^c mutations were induced by FB-elements.[99,100] These mutations are characterized by a frequent formation of several chromosomal rearrangements (Df, In, T). Thus, FB-elements create hot spots for recombination events. They also may be involved in the initiation of homologous recombination.

FB-elements can mobilize the unique sequences. For example, the control part of a transposon producing the w^{DZL} mutation is a unique sequence normally detected in the 21D region of the second chromosome.[99]

FB-elements participate in mobilization of large pieces of genome. Ising and Block described a large w^a transposon of $\sim$100 kb which was excised with a rate of 10^{-3} and inserted into other sites on chromosomes with a frequency of 10^{-4} to 10^{-5}. At the end of this transposon, an FB-element was found and cloned.[101,102]

Generally speaking, in order to be transposed by transposase the element should possess, at the very least, inverted repeats at its termini, even mismatched ones.

B. Reverse Transcriptase-Dependent Passive Mobile Elements
1. RNA Polymerase II Transcriptional Units

Reverse transcription followed by DNA insertion seems to be an important process for amplification of mdg elements (retroviral proviruses). The source of reverse transcriptase in normal cells is the mdg elements of the genome. The virus-like particles containing copia RNA in the cells of *D. melanogaster,* as well as the intracisternal A-particles present in mouse embryonic cells, contain reverse transcriptase.[59] Full-length copies of mdg (retroviral) RNA contain special sequences for priming efficient reverse transcription. However, many other RNAs may also be used by reverse transcriptase, though with a lower efficiency. First, these are the RNA polymerase II transcripts, or true mRNAs. It is well known now that the animal genome contains a number of ''pseudogenes'', or copies of genes, which, in contrast to their prototype, lack introns as well as normal flanking sequences. Usually they are silent as they do not contain a 5'-upstream regulatory region. Many pseudogenes are described.[103-106] As a rule, pseudogenes are flanked by short repeats, indicating their insertion nature. The absence of introns and flanking sequences suggests that the pseudogenes are formed through reverse transcription of processed mRNA. Sometimes a short oligo(A) tail is present which binds the oligo(dT) primer in reverse transcription. In some cases, an inactive pseudogene may be activated if, during the genome rearrangements it contacts the promoter region from another area of the genome. This is probably the way of gene mos activation in some mouse tumors.[107-108]

Another source of RNA for reverse transcription is small nuclear RNA (snRNA). Indeed, a number of pseudogenes were found for the genes responsible for the synthesis of snRNA, which are also known to be transcribed by RNA-polymerase II. Again, these pseudogenes are flanked by short direct repeats, i.e., represent the insertion sequences.[109,110] In several cases, the mechanism of reverse transcription seems to be very clear. For example, the pseudogenes of U12 and U3 RNAs are often truncated, lacking the 3' end. It is known that U3 RNA possesses a secondary structure forming a hairpin involving its 3' end. Therefore, its 3' end may be used as a primer for reverse transcription of the 5' part of the molecule not involved in hairpin formation. The borders of DNA synthesized in vitro by reverse transcriptase with a free U3 RNA template or with an RNP particle containing U3 RNA were exactly the same as the borders of U3 RNA pseudogenes.[110a] Thus, the conclusion about the involvement of reverse transcriptase in pseudogene insertion seems to be well proved.

2. RNA-Polymerase III Transcriptional Units

a. Ubiquitous, Dispersed Short Repetitive Sequences

The most interesting example of transposable elements which may be mobilized through reverse transcription is the ubiquitous, dispersed repeats. These are the B1 and B2 sequences of mouse (rodent) genome and the Alu sequences of the human genome.[29-33,111,111a]

The B1 and B2 sequences are present in the mouse genome, about 10^5 copies each. They are dispersed throughout the whole genome and can occur in nearly every cloned piece of DNA 5 to 10 kb long. The same is true of the Alu sequence though it is even more abundant ($\sim 3 \times 10^5$ copies per haploid human genome).

B1, B2, and Alu are actively transcribed, giving rise to about 1.5 to 2% of all hnRNA. Both strands of B1 and B2 are present in hnRNA and, as a result, dsRNA of 100 to 200 bp in length, or dsRNA-B, is formed. In fact, all dsRNA-B is transcribed from B1 and B2 sequences.[29,112,113] B1, B2, and Alu repeats have been sequenced, and their sizes are 130, 190, and 300 bp; Alu sequences correspond to a partially duplicated B1. All three sequences have been found to be flanked by direct repeats 10 to 20 bp long, which seem to originate from duplication of a target sequence during insertion. Therefore, some of these sequences have been proposed to move around the genome. While studying the polymorphic repetititve sequence (PR1), originally found in the spacer region of mouse ribosomal RNA genes, Kominami et al.[114] have found that B2 sequences (designated in their work as M2) are inserted within the PR1 sequence and flanked by a short direct repeat at both ends. This PR1 segment containing B2 is detected only in the BALB/c strain of laboratory mice. This implies that the B2 sequence has been inserted into the PR1 segment relatively recently during the evolution of mouse strains. Also, Grimaldi and Singer[115] describe an African green monkey-satellite sequence interrupted by an Alu element.

Small polydispersed circular DNA containing Alu sequences has been isolated from monkey cells grown in culture.[116] These circular DNAs seem to represent intermediates in the movement of the Alu-type genetic elements between chromosomal sites.[116,117] These facts support the idea that at least some of the B-type and Alu sequences are mobile in the genome.

An important common feature of these sequences is the presence of the regions reminiscent of the split RNA polymerase III promoter. The consensus sequence for the latter is deduced from the analysis of several RNA-polymerase III genes.[118-125] The location of a possible promoter within short repeats is such that the transcription should start from the beginning of the sequence.[126-128] At the opposite end, the short repeats contain an A-rich region. In addition, B2 contains the signals for RNA-polymerase III termination and several poly-adenylation signals, AATAAA.[31]

Considering these properties of short dispersed repeats, several authors postulated that their transposition should pass through the following steps: (1) transcription by RNA-polymerase III leading to the formation of small RNA with an A-rich 3' end; (2) reverse transcription of small RNA with a d(T) or a U-rich primer; and (3) insertion of the reverse transcripts into a novel position in chromosome with target duplication.[34]

By analogy with the properties of retroviruses, the DNAs generated by reverse transcription may function as intermediates in transposition. If that is the case, small circular DNAs containing mobile, dispersed genetic elements could represent transposable analogs of pro-viral DNAs.

Such a scheme suggests the existence of corresponding small RNAs. They can be easily obtained in vitro by transcription of a cloned sequence with RNA-polymerase III.[128] Recently, small $B2^+$ and $B1^+$ polyadenylated RNAs have been discovered in mouse cells in vivo.[128,129] These small RNAs consist of a B-repeat sequence and a poly(A) tail added post-transcriptionally. Several inhibitor tests show small $poly(A)^+ B2^+$ RNA to be transcribed by RNA polymerase III. It seems very probable that small $B2^+$ RNAs are intermediates in the B-type repeat transposition.

Small B^+ RNA comprises only a minor fraction among the transcripts containing B1 and B2 sequences. Small $B2^+$ RNA is much more abundant than small $B1^+$ RNA and their content in tumor cells is higher than that of normal tissues.

It should be added that in a sequence of short dispersed repeats one can find some homology to other functionally significant elements of the eukaryotic genome, such as consensus exon-introns, intron-exon splicing sites, and replication origins of papova viruses.[31,111]

Short, dispersed repeats possess certain species specificity. Although Alu and B1 are homologous, they clearly differ one from another. The human genome does not contain, or contains only few, copies of B2 homolog.

Among rodents, B1 does not always change in accordance with the evolutionary tree. For example, rat B1 is different from mouse B1, while in more distant species such as hamster and mole, B1 is quite similar to the mouse B1.[130] It is important that within the species the B-type sequences diverge rather slightly (3 to 5% of bases from the consensus sequence). This suggests the existence of certain mechanisms for B sequence correction or conversion. One possibility is the correction by recombination between genomic B sequences and extrachromosomal cDNA copies containing B-repeats. In this case, the sequence transcribed and reverse transcribed more efficiently would have advantages in the "fight for survival" and would create the "consensus sequence", which may differ from one species to another.

b. Large RNA-Polymerase III Transcriptional Units

We carried out the experiments on transcription inhibition using α-amanitin. Under those conditions when the transcription of small nuclear RNAs U1, U2, and U3 (known as being transcribed by RNA polymerase II) was completely blocked, some part of high molecular weight nonribosomal RNA still continued to incorporate label precursor.[131] Its content was estimated as one fourth to one third of total newly synthesized hnRNA and cytoplasmic poly(A)$^+$RNA. One possible explanation of this result is that such high molecular weight transcripts are synthesized by RNA polymerase III from promoters of B-type repeats scattered throughout the genome. If these repeats have lost RNA polymerase III termination signals (by deletion or mutation), RNA polymerase III-mediated transcription would continue from the B-type element through neighboring genomic sequences, giving rise to high molecular weight transcripts. Such long RNAs may sometimes serve as templates in reverse transcription, and thus, may induce transposition or any genomic element including regulatory and other noncoding sequences.

One can conclude from this section that many different elements can play the role of passive transposons, mobilized either by transposase or by reverse transcriptase. For the first type of transposition, the inverted repeats are important. For the second type, the transcription of the element (existence of promoter) and the presence of a sequence appropriate for reverse transcriptase primer binding are required.

IV. EFFECT OF MOBILE ELEMENTS ON GENOME FUNCTIONING

A. Mutation

1. Insertion Mutation

The most obvious result of mobile element insertion into the gene is gene mutation. If the insertion into a coding region took place, the gene would be inactivated with a high probability. For example, mdg 4 insertions into certain parts of the cut locus (III complementation group) led to formation of ct-lethals.[93,94]

If the mobile element is incorporated into the intron or into the regulatory region of the gene, the consequence of this event may be different depending on the exact position of the transposon. The insertion of a small transposon into an intron can remain without visible effect. For example, B1, B2, and Alu sequences can be found within the introns of several normally expressed genes.[132-134] The insertion of a longer transposon may influence the expression of the target gene. Incorporation of the IAP gene into an immunoglobulin gene intron led to a drastic decrease in its expression.[135,136] An intermediate situation can obviously exist (Table 1).

The appearance of foreign sequences in the regulatory region may strongly modify gene expression. Many visible mutations of *D. melanogaster* may possibly be explained in this way. For example, in the cut locus the II group of complementation is considered as a regulatory region.[137] We found that the mdg 4 insertion could induce a number of different visible mutations which were located in this area.[94] Similar data were obtained with other genes of *D. melanogaster*.[100,138]

Unstable mutation induced by transposon insertion can revert to the wild type,[10,11,93,139] sometimes without loss of transposon.[95] This phenomenon may depend on the change in the location of the mobile element within the gene or on the change of its orientation. Such reversions are also unstable.

Mobile elements can influence gene activity, even when located outside the genetic locus (position effects).[138] For example, the w^{DZL} mutation depends on the presence of a 13 kb insertion, including the FB-element to the right from the white locus.[140] This phenomenon may depend on direct or indirect interaction with the regulatory region of corresponding genes. In fact, such insertions can hardly be differentiated from those of the previous group because regulatory regions in eukaryotic genes are not yet well defined.

The influence of mobile elements inserted into flanking regions of genes is well illustrated in the case of yeasts. Ty1 insertion near the genetic locus may dramatically change its expression. Both up and down mutations were observed as well as changes in regulation.[55,141-144]

The question arises: what is the specific importance of transpositions in overall mutagenesis? It is not easy to differentiate between the mutations induced by insertion or by base change. The reversion rate in stable strains is similar in both cases. The most clear cut way to discriminate between two possibilities is a molecular analysis which is, of course, laborious. Less conclusive but much simpler is the genetic approach using the cross of M-females carrying the mutation with P-males. A drastic increase of the reversion rate strongly suggests the transposon nature of the mutation. For instance, stable strain ct^6g^2 could be destabilized by the cross with the MRh12/Cy strain. As a result, the reversion rate ($ct^6 \rightarrow$ ct^+ and $g^2 \rightarrow g^+$) reaches 10^{-3}.[94,95] Thus, both mutations seem to be transposon insertions. The dependence of ct^6 on the mdg 4 insertion has been shown.[47]

The analysis of different mutations in yeast showed that they were also frequently induced by Ty1 element insertion into the genes tested.[55]

Many of the well-known spontaneous mutations in *D. melanogaster* were induced by insertions in the white locus.[140] Several stable spontaneous y, f, bx, and ct-mutations were found to depend on the presence of gypsy (mdg 4) in a corresponding locus (Table 1).[47]

The majority of mutations at a singed locus isolated from natural populations were unstable.[145] A wide distribution of P-factor among flies in natural populations suggests that these mutations were induced by integration of either the P-element, as in the case of several sn and w-mutations,[89] or of other movable elements mobilized by hybrid dysgenesis.

Special analysis of variations in the heat shock gene among the wild population of *D. melanogaster* showed that the rate of insertion is ten times as high as that of base changes. Among several insertions, the P-element and mdg 4 (gypsy) were found.[146]

Mutations other than spontaneous mutations may be induced by movable elements. Several X-ray induced mutations are also transposon-dependent. For example, two fifths of all mutations suppressible by suppressor Hairy-wing which are known to depend on gypsy (mdg 4) insertion are induced by X-ray irradiation.[47] The irradiation of unstable strains by X-rays enhances the mutation rate more efficiently than irradiation of stable strains.[147] Irradiation

Table 1

MUTATIONS IN DIFFERENT LOCI OF *D. MELANOGASTER* AND THEIR RELATIONSHIP WITH MOVABLE ELEMENTS

Locus	Alleles	Transposon	Ref.
Yellow (1A)	y^2	Gypsy (mdg 4)	47
Hairy wing (1B)	Hw^1		47
Scute (1B)	sc^{D1}		47
	sc^{D2}		47
White (3C)	w^a	Copia	154
	$w^{hd81b11}$	Copia	35, 89
	$w^{\pi6}$	P	35, 89
	$w^{\pi12}$	P	35, 89
	$w^{hd80k17}$	P	35, 89
	w^{hd8169}	P	35, 89
	w^{hd81c2}	P	35, 89
	w^c	FB	99
	w^{DZL}	FB	100
	w^{TE}	FB	102
	w^{bf1}	Unknown transposon	140
	w^{sp}		140
	w^1		140
	w^{Zm}		140
	w^{IR1}	I factor	190
	w^{IR3}		190
	w^{a4}	BEL	160
Diminutive (3D)	dm^1	Gypsy (mdg 4)	47
Cut (7B)	ct^6		47
	ct^k		47
	ct^{MR2}	mdg 4	94, 95, 188
	ct^{MRpN}		94, 95, 188
	ct^{MRn}		94, 95, 188
	ct^{MRwR}		94, 95, 188
40 different lethals	ct^{MR1}		94, 95, 188
Singed	sn^w	P	35, 89
	sn^{MR2}		94, 95, 188
	sn^{MR17}		94, 95, 188
	sn^{MR110}		94, 95, 188
	sn^{MR12}		94, 95, 188
Lozenge (8D)		Gypsy (mdg 4)	47
Raspberry (9E)		P	198
RNA polymerase III lo-cus (10C)			199
Garnet (12BC)	g^{MR1}	Copia	188
Forked (15F)	f^1	Gypsy (mdg 4)	47
	f^5		47
	f^{kuhn}		47
Beadex (17AC)	B^2x		47
Cuticle protein gene (44D)		H,M,S, Beagle	189
Hairy (66D)	h^1	Gypsy (mdg 4)	47
	h^{D4}	P	200
Heat shock gene (87A7)		mdg 4 (gypsy), P	146
			146
		Unknown transposon	146
Bithorax (89E)	bx^3	Gypsy (mdg 4)	47
	bx^{341}		47
	bxd^1		47
	bxd^{551}		47
	bxd^{kuhn}		47

of the strain containing an unstable duplication in the white locus leads to w-transposition to the second chromosome.[148] EMS and γ-irradiation also increase the yield of unstable mutations 10 to 20 times reminiscent of viral and phage induction by mutagens.[149]

The role of transposition mutagenesis in higher eukaryotes is less clear. Most mutations in globin gene are found to be induced by base changes and deletions.[150,151] It may be, however, that in this particular case the selection favors these types of mutations as compared to insertions which should be lethal. The down mutation in the immunoglobin genes induced by the IAP gene insertion[135,136] has already been mentioned. More data are necessary for a final conclusion.

2. Suppression of Insertion Mutations

An interesting feature of several mutations in *D. melanogaster* induced by mobile elements is that they can be suppressed. Recessive suppressor su(Hw) suppresses several alleles in different loci.[152] Among 19 suppressible alleles in 10 elements, gypsy (mdg 4) is detected.[47] Thus, the mutations in su(Hw) eliminate the inactivating effect of mdg 4 on several genes.

Besides su(Hw), a number of different suppressors are found in *D. melanogaster* which act on the mutations in several loci. These are su(w^a), su(f), su(pr), and su(S).[153] It seems that in the case of su(w^a) the suppression depends on interaction with the mobile element copia, which is responsible for w^a mutation.[154]

The mechanism of suppressor action remains obscure. It is clear that the mobile elements are not removed from their place upon the suppressor.

Other types of controlling elements are those inhibiting the transposition process. For example, MR-factors inducing the chromosome cleavage, male recombination, and unstable mutagenesis are suppressed by a number of suppressors.[155] The strains inducing male recombination contain a high number of P-elements, possibly needed to overcome the suppressor effects. Thus, both the transposition and the mutation produced by the latter may be, to some extent, controlled by different genetic elements.

3. Reversions and Deletions

The excision of the transposable element by a transposase may be rather precise and in this case it leads to reversion to the wild type. The P-element, FB-sequences, and mdg elements may serve as substrates for such precise excision. The elements lacking inverted repeats at their termini probably cannot be excised in such a way.

Precise excision of the P-element, including a duplicated target sequence, has been shown by Rubin et al.[35] The P-factor-dependent transposase seems to be involved.[91] Precise excision of the FB-element, also eliminating the end duplication, occurred at a high rate. Insertion mutation w containing a 10 kb FB transposable element reverts, in the homozygous strain, in the absence of hybrid dysgenesis and in the absence of P-element. Thus, the precise excision of 10 kb FB-element[100] is induced by a cellular transposase different from the P-element-induced enzyme. The high rate of precise excision of transposable elements in *Drosophila* contrasts to the situation in prokaryotes where the precise excision of transposons occurs much more rarely than other events.[156]

4. The Excision of Transposons is Not Always Precise

Several cases of nonprecise excision should be considered. First, one of LTRs of mdg may not be removed. Solo LTRs are usually absent from the *D. melanogaster* genome. However, in a few cases the one LTR of mdg 4 remained in the genome after the removal of the element itself.[47] In contrast, such events should happen quite frequently in yeast where a number of solo LTRs (solo sequences) are scattered throughout the genome.[19,55] Second, only partial excision of a transposable element may take place. As an example, w^{DZL} mutation which resulted from insertion of two FB elements can be considered. The excision of the sequence between two repeats oriented in the same direction usually takes place and as a result, the eye color unstable revertants are formed with the frequency of 10^{-3}. These still contain different parts of the original 13 kb insertion.[138,140,15]

Finally, nonprecise excision may lead to the formation of deletions in genetic loci where mobile elements had been located. For example, in unstable w and sn alleles deletions of various lengths ranging from intrageneous to long (covering several loci) were generated. One break was always placed in the region of the mobile element.[7,158,159] Thus, many deletions may be explained by nonprecise excision of mobile elements.

If the existence of inverted repeats at the termini is important for excision, one should consider the case where two oppositely oriented short ubiquitous repeats (B1, B2, or Alu) may be inserted close to each other. In this case, they may create, together with the sequence located in between, an element potentially excisible by transposase. Recently, genome instability in the region of human DNA enriched in Alu repeat sequences has been clearly demonstrated.[117] In this case, DNA rearrangements, involving restriction fragment length polymorphism and variations in copy number, are detected in the human genome by blot hybridization with a cloned segment of human DNA initially present in a cluster of Alu repeat sequences. These rearrangements involve both extrachromosomal, circular duplex DNAs and integrated sequences, indicating the presence of transposable elements in human cells.

In some cases, the transposon-dependent deletions can appear as a result of unequal crossing over rather than of nonprecise excision. One may suggest that if two similar transposons are located in the different but close sites on the chromosome, the unequal crossing over would occur leading to small deletions or tandem duplications. Unequal crossing over was demonstrated in the case of mdg BEL[160] and for the FB-element[161] inserted into the white locus, as well as in the case of Ty1 in the yeast.[162] Unequal crossing over is known to play an important role in the initial stages of evolution. More frequently the excised transposon is then lost, but sometimes it may be inserted at another site on chromosomes, inducing further genome reorganization.[7,35,101,148]

B. Gene Activation

1. Mdg Elements

As was mentioned above, some mobile elements, in particular mdg elements and retroviral proviruses, contain the sequences initiating transcription located within long terminal repeats (LTRs). A typical LTR element contains a TATA-box, the signal for transcription initiation; an AATAAA sequence, the signal for RNA polyadenylation; and a special control sequence, or enhancer which in some way remarkably increases the transcription level.[27] These elements are responsible for active transcription of the mdg element. As both LTRs are identical, one can expect that the right-handed LTR should induce the transcription extending to the flanking host sequence. In this way, the insertion of the mdg element may activate the adjacent gene. The enhancers are known to be elements acting at a rather long distance and in both directions. If LTR carries a powerful enhancer, the mdg insertion could activate the transcription of adjacent genes from their own promoters (TATA boxes). The LTRs of numerous mdgs (retroviruses) possess quite different enhancing potentials and so their effect should vary over a wide range.[163,164]

An interesting point is that the transcription from LTR elements can be controlled by several regulatory factors. The most striking example is the LTR of mouse mammary tumor virus (MMTV): the transcription beginning in its LTR elements is under the strict control of steroid hormones.[165,166] The enhancers of MMTV contain hormone binding sites.

Some yet unknown control factors trigger the LTRs of IAP genes in early embryogenesis and switch them off in the tissues of adults.[83]

In the yeast, the expression of several genes changes from inducible to constitutive as a result of Ty1 insertion close to these genes.[167-169] The orientation of Ty1 and the gene is always head-to-head.

The Ty1 LTR sequence is not detected within the transcript. Possibly LTR-induced activation depends on the enhancer present in LTR.

With *D. melanogaster* mdg elements, the transcripts containing LTR sequences and non-mdg sequences could be detected by hybridization of Northern blots with the separated strands of LTR elements and other mdg sequences.[170]

Numerous cases of adjacent activation have been obtained with retroviral proviruses. The insertion of ALV near the protooncogene c-myc is shown to activate it, resulting in the development of lymphomas. The orientation of the virus and the gene may be different. The transcription of c-myc could start from LTR or from its own myc promoters.[171] Thus, the enhancer function of LTR is most important.

Several other examples of oncogene activation by transposed mdg-like elements were presented. For instance, one mouse lymphoma is induced by insertion of a copy of the IAP gene close to the gene c-mos.[107,108]

A number of artificial constructions consisting of LTR elements linked to other structural genes were prepared and found to be very efficient in cell transformation. One example is a chimera consisting of mdg 1 LTR and the gene v-src, which can efficiently induce oncogenic transformation upon transfection of 3T3 cells.[172]

Thus, one may conclude that mdg elements can act as strong activators of gene expression, sometimes in a controllable fashion.

2. Other Mobile Sequences

It is possible that some other mobile elements may also activate gene expression. Short, dispersed sequences of the B-type contain RNA polymerase III promoters (see above), but it is not clear whether long transcripts beginning from their promoters really exist and whether they may be used further for translation.

Sequences of the P-element type may also be active inducers of transcription, though the data are not yet available.

C. Gene Modification

Dispersed repeats of the B-type contain a number of signal sequences within their short nucleotide sequence. The insertion of such elements into a certain gene may introduce functional control elements which change the gene.

We know that many transcriptional units contain B-type sequences which are transcribed.[112,113] The content of B1 and B2 sequences in mouse hnRNA is very high. Most of these are located within the introns and destroyed during processing. However, a small but significant part of B1 and B2 survive the processing and may be detected within mature mRNA molecules.[129,173-176] cDNA clones containing B1 and B2 sequences have been selected and some of them have been sequenced.[176] A full-length B2 sequence is detected at the 3' end of some mRNAs. In B2 sequences AATAAA is located before poly(A), thus playing the role of the polyadenylation signal.[177] A similar result has been recently obtained with B1-containing mRNA.[178] It is possible that the incorporation of B2 into an exon generates a new polyadenylation signal, separating the gene from its downstream part.

It is interesting that all B2 sequences found in mRNA are oriented in the same way, i.e., only one strand that contains the polyadenylation signal survives the hnRNA processing.[177]

The presence of homologies to exon-intron and intron-exon junctions in Alu, B1, B2, and several other repetitive sequences of different organisms may be functionally important.[30-34,111] It has been suggested that the insertion of such repeats induces the appearance of novel splicing sites which modifies the general splicing scheme for a certain gene. The experimental data, however, are absent.

Movable elements may dissect genes. An example of such dissection, though induced not by the movable element but by the chromosome translocation, is breakage of gene c-myc in several mouse lymphomas. The 5' noncoding exon is removed and the promoter-like sequence located in the first intron takes the function of transcription initiation. As a result, shorter mRNA with a smaller 5' noncoding sequence is transcribed. Such RNA is probably more efficient in translation.[179-181]

D. Gross Genome Rearrangements

At least some mobile elements (for example, mdg 1) are preferentially located in the regions of the so-called intercalary heterochromatin.[14,15] These areas of chromosomes are characterized by frequent ectopic pairing, late replication, and under-replication. As a result they often become the sites of chromosome breakage. Chromosomal translocation may take place in these regions. One can speculate that mobile repetitive sequences are directly involved in chromosomal translocations and thus in gross chromosome rearrangements.

The presence of mobile elements in mutable alleles induces chromosome translocations in corresponding sites.[182-184] In P-M dysgenic hybrids obtained with males of strains π^2, a number of chromosomal rearrangements appear.[185] It is found that the X- chromosome of the π^2 strain contains five hot spots for chromosome cleavage.[185,186] At these particular sites the P-elements are detected by *in situ* hybridization.[89]

Another possible gross effect of the mobile element on the genome is dissemination of replication initiation sequences which may be present in B1 and Alu.[45,111] At the moment, it is not clear whether the sequences homologous to papovavirus replication origins present in B1 and Alu can indeed serve for replication initiation in the cell. Only some indirect evidence has been presented. If this is true, the insertion of these elements should change the replication machinery. This could create conditions favoring the amplification of genetic material in certain regions of the genome; however, all these suggestions are still speculative.

E. Creation of Novel Genetic Material

Amplifcation of movable elements may create large amounts of novel genetic material. It is well known that in several species the total content of repetitive sequences (possibly movable elements) is very high. Amplification of mobile elements of the mdg type was observed in *D. melanogaster* cultured cells. A 10- to 15-fold amplification of several mdg families occurred during cell cultivation in vitro.[16,20,36,52] This amplification may give the cells some advantage in the competitive, nonregulated growth in the culture.

Novel genetic material can be created in different ways. We have already mentioned the possibility of large transcript formation using the promoters of mobile elements (B-type sequences). The reverse transcription of such transcripts followed by DNA insertion may lead to dissemination of any sequence present in the genome to new sites. In this way the related genes can appear in different chromosomes.

The third possible mechanism for the increase of genome size is unequal crossing over which may also be induced by repetitive mobile elements located together on the chromosome. In this case, the tandem duplication of a certain part of the genome can be obtained. For example, insertion sequences and tandem repetitions as sources of variation in the MIF-1 dispersed repeat family of the mouse are clearly demonstrated.[187]

V. EVOLUTIONARY SIGNIFICANCE OF MOBILE ELEMENTS

A. Variability: Multiple Mutagenesis Induced by Transpositions and its Evolutionary Significance

1. Variability Induced by Mobile Elements

The data presented above show that movable elements may strongly influence the work of the genome. The genes may be inactivated or activated. Their regulation may be completely altered or only slightly, quantitatively changed. Genes may be deleted or duplicated either tandemly or by being transferred to a distant position. Such effects create variability and polymorphism in the population. It is now clear that the movable elements are one of the major sources of differences between individuals of the same species.

If the changes are selectively advanced, they may be spread through the population. However, in this respect the mutagenesis induced by movable elements does not differ much from other types of mutagenesis.

One of the major difficulties in construction of evolution models is the difficulty of explaining the appearance of new complex features by accumulation of a number of unrelated changes, each of which taken separately, does not add to the fitness of the individual.

The recent data show that mutagenesis induced by transposable elements may help to solve this problem. Actually the mutagenesis induced by transpositions is at least sometimes multiple and possesses a saltatory character.[95,188]

2. Saltatory Character of Transposition and Multiple Mutagenesis

Multiple mutagenesis was observed in a destabilized strain carrying a ct^{MR2} mutation obtained by hybrid dysgenesis. This mutation, induced by mdg 4 insertion, remained unstable for many generations, giving rise to stable and unstable reversions and novel mutations in the cut locus.[94,95] Stable reversions were associated with the mdg 4 removal from the cut locus (7B region of X-chromosome). In other cases mdg 4 remained in 7B, changing its location within the locus. Virtually all derivatives of ct^{MR2} were again unstable, giving rise to reversions and novel mutations. The rate of reversion to ct^+ or mutation changes varied in the range from 10^{-2} to 10^{-4} in different strains. However, in those flies in which the changes in the cut locus took place, several other transposition events occurred at the same time. They resulted in the appearance of novel unstable mutations and in the insertion of the P-element, several mdg elements, and FB-sequences into the novel sites of the X-chromosome (in ct^{MR2} strain P-elements were present only in autosomes). A number of independent transposition events could be observed in one fly, i.e., in one and the same germ cell. Appearance of multiple mutations in clusters showed that the transpositions occurred at the premeiotic stage.

In animals taken at random from the ct^{MR2} strain, the changes in the location of transposons in the X-chromsome are usually not seen, while in revertants to ct^+ they are almost obligatory. Thus, the ct^+ reversion seems to be a marker for transposition processes. In the novel strains obtained, the location of the transposon remains rather stable until the novel "transposition burst" occurs in one or few germ cells.[95,188]

As an example of triple mutation events one could mention the following: $ct^{MR2} \rightarrow ct^+$, y, w, or $ct^{MR2}r^{MR1} \rightarrow cm\ ct^{PN}r^+$. The number of transpositions visible in the X-chromosome upon *in situ* hybridization with copia, mdg 1,2,3,4 P-elements, and FB-elements, constitutes from 12 to 20[188] (Figure 1). Considering that only some families of the mobile elements have been studied and that only one chromsome has been analyzed, it can be concluded that the real number of transpositions should reach several hundred. However, it is not clear whether we can extrapolate the data obtained on *D. melanogaster* to other eukaryotes. Further experiments are necessary.

Obviously, multiple mutagenesis induced by the "transposition bursts" should create organisms differing in many respects from their parents and thus, may serve as a basis for "evolutionary jumps".

3. Specificity of Transposition Resulting in Partially Site-Specific Mutagenesis

Data are being accumulated that indicate that the sites of mobile element insertions are not absolutely random, but display more or less prominent site specificity. It has been shown previously that mdg 1 is usually located in the sites of intercalary heterochromatin. The P-

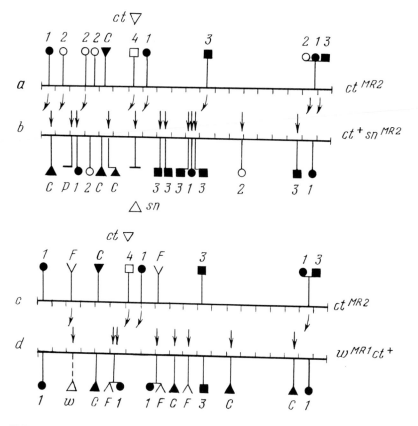

FIGURE 1.   Schematic presentation of the results of *in situ* hybridization of ct[MR2], ct[+]sn[MR2], and w[MR1]ct[+] X-chromosomes with different probes ("transposition bursts"). (a, c) Distribution of hybridization sites along ct[MR2] X-chromosome; (b) that of ct[+]sn[MR2] X-chromosome; and (d) that of w[MR1]ct[+] X-chromsome. In (1, 2) the hybridization was performed with mdg 1 (1), mdg 2 (2), mdg 3 (3), mdg 4 (4), copia (C), and P-element (P); in (3, 4) with mdg 1, mdg 3, mdg 4, copia, P-element (same symbols) and FB-element (F). The triangles (▽) indicate the location of mutations characteristic for the strains analyzed. The arrows show the transposition events: the loss of mobile elements (↙) and their insertion (↓).

element is shown to be located preferentially in the loci singed, white, and a few others.[35,95] mdg HMS Beagle and 297 were usually framed by TATA duplication, and thus are incorporated in TATA-boxes of several genes.[51,189]

Another type of transposition specificity is the appearance of certain pairs or even triads of different events which may be induced by different mobile elements. In the course of studies on unstable ct[MR2] mutation and its derivatives we have encountered a number of such specific combinations. For instance, the y and w mutations have frequently appeared together (at least in one third of total y and in one half of total w mutations). Change ct[MR2] → ct[MRpN] is frequently accompanied by appearance of a cm mutation. In strain ct[MR2]r[MR1], the transition to ct[MRpN]r[+] is always associated with unstable mutations in the carmine locus. Such type specificity may depend upon the relative site-specificity of transposons and on certain control genomic factors which mobilize that specific set of transposable elements.[95]

Partial specificity of transposition may determine to some extent the direction of induced variability.

B. Regulation of Transposition

1. P-Elements and Other Factors Encoding Transposases

In the above-mentioned experiments, the transposition bursts occurred in the genome destabilized by P-M dysgenesis. The role of the P-element that generates the transposase machinery was obvious. The number of P-elements present in the genome was important for the level of the mobility. Also, the strains which lost the transposon mobility could again be activated by a cross with the strain carrying multiple P-factors.

However, the presence of the P-factor is necessary, but not sufficient, for the transposition bursts in the destabilized strains. P-elements are present in all germ cells of a given organism but only in few of them do these bursts occur.[188]

The nature of such "triggers" in destabilized strains is quite obscure. The P-factors themselves seem to encode, in addition to transposase, the proteins with regulatory functions inhibiting transposition. One possibility is that the "transposition triggers" interfere with them. Further studies are necessary.

The P-factor is certainly not the only genetic element encoding the transposase and destabilizing the genome.

At the moment, two hybrid dysgenesis systems in *D. melanogaster* are known. The second systems is the I-R (inducer-receptor) system. The active factor in this system is an I-element, a sequence 5.4 kb long, which has been cloned and is now under investigation.[190] It also possibly encodes transposase activity. Several other elements which act in a similar way may exist in *D. melanogaster* and other eukaryotes. The existence of the strains containing or not containing P-elements or other similar elements creates the possibility of hybrid dysgenesis. Obviously dysgenesis should be especially frequent at the border between different population areals.

2. Transposition Depending on Reverse Transcriptase

Reverse transcriptase genes probably exist in any cell. The retrovirus-like particles of *D. melanogaster* possess the reverse transcriptase activity.[59,191] The same is true for mouse intracisternal A-particles.[83] At least some mdgs (retroviral proviruses) are nondefective and can produce the active enzyme. The expression of such elements varies in a wide range, but at least at certain stages it is high enough. For example, IAP genes are efficiently expressed during early embryogenesis, though adult tissues may allow some IAP-gene transcription. A striking activation of IAP gene expression takes place in many tumors.[83]

In spite of these facts, the reverse transcriptase-induced transpositions again seem to be very rare. For instance, in mouse and chicken strains the endogenous proviruses are very stable and new copies appear in the genome very rarely, at least in germ cells. Possibly some repressing by immune factors exists in the cell. On the other hand, the infection with exogenous virus always leads to integration into novel sites in the genome. The horizontal species-to-species transition of provirus genomes has also been well established.

The factors activating reverse transcriptase-dependent transposition are to be investigated.

3. The Role of Horizontal "Infection"

It seems probable that the appearance in a population of a new active transposon, such as the P-element or exogenous retrovirus, is an important factor in transposition activation leading, in particular, to transposition "bursts". Such phenomena can be observed not only in the laboratory but also in nature. For example, sometimes in the wild population of *D. melanogaster* a certain mutation becomes most abundant.

In the 1930s, a striking increase of mutagenesis in yellow and white loci in different natural populations of *D. melanogaster* was noticed.[192,193] In the late 1960s, a similar mutation burst was observed but this was abnormal abdomen mutation.[194,195] In 1973, the sn mutation became dominant in natural populations which were separated by thousands of kilometers.[10] Such synchronous mutation changes in natural populations were designated as "mutation vogue".[195]

The vogue for sn-mutation may be explained by the expansion of the P-element among natural populations in the 1970s. The locus singed is one of the favorite loci for the P-element insertion.[35,95]

The change of the "mutation fashion" might depend upon the change of the transposon which conquers the population if possessing its own site specificity.

C. Reproductive Isolation and Speciation

Besides multiple variability, the movable elements probably play an important role in the appearance of reproductive isolation, which is a prerequisite for the formation of new species.

Some simple genomic rearrangements could lead to speciation as it takes place in the case of *D. melanogaster*.[155] The inversions as well as other genomic rearrangements can readily be induced by transposable elements of a different nature.

The accumulation of many changes in the location of movable elements in the course of several "transposition bursts" in the strains with a destabilized genome could also prevent the appropriate meiotic chromosome pairing in the hybrids with the parental strain.

Thus, after one or several "bursts" which were simply due to the appearance of mobile elements at new sites or because of gross genomic rearrangements, a new strain could appear which is unable to mate with the parental. If the genetic changes provide this strain with selective advantages, it survives and gives rise to a new species.

The difference between movable elements of various species may also contribute to their reproductive isolation. Such differences may appear, for example, instantly as a result of the infection of a particular population with the novel transposable element such as retrovirus or the P-factor, or the change of mobile element may be the result of its own evolution passing through a number of mutation-conversion steps. In this case, in a separate group of animals a changed mobile element may appear.

The crosses between the groups with diverged mobile elements may encounter several difficulties. First, crossing over may be blocked. Second, the absence of appropriate immune factors for a new transposon may induce a high level of possibly lethal transpositions and the appearance of sterility (one of the major features of hybrid dysgenesis).

D. Formation of Novel Genetic Material

Amplification of mobile elements themselves increases the amount of genetic material in the genome. This extra material may be used for the formation of new genes in evolution. We have already mentioned that almost any part of the genome may act as a passive transposable element mobilized through reverse transcription of RNA-polymerase II or III transcripts. Such novel construction may appear in any region of the genome. In this way, the complete genes, their parts, larger sequences of DNA, or small signal sequences can be spread throughout the genome. As a result, new genes or some novel constructions consisting of parts of different genes or of their regulatory elements can appear.

Finally, the movable elements may become the sites of unequal crossing over that can lead to the tandem duplication of genetic material.[160] In several cases, gene duplication may have proceeded in this way.

VI. CONCLUSION

The data presented in this chapter allow us to conclude that the movable elements play an important role in evolution and species formation. First, they are responsible for genetic variability among the representatives of the same species, providing material for natural selection. Second, transposition bursts could give rise to multiple changes in organisms leading up to "evolutionary jumps". Third, movable elements may immediately induce the formation of novel genes in the course of transposition or by providing the material for evolution. Fourth, the mobile elements may be responsible for reproductive isolation and speciation.

Movable genetic elements are usually considered as selfish DNA or genomic parasites.[196,197] In several cases, they may fulfill some useful functions in the work of the genome (as was mentioned previously in the chapter), and therefore seem to represent symbionts rather than parasites.

Probably more important is that they make the eukaryotic genome very flexible and therefore able to evolve with a rather high speed. Therefore, the organisms with mobile elements should gain an advantage over those lacking such elements in changing conditions of environment. This may be one of the reasons that these elements are so numerous and widespread among living creatures.

REFERENCES

1. **Demerec, M. and Reddish, G.,** A frequently mutating character in *Drosophila virilis, Proc. Natl. Acad. Sci. U.S.A.,* 12, 11, 1926.
2. **Demerec, M.,** Miniature-alpha — a second frequently mutating character in *Drosophila virilis, Proc. Natl. Acad. Sci. U.S.A.,* 13, 249, 1926.
3. **Demerec, M.,** The behaviour of mutable gene, Proc. V. Int. Congr. Genet., Berlin, 1927, 183.
4. **McClintock, B.,** Controlling elements and the gene, *Cold Spring Harbor Symp. Quant. Biol.,* 21, 197, 1956.
5. **McClintock, B.,** The control of gene action in maize, *Brookhaven Symp. Biol.,* 18, 162, 1965.
6. **Starlinger, P. and Saedler, K.,** Insertion mutations in microorganisms, *Biochemie,* 54, 177, 1983.
7. **Green, M. M.,** The genetics of mutable gene at the *white* locus of *Drosophila melanogaster, Genetics,* 56, 467, 1967.
8. **Green, M. M.,** Some observations and comment on mutable and mutator genes in *Drosophila, Genetics,* 73 (Suppl.), 187, 1973.
9. **Green, M. M.,** Genetic instability in *Drosophila melanogaster:* mutable miniature, *Mutation Res.,* 29, 77, 1975.
10. **Golubovsky, M. D.,** Instability of *singed* locus in *Drosophila melanogaster:* phenotypically mutant and normal alleles mutating in accordance with the rule "all or none", *Genetics (USSR),* 13, 845, 1977.
11. **Golubovsky, M. D.,** Paramutation phenomenon in *Drosophila, Genetics (USSR),* 13, 1605, 1977.
12. **Golubovsky, M. D., Ivanov, Yu, M., and Green, M. M.,** Genetic instability in *Drosophila melanogaster:* putative multiple insertion mutants at the *singed* bristle locus, *Proc. Natl. Acad. Sci. U.S.A.,* 74, 2973, 1977.
13. **Finnegan, D. J., Rubin, G. M., Young, H. W., and Hogness, D. S.,** Repeated gene families in *Drosophila melanogaster, Cold Spring Harbor Symp. Quant. Biol.,* 42, 1053, 1978.
14. **Ilyin, Y. V., Tchurikov, N. A., Ananiev, E. V., Ryskov, A. P., Yenikolopov, G. N., Limborska, S. A., Maleeva, N. E., Gvozdev, V. A., and Georgiev, G. P.,** Isolation and characterization of eukaryotic DNA fragments containing structural genes and adjacent sequences, *Cold Spring Harbor Symp. Quant. Biol.,* 42, 959, 1978.
15. **Ananiev, E. V., Gvozdev, V. A., Ilyin, Y. V., Tchurikov, N. A., and Georgiev, G. P.,** Reiterated genes with varying location in intercalary heterochromatin of *Drosophila melanogaster, Chromosoma,* 70, 1, 1978.
16. **Potter, S. S., Brorein, W. J., Dunsmuir, P., and Rubin, G. M.,** Transposition of element of the 412, *copia* and 297 dispersed repeated gene families in Drosophila, *Cell,* 17, 415, 1979.
17. **Strobel, E., Dunsmuir, P., and Rubin, G. M.,** Polymorphism in the chromosomal locations of element of the 412, *copia* and 297 dispersed repeated gene families in *Drosophila, Cell,* 17, 429, 1979.
18. **Young, M. W.,** Middle repetitive DNA: a fluid component of the *Drosophila* genome, *Proc. Natl. Acad. Sci. U.S.A.,* 76, 6274, 1979.
19. **Cameron, J. R., Loh, E. Y., and Davis, R. W.,** Evidence for transposition of dispersed repetitive DNA families in yeast, *Cell,* 16, 739, 1979.
20. **Ilyin, Y. V., Chmeliauskaite, V. G., Ananiev, E. V., and Georgiev, G. P.,** Isolation and characterization of a new family of mobile dispersed genetic element, mdg3, in *Drosophila melanogaster, Chromosoma,* 81, 27, 1980.
21. **Spradling, A. C. and Rubin, G. M.,** Drosophila genome organization: conserved and dynamic aspects, *Annu. Rev. Genet.,* 15, 219, 1981.
22. **Pierce, D. A. and Lucchesi, J. C.,** Analysis of a dispersed repetitive DNA sequence in isogenic lines of Drosophila, *Chromosoma,* 82, 471, 1981.

23. **Rubin, G. M.,** Dispersed repetitive DNAs in Drosophila, in *Mobile Genetic Elements,* Shapiro, J. A., Ed., Academic Press, New York, 1983, 329.

24. **Georgiev, G. P., Ilyin, Y. V., Ryskov, A. P., and Kramerov, D. A.,** Mobile dispersed genetic elements and their possible relation to carcinogenesis, *Mol. Biol. Rep.,* 6, 249, 1980.

25. **Finnegan, D. J.,** Transposable elements and proviruses, *Nature (London),* 292, 800, 1981.

26. **Kugimiya, W., Ikenaga, H., and Saigo, K.,** Close relationship between the long terminal repeats of avian leukosis-sarcoma virus and copia-like movable genetic elements of Drosophila, *Proc. Natl. Acad. Sci. U.S.A.,* 80, 3139, 1983.

27. **Varmus, H. E.,** Retrovirus, in *Mobile Genetic Elements,* Shapiro, J. A., Ed., Academic Press, New York, 1983, 411.

28. **Calos, M. and Miller, J. H.,** Transposable elements, *Cell,* 20, 579, 1980.

29. **Kramerov, D. A., Grigoryan, A. A., Ryskov, A. P., and Georgiev, G. P.,** Long double-stranded sequences of nuclear pre-mRNA (ds-RNA-B) consist of a few highly abundant classes of sequences: evidence from DNA cloning experiments, *Nucleic Acids Res.,* 6, 697, 1979.

30. **Krayev, A. S., Kramerov, D. A., Skryabin, K. G., Ryskov, A. P., Bayev, A. A., and Georgiev, G. P.,** The nucleotide sequence of the ubiquitous repetitive DNA sequence B1 complementary to the most abundant class of mouse fold-back RNA, *Nucleic Acids Res.,* 8, 1201, 1980.

31. **Krayev, A. S., Marusheva, T. A., Kramerov, D. A., Ryskov, A. P., Skryabin, K. G., Bayev, A. A., and Georgiev, G. P.,** Ubiquitous transposon-like repeats B1 and B2 of the mouse genome: B2 sequencing, *Nucleic Acids Res.,* 10, 7461, 1982.

32. **Houck, C. M., Rinehart, F. P., and Schmid, C. W.,** A ubiquitous family of repeated DNA sequences in the human genome, *J. Mol. Biol.,* 132, 289, 1979.

33. **Rubin, C. M., Houck, C. M., Deininger, P. L., Friedmann, T., and Schmid, C. W.,** Partial nucleotide sequence of the 300-nucleotide interspersed repeated human DNA sequences, *Nature (London),* 284, 372, 1980.

34. **Jagadeeswaran, P., Forget, B. C., and Weissman, S. M.,** Short interspersed repetitive DNA elements in eucaryotes: transposable DNA elements generated by reverse transcription of RNA polIII transcripts? *Cell,* 26, 141, 1981.

35. **Rubin, G. M., Kidwell, M. G., and Bingham, P. M.,** The molecular basis of P-M hybrid dysgenesis: the nature of induced mutations, *Cell,* 29, 987, 1982.

36. **Ilyin, Y. V., Chmeliauskaite, V. G., Ananiev, E. V., Lyubomirskaya, N. V., Kulguskin, V. V., Bayev, A. A., Jr., and Georgiev, G. P.,** Mobile dispersed element MDG1 of *Drosophila melanogaster:* structural organization, *Nucleic Acids Res.,* 8, 5333, 1980.

37. **Ilyin, Y. V., Chmeliauskaite, V. G., Kulguskin, V. V., and Georgiev, G. P.,** Mobile dispersed genetic element mdg1 of *Drosophila melanogaster:* transcription pattern, *Nucleic Acids Res.,* 8, 5347, 1980.

38. **Kulguskin, V. V., Ilyin, Y. V., and Georgiev, G. P.,** Mobile dispersed genetic element mdg1 of *Drosophila melanogaster:* nucleotide sequence of long terminal repeats, *Nucleic Acids Res.,* 9, 3451, 1981.

39. **Kulguskin, V. V., Ilyin, Y. V., and Georgiev, G. P.,** Mobile dispersed genetic elements mdg1: nucleotide sequence of long terminal repeats, *Genetics (USSR),* 19, 869, 1982.

40. **Bayev, A. A., Krayev, A. S., Lyubomirskaya, N. V., Ilyin, Y. V., Skryabin, K. G., and Georgiev, G. P.,** The transposable element mdg3 in *Drosophila melanogaster* is flanked with the perfect and mismatched inverted repeats, *Nucleic Acids Res.,* 8, 3263, 1980.

41. **Levis, R., Dunsmuir, P., and Rubin, G. M.,** Terminal repeats of the Drosophila transposable element *copia:* nucleotide sequence and genomic organization, *Cell,* 21, 581, 1980.

42. **Dunsmuir, P., Brorein, W. J., Jr., Simon, M. A., and Rubin, G. M.,** Insertion of the Drosophila transposable element *copia* generates a 5 base pair duplication, *Cell,* 21, 575, 1980.

43. **Flavell, A. J., Ruby, S. W., Toole, J. J., Roberts, B. E., and Rubin, G. M.,** Translation and developmental regulation of RNA encoded by the eukaryotic transposable element *copia, Proc. Natl. Acad. Sci. U.S.A.,* 77, 7107, 1980.

44. **Flavell, A. J., Levis, R., Simon, M., and Rubin, G. M.,** The 5'-termini of RNAs encoded by the transposable element *copia, Nucleic Acids Res.,* 9, 6279, 1981.

45. **Joung, M. V. and Schwartz, H. E.,** Nomadic gene families in Drosophila, *Cold Spring Harbor Symp. Quant. Biol.,* 45, 629, 1981.

46. **Will, B. M., Bayev, A. A., and Finnegan, D. J.,** Nucleotide sequence of terminal repeats of 412 transposable elements of *Drosophila melanogaster.* A similarity to proviral long terminal repeats and its implications for the mechanism of transposition, *J. Mol. Biol.,* 153, 897, 1981.

47. **Modolell, J., Bender, W., and Meselson, M.,** *Drosophila melanogaster* mutations suppressible by the suppressor *Hairy*-wing are insertions of a 7.3-kilobase movable element, *Proc. Natl. Acad. Sci. U.S.A.,* 80, 1678, 1983.

48. **Bayev, A. A., Lyubomirskaya, N. V., Dzhumagaliev, E. B., Ananiev, E. V., Amiantova, I. G., and Ilyin, Y. V.,** Structural organization of transposable element mdg4 from *Drosophila melanogaster* and nucleotide sequence of its long terminal repeats, *Nucl. Acids Res.,* 12, 3707, 1984.

49. **Dzhumagaliev, E. V., Bayev, A. A., Jr., and Ilyin, Y. V.,** Primary structure of long terminal repeats and adjacent portions of mobile dispersed gene MDG4 in *Drosophila melanogaster, Dokl. Acad. Nauk U.S.S.R.,* 273, 214, 1983.

50. **Scherer, G., Tschudi, C., Perera, J., Delius, H., and Pirrota, V.,** B104, a new dispersed repeated gene family in *Drosophila melanogaster* and its analogy with retroviruses, *J. Mol. Biol.,* 157, 435, 1982.

51. **Ikenaga, H. and Saigo, K.,** Insertion of a movable genetic element, 297, into the T-A-T-A box for the H3 histone gene in *Drosophila melanogaster, Proc. Natl. Acad. Sci. U.S.A.,* 79, 4143, 1982.

52. **Tchurikov, N. A., Ilyin, Y. V., Skryabin, K. G., Ananiev, E. V., Bayev, A. A., Jr., Krayev, A. S., Zelentsova, E. S., Kulguskin, V. V., Lyubomirskaya, N. V., and Georgiev, G. P.,** General properties of mobile dispersed genetic elements in *Drosophila* melanogaster, *Cold Spring Harbor Symp. Quant. Biol.,* 45, 655, 1981.

53. **Ilyin, Y. V., Chmeliauskaite, V. G., and Georgiev, G. P.,** Double-stranded sequences in RNA *Drosophila melanogaster:* relation to mobile dispersed genes, *Nucleic Acids Res.,* 8, 3439, 1980.

54. **Georgiev, G. P., Ilyin, Y. V., Chmeliauskaite, V. G., Ryskov, A. P., Kramerov, D. A., Skryabin, K. G., Krayev, A. S., Lukanidin, E. M., and Grigoryan, M. S.,** Mobile dispersed genetic elements and other middle repetitive DNA sequences in the genomes of *Drosophila* and mouse: transcription and biological significance, *Cold Spring Harbor Symp. Quant. Biol.,* 45, 641, 1981.

55. **Roeder, G. S. and Fink, G. R.,** Transposable elements in yeast, in *Mobile Genetic Elements,* Shapiro, J. A., Ed., Academic Press, New York, 1983, 300.

56. **Gafner, J. and Philippsen, P.,** The yeast transposon Ty1 generates duplications of target DNA upon insertion, *Nature (London),* 286, 414, 1980.

57. **Farabaugh, P. J. and Fink, G. R.,** Insertion of the eukaryotic transposable element Ty1 creates a 5-base pair duplication, *Nature (London),* 286, 352, 1980.

58. **Weiss, R. A., Teich, N., Varmus, H. E., and Coffin, J. M., Eds.,** Molecular Biology of Tumor Viruses: RNA Tumor Viruses, Cold Spring Harbor Laboratory, Cold Spring Harbor, N.Y., 1982.

59. **Shiba, T. and Saigo, K.,** Retrovirus-like particles containing RNA homologous to the transposable element *copia* in *Drosophila melanogaster, Nature (London),* 302, 109, 1983.

60. **Ono, M., Cole, M. D., White, A. T., and Huang, R. C. C.,** Sequence organization of cloned intracisternal A particle genes, *Cell,* 21, 465, 1980.

61. **Kuff, E. L., Smith, L. A., and Lueders, K. K.,** Intracisternal A-particle genes in Mus musculus: a conserved family of retrovirus-like elements, *Mol. Cell Biol.,* 1, 216, 1981.

62. **Keshet, E., Shaul, Y., Kaminchik, J., and Aviv, H.,** Heterogeneity of ''virus-like'' genes encoding retrovirus-associated 30S RNA and their organization within the mouse genome, *Cell,* 20, 431, 1980.

63. **Dolberg, D. S., Bacheler, L. T., and Fan, H.,** Endogenous type C retroviral sequences of mice are organized in a small number of virus-like classes and have been acquired recently, *J. Virol.,* 40, 96, 1981.

64. **Coffin, J.,** Structure of the retroviral genomes, in *Molecular Biology of Tumor viruses: RNA Tumor Viruses,* Weiss, R. A., Teich, N., Varmus, H. E., and Coffin, J. M., Eds., Cold Spring Harbor Laboratory, Cold Spring Harbor, N.Y., 1982, 261.

65. **Georgiev, G. P., Kramerov, D. A., Ryskov, A. P., Skryabin, K. G., and Lukanidin, E. M.,** Dispersed repetitive sequences in eukaryotic genomes and their possible biological significance, *Cold Spring Harbor Symp. Quant. Biol.,* 47, 1109, 1983.

66. **Coffin, J.,** Endogenous proviruses, in *Molecular Biology of Tumor Viruses: RNA Tumor Viruses,* Weiss, R. A., Teich, N., Varmus, H. E., and Coffin, J. M., Eds., Cold Spring Harbor Laboratory, Cold Spring Harbor, N.Y., 1982, 1109.

67. **Teich, N.,** Taxonomy of retroviruses, in *Molecular Biology of Tumor Viruses: RNA Tumor Viruses,* Weiss, R. A., Teich, N., Varmus, H. E., and Coffin, J. M., Eds., Cold Spring Harbor Laboratory, Cold Spring Harbor, N.Y., 1982, 25.

68. **Martin, M. A., Bryan, T., Rasheed, S., and Kahn, A. S.,** Identification and cloning of endogenous retroviral sequences present in human DNA, *Proc. Natl. Acad. Sci. U.S.A.,* 78, 4892, 1981.

69. **Fanning, T. G.,** Characterization of a highly repetitive family of DNA sequences in the mouse, *Nucleic Acids Res.,* 10, 5003, 1982.

70. **Gebhard, W., Meitinger, T., Höchtl, J., and Zachau, H. G.,** A new family of interspersed repititive DNA sequences in the mouse genome, *J. Mol. Biol.,* 157, 453, 1982.

71. **Wilson, R. and Storb, Y.,** Association of two different repetitive DNA elements near immunoglobulin light chain genes, *Nucleic Acids Res.,* 11, 1803, 1983.

72. **Noda, M., Kurihara, M., and Takano, T.,** Retrovirus-related sequences in human DNA: detection and cloning of sequences which hybridize with the long terminal repeat of baboon endogenous virus, *Nucleic Acids Res.,* 10, 2805, 1982.

73. **Meunier-Rotival, M., Soriano, P., Cuny, G., Strauss, F., and Bernardi, G.,** Sequence organization and genomic distribution of the major family of interspersed repeats of mouse DNA, *Proc. Natl. Acad. Sci. U.S.A.,* 79, 355, 1982.

74. **Repaske, R., O'Neill, R. R., Steele, P. E., and Martin, M. A.,** Characterization and partial nucleotide sequences of endogenous type C retrovirus segments in human chromosmal DNA, *Proc. Natl. Acad. Sci. U.S.A.,* 80, 678, 1983.

75. **Kole, L. B., Haynes, S. R., and Jelinek, W. R.,** Discrete and heterogeneous high molecular weight RNAs complementary to a long dispersed repeat family (a possible transposon) of human DNA, *J. Mol. Biol.,* 165, 257, 1983.

76. **Gilboa, E., Mitra, S. W., Goff, S., and Baltimore, D.,** A detailed model of reverse transcription and tests of crucial aspects, *Cell,* 18, 93, 1979.

77. **Varmus, H. E. and Swanstrom, R.,** Replication of retroviruses, in *Molecular Biology of Tumor Viruses: RNA Tumor Viruses,* Weiss, R. A., Teich, N., Varmus, H. E., and Coffin, J. M., Eds., Cold Spring Harbor Laboratory, Cold Spring Harbor, N.Y., 1982, 369.

78. **Tulchinskyi, E., Grigoryan, M. S., Kramurov, D. A., Tulchinsky, E. M., Revasova, E. S., and Lukanidin, E. M.,** Activation of putative transposition in intermediate formation in tumor cells, *EMBO J.,* 4, 2209, 1985.

79. **Flavell, A. J. and Ish-Horowicz, D.,** Extrachromosomal circular copies of the eukaryotic transposable element *copia* in cultured Drosophila cells, *Nature (London),* 292, 591, 1981.

80. **Flavell, A. J. and Ish-Horowicz, D.,** The origin of extrachromosomal circular *copia* elements, *Cell,* 34, 415, 1983.

81. **Lyubomirskaya, N. V., Gorelova, T. V., Ilyin, Y. V., and Schuppe, N. G.,** Circular DNAs corresponding to mobile dispersed genes in *Drosophila melanogaster* culture cells, *Dokl. Acad. Nauk USSR,* 276, 246, 1984.

82. **Ilyin, Y. V., Schuppe, N. G., Lyubomirskaya, N. V., Gorelova, T. V., and Arkhipova, I. R.,** Circular copies of mobile dispersed genetic elements in cultured *Drosophila melanogaster* cells, *Nucleic Acids Res.,* 12, 7517, 1984.

83. **Kelly, F. and Condamine, H.,** Tumor viruses and early mouse embryos, *Biochim. Biophys. Acta,* 651, 105, 1982.

84. **Kuff, E. L., Leuder, K. K., and Scolnick, E. M.,** Nucleotide sequence relationship between intracisternal type A particles of Mus musculus and an endogenous retrovirus (M432) of Mus cervicolar, *J. Virol.,* 28, 66, 1978.

85. **Dowsett, A. and Young, M. W.,** Differing levels of dispersed repetitive DNA among closely related species of *Drosophila, Proc. Natl. Acad. Sci. U.S.A.,* 79, 4570, 1982.

86. **Dowsett, A. P.,** Closely related species of *Drosophila* can contain different libraries of middle repetitive DNA sequences, *Chromosoma,* 88, 104, 1983.

87. **Evgen'ev, M. B., Yenikolopov, G. N., Peunova, N. I., and Ilyin, Y. V.,** Transposition of mobile genetic elements in interspecific hybrids of Drosophila, *Chromosoma,* 85, 375, 1982.

88. **Bingham, P. M., Levis, R., and Rubin, G. M.,** Cloning of DNA sequences from the *white* locus of *D. melanogaster* by a novel and general method, *Cell,* 25, 693, 1981.

89. **Bingham, P. M., Kidwell, M. G., and Rubin, G. M.,** The molecular basis of P-M hybrid dysgenesis: the role of the P-element, a P-strain-specific transposon family, *Cell,* 29, 995, 1982.

90. **O'Hare, K. O. and Rubin, G. M.,** Structures of P-transposable elements and their sites of insertion and excision in the *Drosophila melanogaster* genome, *Cell,* 34, 25, 1983.

91. **Spradling, A. C. and Rubin, G. M.,** Transposition of cloned P-elements into *Drosophila* germ line chromsomes, *Science,* 218, 341, 1981.

92. **Rubin, G. M. and Spradling, A. C.,** Genetic transformation of *Drosophila* with transposable element vectors, *Science,* 218, 348, 1982.

93. **Gerasimova, T. I.,** Genetic instability of the *cut* locus of *Drosophila melanogaster* induced by the *MRh*12 chromosome, *Mol. Gen. Genet.,* 184, 544, 1981.

94. **Gerasimova, T. I., Ilyin, Y. V., Mizrokhi, L. J., Semjonova, L. V., and Georgiev, G. P.,** Mobilization of transposable element mdg4 by hybrid dysgenesis generates a family of unstable *cut* mutations in *Drosophila melanogaster, Mol. Gen. Genet.,* 193, 488, 1984.

95. **Gerasimova, T. I., Matunina, L. V., Ilyin, Y. V., and Georgiev, G. P.,** Simultaneous transposition of different mobile elements: relation to multiple unstable mutagenesis in *Drosophila melanogaster, Mol. Gen. Genet.,* 194, 517, 1984.

96. **Potter, S. S., Truett, M., Phillips, M., and Moher, A.,** Eukaryotic transposable elements with inverted terminal repeats, *Cell,* 20, 639, 1980.

97. **Truett, M. A., Jones, R. S., and Potter, S. S.,** Unusual structure of the FB family of transposable elements in *Drosophila Cell,* 24, 253, 1981.

98. **Potter, S. S.,** DNA sequence of a fold back transposable element in *Drosophila, Nature (London),* 297, 201, 1982.

99. **Levis, R., Collins, M., and Rubin, G.,** FB elements are the common basis for the instability of the w^{DZL} and w^c *Drosophila* mutations, *Cell,* 30, 551, 1982.

100. **Collins, M. and Rubin, G.,** High frequency precise excision of the *Drosophila* fold back transposable element, *Science,* 303, 259, 1983.

101. **Ising, G. and Block, K.,** Derivation-dependent distribution of insertion sites for a *Drosophila* transposon, *Cold Spring Harbor Symp. Quant. Biol.,* 45, 527, 1981.

102. **Goldberg, M. L., Paro, R., and Gehring, W. L.,** Molecular cloning of the *white* locus region of *Drosophila melanogaster* using a large transposable element, *EMBO J.,* 1, 93, 1982.

103. **Vanin, E. F., Goldberg, G. I., Tucker, P. W., and Smithies, O.,** A mouse α-globin-related pseudogene lacking intervening sequences, *Nature (London),* 286, 222, 1980.

103a. **Nishioka, Y., Leder, A., and Leder, P.,** Unusual α-globin-like gene that has clearly lost both globin intervening sequences, *Proc. Natl. Acad. Sci. U.S.A.,* 77, 2806, 1980.

104. **Hollis, G. F., Hieter, P. A., McBride, O. W., Swan, D., and Leder, P.,** Processed genes: a dispersed human immunoglobulin gene bearing evidence of RNA-type processing, *Nature (London),* 296, 321, 1982.

105. **Karin, M. and Richards, R. J.,** Human metallothionein genes — primary structure of the metallothionein-II gene and a related processed gene, *Nature (London),* 299, 797, 1982.

106. **Lemischka, I. and Sharp, P. A.,** The sequences of an expressed rat α-tubulin gene and a pseudogene with an inserted repetitive element, *Nature (London),* 300, 330, 1982.

107. **Rechavi, G., Givol, D., and Canaani, E.,** Activation of a cellular oncogene by DNA rearrangement: possible involvement of an IS-like element, *Nature (London),* 300, 607, 1982.

108. **Kuff, E., Feenstra, A., Lueders, K., Rechavi, G., Givol, D., and Canaani, E.,** Homology between an endogenous viral LTR and sequences inserted in an activated cellular oncogene, *Nature (London),* 302, 547, 1983.

109. **Van Arsdell, S. W., Danison, R. A., Bernstein, L. B., Weiner, A. M., Manser, T., and Gesteland, R. F.,** Direct repeats flank three small nuclear RNA pseudogenes in the human genome, *Cell,* 26, 11, 1981.

110. **Denison, R. A., Van Arsdell, S. W., Bernstein, L. B., and Weiner, A. M.,** Abundant pseudogenes for small nuclear RNAs are dispersed in the human genome, *Proc. Natl. Acad. Sci. U.S.A.,* 78, 810, 1981.

110a. **Bernstein, L. B., Mount, S. M., and Weiner, A. M.,** Pseudogenes for human small nuclear RNA U3 appear to arise by integration of self-primed reverse transcripts of the RNA into new chromosomal sites, *Cell,* 32, 461, 1983.

111. **Jelinek, W. R., Toomey, T. P., Leinwand, L., Duncan, C. H., Choudary, P. V., Biro, P. A., Weissman, S. M., Rubin, C. M., Houck, C. M., Deininger, P. L., and Schmid, C. W.,** Ubiquitous, interspersed repeated sequences in mammalian genomes, *Proc. Natl. Acad. Sci. U.S.A.,* 77, 1398, 1980.

111a. **Haynes, S. R., Tomey, T. P., Leinwand, L., and Jelinek, W. R.,** The Chinese hamster Alu equivalent sequence: a conserved highly repetitious, interspersed deoxyribonucleic acid sequence in mammals has a structure suggestive of a transposable element, *Mol. Cell. Biol.,* 1, 573, 1981.

112. **Ryskov, A. P.,** Double-stranded structures and complementary sequences in nuclear precursors of mRNA, *Adv. Mod. Biol. (U.S.S.R.),* 86, 163, 1978.

113. **Georgiev, G. P. and Ryskov, A. P.,** The structure of transcription and the regulation of transcription, in *Eukaryotic Gene Regulation,* Vol. 1, Kolodny, G. M., Ed., CRC Press, Boca Raton, Fla., 1980, 33.

114. **Kominami, R., Muramatsu, M., and Moriwaki, K.,** A mouse type 2 Alu sequence (M2) is mobile in the genome, *Nature (London),* 301, 87, 1983.

115. **Grimaldi, G. and Singer, M. F.,** A monkey Alu sequence is flanked by 13-base pair direct repeats of an interrupted α-satellite DNA sequence, *Proc. Natl. Acad. Sci. U.S.A.,* 79, 1497, 1982.

116. **Krolewski, J. J., Bertelsen, A. H., Humayun, M. Z., and Rush, M. G.,** Members of the Alu family of interspersed, repetitive DNA sequences are in the small circular DNA population of monkey cell grown in culture, *J. Mol. Biol.,* 154, 399, 1982.

117. **Calabretta, B., Robberson, D. L., Barrera-Saldana, H. A., Lambrou, T. P., and Saunders, G. F.,** Genome instability in a region of human DNA enriched in Alu repeat sequences, *Nature (London),* 296, 219, 1982.

118. **Kressmann, A., Hofstetter, H. D., Capua, E., Grosschedl, R., and Birnstiel, M. L.,** A tRNA gene of Xenopus laevis contains at least two sites promoting transcription, *Nucleic Acids Res.,* 7, 1749, 1979.

119. **Bogenhagen, D. F., Sakonju, S., and Brown, D. D.,** A control region in the center of the 5S RNA gene directs specific initiation of transcription. II. The 3' border of the region, *Cell,* 19, 27, 1980.

120. **Fowlkes, D. M. and Shenk, T.,** Transcriptional control regions of the adenovirus VA1 RNA gene, *Cell,* 22, 405, 1980.

121. **Sakonju, S., Bogenhagen, D. F., and Brown, D. D.,** A control region in the center of the 5S RNA gene directs specific initiation of transcription. I. The 5' border of the region, *Cell,* 19, 13, 1980.

122. **Sharp, S., DeFranco, D., Dingermann, T., Farrell, P., and Söll, D.,** Internal control regions for transcription of eukaryotic tRNA genes, *Proc. Natl. Acad. Sci. U.S.A.,* 78, 6657, 1981.

123. **Galli, G., Hofstetter, H., and Birnstiel, M. L.,** Two conserved sequence blocks within eukaryotic tRNA genes are major promoter elements, *Nature (London),* 294, 626, 1981.

124. **Hofstetter, H., Kressmann, A., and Birnstiel, M. L.,** A split promoter for a eucaryotic tRNA gene, *Cell,* 24, 573, 1981.

125. **Korn, L. J.,** Transcription of *Xenopus* 5S ribosomal RNA genes, *Nature (London),* 295, 101, 1982.

126. **Duncan, C., Biro, P. A., Choundary, P. V., Elder, J. T., Wang, R. R. C., Forget, B. G., de Riel, J. K., and Weissmann, S. M.,** RNA polymerase III transcription units are interspersed among human non α-globin genes, *Proc. Natl. Acad. Sci. U.S.A.,* 76, 5095, 1979.

127. **Elder, J. T., Pan, J., Duncan, C. H., and Weissmann, S. M.,** Transcriptional analysis of interspersed repetitive polymerase III transcription units in human DNA, *Nucl. Acid Res.,* 9, 1171, 1981.

128. **Haynes, S. R. and Jelinek, W. R.,** Low molecular weight RNAs transcribed *in vitro* by RNA polymerase III from Alu-type dispersed repeats in Chinese hamster DNA are also found *in vivo, Proc. Natl. Acad. Sci. U.S.A.,* 78, 6130, 1981.

129. **Kramerov, D. A., Lekakh, I. V., Samarina, O. P., and Ryskov, A. P.,** The sequences homologous to major interspersed repeats B1 and B2 of mouse genome are present in mRNA and small cytoplasmic poly(A)$^+$RNA, *Nucleic Acids Res.,* 10, 7477, 1982.

130. **Kramerov, D. A. and Lomidze, N. V.,** "Paradoxical" evolution of the individual transcribed repeated sequence DNA (B1) in rodents, *Proc. Acad. Sci. U.S.S.R.,* 255, 1005, 1980.

131. **Kramerov, D. A., Tillib, S. V., Lekakh, I. V., Ryskov, A. P., and Georgiev, G. P.,** Small poly(A)-containing B2 RNA: biosynthesis and distribution in cytoplasma, *J. Mol. Biol.,* submitted.

132. **Barta, A., Richards, R. J., Baxter, L. D., and Shine, J.,** Primary structure and evolution of rat growth hormone gene, *Proc. Natl. Acad. Sci. U.S.A.,* 78, 4867, 1981.

133. **Kioussis, D., Eiferman, F., Van de Rijn, P., Gorin, M. B., Ingram, R. S., and Tilgham, S. M.,** The evolution of α-fetoprotein and albumin. II. The structures of the α-fetoprotein and albumin genes in the mouse, *J. Biol. Chem.,* 256, 1960, 1981.

134. **Tsukada, T., Watanabe, Y., Nakai, Y., Imura, H., Nakanishi, S., and Numa, S.,** Repetitive DNA sequences in the human corticotropin-β-lipotropin precursor gene region: Alu family members, *Nucleic Acids Res.,* 10, 1471, 1982.

135. **Hawley, R., Shulman, M., Murialdo, H., Gilson, D., and Hozumi, N.,** Mutant immunoglobulin genes have repetitive DNA elements inserted into their intervening sequences, *Proc. Natl. Acad. Sci. U.S.A.,* 79, 7425, 1982.

136. **Kuff, E., Feenstra, A., Lueders, K., Smith, L., Hawley, R., Hozumi, N., and Shulman, M.,** Intra-cisternal A-particle genes as movable elements in the mouse genome, *Proc. Natl. Acad. Sci. U.S.A.,* 80, 1992, 1983.

137. **Johnson, T. K. and Judd, B. H.,** Analysis of the *cut* locus of *Drosophila melanogaster, Genetics,* 92, 485, 1979.

138. **Levis, R. and Rubin, G. M.,** The unstable w^{DZL} mutation of *Drosophila* is caused by a 13 kilobase insertion that is imprecisely excised in phenotypic revertants, *Cell,* 30, 543, 1982.

139. **Rasmuson, B. and Green, M. M.,** Genetic instability in *Drosophila melanogaster:* a mutable tandem duplication, *Mol. Gen. Genet.,* 133, 249, 1974.

140. **Zachar, Z. and Bingham, P. M.,** Regulation of *white* locus expression: the structure of mutant alleles at the *white* locus of *D. melanogaster, Cell,* 30, 529, 1982.

141. **Roeder, G. S. and Fink, G. R.,** DNA rearrangements associated with a transposable element in yeast, *Cell,* 21, 239, 1980.

142. **Roeder, G. S., Farabauph, P. J., Chaleff, D. T., and Fink, G. R.,** The origins of gene instability in yeast, *Science,* 209, 1375, 1980.

143. **Williamson, V. M., Young, E. T., and Griacy, M.,** Transposable elements associated with constitutive expression of alcohol dehydrogenase II, *Cell,* 23, 605, 1981.

144. **Errede, B., Cardillo, T. S., Wever, G., and Sherman, F.,** Studies on transposable elements in yeast. I. ROAM mutations causing increased expression of yeast genes: their activation by signals directed toward conjugation functions and their formation by insertion of Ty1 repetitive elements, *Cold Spring Harbor Symp. Quant. Biol.,* 45, 593, 1981.

145. **Ivanov, Y. M. and Golubovsky, M. D.,** Increase of mutation rate and appearance of unstable alleles of the *singed* gene in natural populations of *D. melanogaster, Genetics (U.S.S.R.),* 13, 655, 1974.

146. **Leigh Brown, A. J.,** Variation at the 87A heat shock locus in *Drosophila melanogaster, Proc. Natl. Acad. Sci. U.S.A.,* 80, 5330, 1983.

147. **Sohels, F. H. and Eeken, J. C. J.,** Influence of the MR (mutator) factor on X-ray induced genetic damage, *Mutat. Res.,* 83, 201, 1981.

148. **Rasmuson, B., Montell, I., Rasmuson, A., Svahlin, H., and Westerberg, B. M.,** Genetic instability in *Drosphila melanogaster.* Evidence for regulation excision and transposition at the *white* locus, *Mol. Gen. Genet.,* 177, 567, 1980.

149. **Rasmuson, B., Svahlin, H., Rasmuson, A., Montell, I., and Olafson, H.,** The use of a mutationally unstable X-chromosome in *D. melanogaster* for mutagenicity testing, *Mutat. Res.,* 54, 33, 1978.

150. **Weatherall, D. J. and Clegg, J. B.,** Thalassemia revisited, *Cell,* 29, 7, 1982.
151. **Limborska, S. A.,** Globin gene systems, in *Molecular Biology,* Vol. 19, Kiselev, L. L., Eds., VINITI, Moscow, 1982, 84.
152. **Lewis, E. B.,** *Drosophila Inf. Serv.,* 23, 59, 1949.
153. **Lindsley, D. L. and Grell, E. H.,** Genetic variations of *Drosophila melanogaster,* in Carnegie Institute of Washington Publication, No627, Washington, D.C., 1968.
154. **Bingham, P. M. and Judd, B. H.,** A copy of the *copia* transposable element is very tightly linked to the w^a allele at the *white* locus of *Drosophila melanogaster, Cell,* 25, 705, 1981.
155. **Thompson, J. N. and Woodruff, R. C.,** Mutator genes: pace makers of evolution, *Nature (London),* 274, 317, 1978.
156. **Kleckner, N. A.,** Transposable elements in prokaryotes, *Cell,* 11, 11, 1977.
157. **Levis, R., Bingham, P. M., and Rubin, G. M.,** Physical map of the *white* locus of Drosophila melanogaster, *Proc. Natl. Acad. Sci. U.S.A.,* 79, 564, 1982.
158. **Simmons, M. J. and Lim, J. K.,** Site specificity of mutations arising in dysgenic hybrids of *D. melanogaster, Genetics,* 77, 6042, 1980.
159. **Green, M. M.,** Genetic instability in *Drosophila melanogaster:* deletion induction by insertion sequence, *Proc. Natl. Acad. Sci. U.S.A.,* 79, 5367, 1982.
160. **Goldberg, M. L., Sheen, J. Yu., Gehring, W. J., and Green, M. M.,** Unequal crossing-over associated with a symmetrical synapsis between nomadic elements in the Drosophila, *Proc. Natl. Acad. Sci. U.S.A.,* 80, 5017, 1983.
161. **Karess, R. E. and Rubin, G. M.,** A small tandem duplication is responsible for the unstable white-ivory mutation in *Drosophila, Cell,* 30, 63, 1983.
162. **Roeder, G. S.,** Unequal crossing-over between yeast transposable element, *Mol. Gen. Genet.,* 190, 117, 1983.
163. **Khoury, G. and Gruss, R.,** Enhancer elements, *Cell,* 33, 313, 1983.
164. **Tsichlis, P. N. and Coffin, J. M.,** Recombinants between endogenous and exogenous avian tumor viruses: role of the C region and other portions of the genome in the control of replication and transformation, *J. Virol.,* 33, 238, 1980.
165. **Govindan, M. V., Spiess, E., and Majors, J. E.,** Purified glucocorticoid receptor-hormone complex from rat liver cytosol binds specifically to cloned mouse mammary tumor virus long terminal repeats *in vitro, Proc. Natl. Acad. Sci. U.S.A.,* 79, 5157, 1982.
166. **Chandler, V. L., Maler, B. A., and Yamamoto, K.,** DNA sequences bound specifically by glucocorticoid receptor *in vitro* render a heterologous promoter hormone responsive *in vivo, Cell,* 33, 489, 1983.
167. **Williamson, V. M., Young, E. T., and Ciriacy, M.,** Transposable elements associated with constitutive expression of yeast alcohol dehydrogenase II, *Cell,* 23, 605, 1981.
168. **Errede, B., Cardillo, T. S., Sherman, F., Dubois, E., Deschamps, J., and Wiame, J.-M.,** Mating signals control expression of mutations resulting from insertion of transposable repetitive element adjacent to diverse yeast genes, *Cell,* 22, 427, 1980.
169. **Errede, B., Cardillo, T. S., Wever, G., and Sherman, F.,** Studies of transposable elements in yeast. I. ROAM mutations causing increased expression of yeast genes: their activation by signals directed toward conjugation functions and their formation by insertion of Ty1 repetitive elements, *Cold Spring Harbor Symp. Quant. Biol.,* 45, 593, 1980.
170. **Ilyin, Y. V.,** unpublished data, 1983.
171. **Payne, G. S., Bishop, J. M., and Varmus, H. E.,** Multiple arrangements of viral DNA and an activated host oncogene inbursal lymphomas, *Nature (London),* 295, 209, 1982.
172. **Ninkina, N.,** personal communication, 1983.
173. **Ryskov, A. P., Limborska, S. A., and Georgiev, G. P.,** Hybridization of mRNA and pre-mRNA with the sequences forming double-stranded structures in pre-mRNA, *Mol. Biol. Rep.,* 1, 215, 1973.
174. **Naora, H. and Whitelam, J. M.,** Presence of sequences hybridizable to dsRNA in cytoplasmic mRNA molecules, *Nature (London),* 256, 756, 1975.
175. **Ryskov, A. P., Tokarskaya, O. V., and Georgiev, G. P.,** Direct demonstration of a complementarity between mRNA and double-stranded sequences of pre-mRNA, *Mol. Biol. Rep.,* 2, 353, 1976.
176. **Lalanne, J. L., Bregegere, F., Delarbre, C., Abastado, J. P., Gachelin, G., and Kourilsky, P.,** Comparison of nucleotide sequences of mRNAs belonging to the mouse H-2 multigene family, *Nucleic Acids Res.,* 10, 1039, 1982.
177. **Ryskov, A. P., Ivanov, P. L., Kramerov, D. A., and Georgiev, G. P.,** Mouse ubiquitous B2 repeat in polysomal and cytoplasmic poly(A) $^+$RNAs: unidirectional orientation and 3'-end localization, *Nucleic Acids Res.,* 11, 6541, 1983.
178. **Ryskov, A. P.,** unpublished data, 1983.
179. **Hamlyn, P. H. and Rabbits, T. H.,** Translocation joining c-myc and immunoglobulin 1 genes in a Burkitt lymphoma revealing a third exon in the c-myc oncogene, *Nature (London),* 304, 135, 1983.

180. **Maguire, R. T., Robins, T. S., Thorgeirsson, S. S., and Heilman, C. A.,** Expression of cellular myc and mos genes in undifferentiated B cell lymphomas of Burkitt and non-Burkitt types, *Proc. Natl. Acad. Sci. U.S.A.,* 80, 1947, 1983.

181. **Adams, J. M., Gerondakis, S., Webb, E., Corcoran, L. M., and Cory, S.,** Cellular myc oncogene is altered by chromsome translocation to an immunoglobulin locus in murine plasmacytoma and its rearranged similarity in human Burkitt lymphomas, *Proc. Natl. Acad. Sci. U.S.A.,* 80, 1982, 1983.

182. **Lefevre, G., Jr. and Green, M. M.,** Genetic duplication in the *white*-split interval of the X-chromosome in *Drosophila melanogaster, Chromsoma,* 36, 391, 1972.

183. **Bingham, P. M.,** A novel dominant mutant allele at the *white* locus of *Drosophila melanogaster* is mutable, *Cold Spring Harbor Symp. Quant. Biol.,* 45, 519, 1981.

184. **Lim, J. K.,** Site-specific intrachromsomal rearrangements in *Drosophila melanogaster:* cytological evidence for transposable elements, *Cold Spring Harbor Symp. Quant. Biol.,* 45, 553, 1981.

185. **Berg, R. L., Engels, W. R., and Kreher, R. A.,** Site specific X-chromosome rearrangements from hybrid dysgenesis in *Drosophila melanogaster, Science,* 210, 427, 1980.

186. **Engels, W. R. and Preston, C. R.,** Identifying P factors in *Drosophila* by means of chromsome breakage hotspots, *Cell,* 26, 421, 1981.

187. **Brown, S. D. M. and Piechaczyk, M.,** Insertion sequences and tandem repetitions as sources of variation in a dispersed repeat family, *J. Mol. Biol.,* 165, 249, 1983.

188. **Gerasimova, T. I., Mizrokhi, L. Yu., and Georgiev, G. P.,** Transposition bursts in genetically unstable *Drosophila melanogaster, Nature (London),* 309, 714, 1984.

189. **Snyder, M. P., Kimbrell, D., Hunkapiller, M., Hill, R., Fistrom, J., and Davidson, N.,** A transposable element that splits the promoter region inactivates a *Drosophila* cuticle protein gene, *Proc. Natl. Acad. Sci. U.S.A.,* 79, 7430, 1981.

190. **Bucheton, A.,** The IR system of hybrid dysgenesis in *D. melanogaster,* Molecular study of I factor, 8th Eur. *Drosophila* Res. Conf., Cambridge, England, September, 1983.

191. **Heine, H. W., Kelly, D. C., and Avery, R. J.,** The detection of intracellular retrovirus-like entities in *Drosophila melanogaster* cell culture, *J. Gen. Virol.,* 49, 385, 1980.

192. **Berg, R. L.,** On the relationship between mutability and selection in natural population of *Drosophila melanogaster, J. Gen. Biol. (U.S.S.R.),* 9, 299, 1948.

193. **Duseeva, N. D.,** On the specificity and periodicity of the mutability in natural population of *Drosophila melanogaster, Dokl. Acad. Nauk U.S.S.R.,* 60, 665, 1948.

194. **Berg, R. L.,** A sudden and synchronous increase in the frequency of abnormal abdomen in the geographically isolated populations of *D. melanogaster, Drosophila Inf. Serv.,* 48, 94, 1972.

195. **Golubovsky, M. D., Ivanov, Y. N., Zakharov, I. K., and Berg, R. L.,** Synchronous and parallel changes in the gene pool at natural populations of the trait fly *Drosophila melanogaster, Genetika (U.S.S.R.),* 10, 72, 1974.

196. **Orgel, L. E. and Crick, F. H. C.,** Selfish DNA: the ultimate parasite, *Nature (London),* 284, 604, 1980.

197. **Doolittle, W. F. and Sapienza, C.,** Selfish genes, the phenotype paradigm and genome evolution, *Nature (London),* 284, 601, 1980.

198. **Eeken, J. C. J.,** The nature of MR-mediated site-specific lesions in P-element free and P-element carrying X-chromosomes in *Drosophila melanogaster,* 8th Eur. *Drosophila* Res. Conf., Cambridge, England, September, 1983.

199. **Searles, L. L., Jokerst, R. S., Bingham, P. M., Voelker, R. A., and Greenleaf, A. L.,** Molecular cloning of sequences from a *Drosophila* RNA polymerase II locus by P-element transposon tagging, *Cell,* 31, 585, 1982.

200. **Ish-Horowicz, D., Howard, K. R., Ingham, P. W., Leigh Brown, A. J., and Pinchin, S. M.,** Molecular and genetic studies of the *hairy* locus, a gene required for embryonic segmentation, 8th Eur. *Drosophila* Res. Conf., Cambridge, England, September 1983.

Chapter 3

RIBOSOMAL RNA PROCESSING SITES

Robert J. Crouch and Jean-Pierre Bachellerie

TABLE OF CONTENTS

I. INTRODUCTION

Synthesis of equimolar quantities of three of the RNA molecules found in ribosomes is generally achieved via the production of a single rRNA precursor molecule.[1-7] In prokaryotes, in particular *Escherichia coli,* the precursor contains 16S, 23S, and 5S. The prokaryotic pre-rRNA sometimes contains one or more tRNA molecules. In contrast, eukaryotic pre-rRNA molecules include a 5.8S rRNA molecule, do not contain 5S rRNA sequences, and only a few contain tRNA sequences. The general organization of prokaryotic and nuclear eukaryotic rRNA genes is depicted in Figure 1. Mature rRNA species are produced by a series of cleavages which are collectively known as rRNA processing. Recent reviews on rRNA processing in *E. coli,*[4] mitochondria,[5,6] and eukaryotic organisms[1,2,7] have appeared. The details of rRNA processing are relatively well understood for *E. coli* but are less well known for eukaryotic rRNA. In *E. coli,* the processing pathway has been analyzed both biochemically and genetically, whereas in eukaryotes most studies have been of a biochemical nature. For several years it was felt that a general unique pathway for processing of rRNA existed in each organism. However, recent information[8,9] suggests that there may be several alternate pathways within a given cell. It must be pointed out that many of the "interme-diates" have not been shown to be precursors by pulse-chase experiments. The discussion of a processing site in terms of evolutionary conservation or comparison is still somewhat tenuous since in several instances it is not known with absolute certainty whether the "intermediate" that defines the processing site is a true intermediate. We can define proc-essing sites at the ends of mature rRNA species in the sense that these ends are produced via processing, but we are unable to make any general conclusion concerning the nature of the enzymatic process or processes that are involved in making up the processing site. Indeed, with the diversity that biological systems exhibit in their synthesis and make-up of rRNAs, it may be that diverse mechanisms and processing sites will be typical rather than exceptional.

As a case in point, a few organisms have interruptions in their large ribosomal RNA. *Drosophila* 28S rRNA genes exist in both an interrupted and uninterrupted form.[10-13] In some *Tetrahymena* species,[15,16] all copies of their large rRNA interrupted, and *Neurospora* mitochondrial rRNA contain an intron. Each organism seems to handle the intron in different ways. In *Drosophila* the 28S genes with introns are not transcribed.[14] In *Tetrahymena* the RNA sequence of the intron is chemically reactive and excision and ligation of the RNA can be carried out in vitro in the absence of any proteins.[16] The reaction to remove the intron of the large rRNA of *Neurospora*[18] has some similarities to the *Tetrahymena* intron removal in that they both add an extra guanosine residue to the intron sequence during excision. Unlike the *Tetrahymena* case, the *Neurospora* mitochondrial RNA seems to require formation

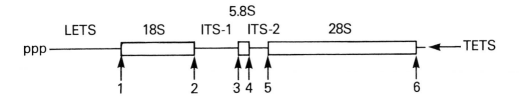

FIGURE 1. Eukaryotic transcription unit. rRNA transcriptional unit for typical eukaryote. LETS = Leader External Transcribed Spacer; ITS = Internal Transcribed Spacer (1 and 2); TETS = Terminal External Transcribed Spacer; 1 through 6 refer to sites at the end of mature rRNA species. Sites other than those at the mature ends are, in this scheme, referred to as fractional numbers. In the text, 18S and 28S are often used in a generic sense for small and large ribosomal rRNA species. LETS is sometimes referred to as ETS.

of a pre-rRNA ribosomal RNA-protein (pre-rRNP) to place the ends of the intron in proper position for the excision reaction to occur. However, a complete set of proteins is not always required for processing since yeast mutants[19] that are devoid of certain pre-ribosomal proteins appear to process rRNA normally.

The material presented in this chapter, by necessity must be limited to a simple comparison of sequences which are known processing sites or are suggested to be processing sites by inferences made by comparative analysis of putative processing sites with homologous regions of rRNA gene sequences from different organisms.

We have chosen to divide this review into sections in which we discuss rRNA genes and processing in related species. Although the divisions are somewhat arbitrary, there are some features of organization or processing that are distinctive for each grouping.

II. BACTERIAL PROCESSING

A. Bacterial rRNA Processing

In wild type *E. coli,* processing of pre-rRNA occurs during the course of transcription, as is evident from the electron micrographs of Miller et al.[20] Detailed studies show that several enzymatic steps are involved in the production of the termini of 16 and 23S rRNAs. RNase III is responsible for generating a precursor 16S rRNA that contains 114 additional nucleotides at the 5′ end of 16S and an additional 33 nucleotides at the 3′ terminus.[21-24] RNase III also produces a precursor to 23S rRNA[25] containing 7 and 8 additional nucleotides at the 5′ and 3′ termini, respectively. Mature 16 and 23S are processed from the RNase III produced molecules by RNase M16 and RNase "M23".[26,27] Evidence supports the requirement for an RNP at these late stages.

In the pre-rRNA, extensive base pairing is possible for sequences flanking 16S rRNA.[24] Such double-stranded RNA regions may then be cleaved by RNase III. Transcripts that terminate in the 5′ half of the 16S rRNA sequences are not cleaved in vitro,[28] even though they possess the primary sequence cleaved by RNase III. This is taken as evidence to suggest that both regions flanking the 16S rRNA are required for RNase III cleavage. Extensive base pairing is also possible between sequences flanking 23S rRNA in pre-rRNA.[25] However, its involvement as a signal for RNase III action remains to be established. Using results obtained from S1 nuclease mapping analysis, King and Schlessinger[29] conclude that the 5′ end of 23S rRNA can be cleaved by RNase III before the 3′ end of the same molecule is synthesized. They suggest two interpretations of these results: (1) the newly synthesized 23S 5′ end could hybridize to the 3′ end of a separate pre-rRNA (i.e., an intermolecular reaction) or (2) RNase III recognition sites are present at each end of the 23S rRNA sequences. It is possible to determine if there are independent sites by making transcripts containing each end separately and asking if RNase III cleaves the transcripts. Intermolecular vs. intramolecular RNase III sites might be more difficult to test.

BOX A SEQUENCES

Nut L	CGCUCUUAAAAAUUAAG
Nut R	CGCUCUUACACAUUCCA
P21	CGCUCUUUAA
P22	UGCUCUUUAA
T7(0.3)	CGCUCUUUAACAAUCUG
16S E.coli	UGCUCUUUAACAAUUUA
23S E.coli	UGCUCUUUAAAAAUCUG
16S B.sub.	UGAUCUUUGAAAACUAA
23S B.sub.	UGUUCUUUGAAAACUAG

FIGURE 2.   Box A sequences. Nut L and Nut R are Box A sequences of phage. P21 and P22 are from phages P21 and P22. T7 (0.3) is a phage T7 Box A sequence near gene 0.3. See text for discussion of B. sub. *(B. subtilis)* and *E. coli* 16 and 23S Box A sequences.

B. Comparison of *Bacillus* and *E. coli* rRNA Processing

The organization of rRNA genes in *Bacillus subtilis* is somewhat different from *Escherichia coli*. Most of the rRNA genes are clustered near the origin of replication and only a few pre-rRNA transcripts contain tRNA molecules in the 16S-23S spacer region.[30-34] The regions flanking 16S rRNA sequences can be base paired to form a structure similar to that proposed for *E. coli*. The same is true for 23S rRNA. An interesting feature of these structures in *B. subtilis* is that there is also some sequence similarity between the two stems. As Loughney et al.[34,35] mentioned, the sequence homologies permit alternative structures in which the 5′ flanking region of 16S can pair with the 3′ flanking region of 23S rRNA and the 3′ flanking region of 16S rRNA can pair with the 5′ flanking region of 23S rRNA. The stability calculated for these latter structures is not as great as that of the former. Comparison of *E. coli* and *B. subtilis* processing sites, by searching for sequence homologies, reveals that there is little homology except for a few nucleotides near the 5′ end of 16S rRNA and about 8 to 10 bp, which are in *B. subtilis* form a part of the stem structure. The latter homology may be of no significance but it is worth mentioning since the sequence is similar to a sequence important in antiterminating transcription: the Box A sequence.

The Box A sequence is a part of an apparatus that affects transcription. Together with other cell factors[36] (and in some cases, phage-encoded factors), the Box A sequence acts to antiterminate transcription. It may be to ensure complete transcription of bacterial rRNA genes, an antitermination mechanism must be triggered. The location of the "Box A-like" sequences upstream from both 16S and 23S rRNA[37] (Figure 2) may reflect yet another common feature of rRNA transcription, not a simple sequence recognized as a processing site (i.e., there may be more information in the primary sequence than mature rRNAs and processing sites).

Generally, both primary and secondary structures may play important roles in RNA processing. Similarities and differences in primary sequence are most frequently examined to ascertain a recognition signal for an enzyme. Because RNase III exhibits a specificity for duplex RNA,[38-40] the secondary structure of pre-RNA has been felt to be significant for processing. It is, of course, possible that both primary and secondary structures are important for cleavage and the sequence homology noted above between *B. subtilis* and *E. coli* pre-rRNAs is a primary sequence recognition element.

III. *TETRAHYMENA* INTRON SPLICING

Recent excitement in the area of RNA processing was the result of work from the laboratory of Cech.[16,41] Selfsplicing of the RNA has forced a reconsideration of "enzymatic" mechanisms. To ascertain what portion(s) of this intron are relevant for the splicing reaction, Engberg[42] determined the nucleotide sequence intron from several different species of *Tetrahymena* (Figure 3). The minor differences noted in the sequences added little to our knowledge of the portions of the intron required for cleavage and ligation.

However, certain portions of the *Tetrahymena* intervening sequences exhibit structural homology with sequences from several yeast mitochondrial introns[43,44] (see Figure 3). The four conserved 10 to 20 nucleotide long stretches which are seen, are able to form two pairs of helical stems.

Intron sequences have been determined for two other lower eukaryotes, *Physarum*[45] and *Chlamydomonas*.[46] There appears to be little homology in sequence to *Tetrahymena* introns. Nomiyama et al.[45] have pointed out that in each instance, a U at the 5′ junction joins to a G at the 3′ junction. The significance of this "rule" remains unknown. The relationship between *Tetrahymena* rRNA genes and mitochondrial mRNA genes suggests that splicing of rRNA introns may be related more to mRNA metabolism than to rRNA processing. Indeed, not all species of *Tetrahymena* have interrupted rRNA genes.

IV. rRNA PROCESSING IN FUNGI

A. Organization

Unlike prokaryotic rDNA gene organization, fungi do not include 5S rRNA in the transcript with the other rRNA species. In *Saccharomyces cerevisiae*,[47] there is a close linkage of 5S rRNA genes with the rRNA transcriptional unit, whereas in *Neurospora crassa*[48] and *S. pombe*,[49,50] the 5S rRNA genes are unlinked to the major rRNA transcriptional unit. The latter situation is true for most other eukaryotic systems. *E. coli* has seven copies of the rRNA gene.[51] In contrast, eukaryotic cells contain many copies of rRNA genes. In yeast[52] and *Neurospora*,[53] as is true in most eukaryotes, more than 100 copies of the rRNA genes are repeated in a tandem array. The repeat units are reasonably homogeneous within a given species. For example, the repeat unit of yeast is 9.2 kb[54] and 8.7 kb for *N. crassa*.[55]

B. Processing

The complete nucleotide sequence for the transcribed portion of the rRNA gene repeat unit of *Saccharomyces carlsbergenesis* is known.[56-58,98] Several intermediates have been described which appear to arise somewhat differently than that described for *E. coli*. It seems that only a single cleavage event is required to generate the ends of the mature rRNA species. Veldman et al.[58,98] have suggested that one or more of the cleavages may occur in a manner similar to that for *E. coli* 16S rRNA (i.e., two regions of the transcript which are widely separated in the primary sequence pair to form a single site which is cleavage at two points). Recent results from Planta's group question the necessity for juxtaposition of two such sites (B1 and B2) in their published model.[59] Veldman and associates also suggested the possibility of concensus recognition sites for processing. Their interesting proposal suggests that 5′ processing sites have one consensus sequence and 3′ processing sites a second consensus sequence (Figure 4). Recognition of these sites almost certainly involves more than just the sequence since the ITS regions are 60 to 64% A + U and many similar sequences exist throughout both ITS regions.

C. Comparison of Fungal rRNA Processing Sites

Portions of two related sequences are available. First, Planta's groups has determined the nucleotide sequence flanking site 3 for *Saccharomyces rosei*.[59] It is clear that the sequence

```
                                                                                    9R'
T.t.  AAATAGCAATATTTACCTTTGGAGGGAAAAGTTATCAGGCATGCACCTGGTAGCTAGTCTTTAAACCAATAGATTGCATCGGTTTAAAAGGCAAGACGGTCAAA
T.p.          T       A                                        C    A                               T
T.h.          T       A                                        C    A                               T
T.c.          T       A                                        C    A                               T
                      A

T.t.  TTGCCGGAAAGGGGTCAACAGCCGTTCAGTACCAAGTCTCAGGGGAAACTTTGAGATGGCCTTGCAAAGGGTATGGTAATAAGCTGACGGACATGGTCCTAACC
T.p.              A                                                          A         G
T.h.              A                                                          A         G
T.c.              A                                                          A         G
        B                                9L                    9R                                            2

T.t.  ACGCAGCCAAGTCCTAAGTCAACAGATCTTCTGTTGATATGGATGCAGTTCACAGACTAAAATGTCCGTCCCGGAAGA-T--GTATTCTTCTCATAAGATATAGT
T.p.                                TT     GG-                               A AG
T.h.                                TT     GG-                               A AG
T.c.                                TT     GG-                               A AG

T.t.  CGGACCTCTCCTTAATGGGAGCTAGCGGATGAAGTGACAACACTGAGCGCGCTGGGAACTAATTTGTATGCGAAAGTATATTGATTAGTTTTGGAGTACTCG
T.p.                                            G       G       CA    C       ---A     C-    A -          C A
T.h.                                            G       G       TA    C       ---A     C-    A -          C A
T.c.                                            G       G       TA    C       ---A     C-    A -          C A
```

FIGURE 3.   Ribosomal RNA introns of four Tetrahymena species. Shaded letters are in mature rRNA sequences flanking the intron. Differences in primary sequences are noted by placing the different nucleotide below the sequence for *Tetrahymena thermophila*. No notation is made for identity. "—" refers to insertions or deletions. Solid areas noted as 9R', A, B, 9L, 9R, and 2 are sequences common among several introns as mentioned in text. T.t. = *Tetrahymena thermophila*; T.p. = *Tetrahymena pigmentosa*; T.h. = *Tetrahymena hyperangulari*; T.c. = *Tetrahymena cosmopolitani*.

5' Processing Sites		3' Processing Sites	
SITE 1	UUUUAAGAUAGUUA	SITE 2	UCAUUA
SITE 3	UUUUAAAAUA-UUA	SITE 4	UCAUUU
SITE 5	UCUUAAAGUU-UGA	SITE 6	UGAUUU
		SITE 2.5	UCAAUA
		SITE 4.5	UCGUUU

FIGURE 4. Potential recognition sequences for yeast RNA processing.[58] Sites 1 through 6 are defined in Figure 1 legend. Site 2.5 is between 2 and 3 (in ITS-1) and 4.5 is between 4 and 5 (in ITS-2). 5′ (or 3′) processing sites refer to cleavages generating the 5′ (or 3′) termini of mature rRNAs or maturation intermediates. These concensus sequences are located immediately upstream of the cleavage sites.

```
                                   ITS-1                               5.8S
S. r.   ATT-ATTATAAACCAGTCAAAACCAA----TTTCGTTA-TG-AAAT-TAAAAATATTTAAAACTTTCAACAACGGATCTCT
        ::: :::: :     ::::::: :::    :::::: : :: :::: : :::::::: ::::::::::::::::::::::::
S. c.   ATTCATTAAATTTTTGTCAAAAACAAGAATTTTCGTAACTGGAAATTTTAAAATATTAAAAACTTTCAACAACGGATCTCT
```

FIGURE 5. Site 3 sequences of two yeasts. S.r. = *Saccharomyces rosei*;[59] S.c. = *Saccharomyces carlsbergensis*.[58] Shaded letters are found in 5.8S rRNA. '':'' indicates identities. ''-'' indicates deletions or insertions.

adjacent to the 5′ end of 5.8S rRNA is highly conserved (Figure 5). Heteroduplexes between rRNA genes from several different yeast species suggest that there is a high degree of conservation of this region in many yeast species.[60] Second, Chambers et al.[55] have recently determined the ITS region nucleotide sequence of *Neurospora crassa* rDNA. Some interesting features arise when *N. crassa* and *S. carlsbergensis* sequences are compared. The variability of size of ITS-1 and ITS-2 from organism to organism is apparent in this comparison. ITS-1 of *S. carlsbergensis* is 363 nucleotides — about 2 times the 186 nucleotides of ITS-1 of *N. crassa*. The length of ITS-2 of *S. carlsbergensis* is also almost twice that of *N. crassa* (235 compared to 145). ITS-1 of *S. carlsbergensis* is 35% G + C; ITS-2 is 39% G + C. ITS-1 and ITS-2 are 55% G + C and 58% G + C, respectively, for *N. crassa*. A comparison of the two sequences using a computer program of Wilbur and Lipman[61] is shown in Figure 6. The alignment presented demonstrates the general lack of homology of the ITS regions of *S. carlsbergensis* and *N. crassa*. Even the sequences immediately adjacent to mature rRNA termini are only poorly conserved. In addition, the processing site reported for *S. carlsbergensis* in ITS-1 is not present in *N. crassa*. There is, however, some small conservation of sequence in ITS-2 for the processing site reported in *S. carlsbergensis* (the 3′ end of ''7S'' rRNA precursor; see Figure 6).

Comparisons of primary nucleotide sequences to determine a ''consensus'' sequence for rRNA processing sites suffer from size differences of regions being compared and from base-compositional differences.

V. INSECT rRNA PROCESSING

A. Organization

Ribosomal RNA genes of insects appear as tandem arrays characteristic of most eukaryotic rRNA cistrons. In addition to the usual NTS—LETS—18S—ITS-1—5.8S—ITS-2—28S order of sequence (Figure 1), insects exhibit two other components:

1. The 5.8S region is split into two segments. In two instances DNA sequence analysis has shown the ''5.8S'' and ''2S'' segments to be separated by 28 nucleotides (*Drosophila melanogaster*)[62] and 22 nucleotides (*S. coprophila*)[63] (Figure 7). In contrast, *Bombyx mori* 5.8S rRNA is not fragmented.[64]

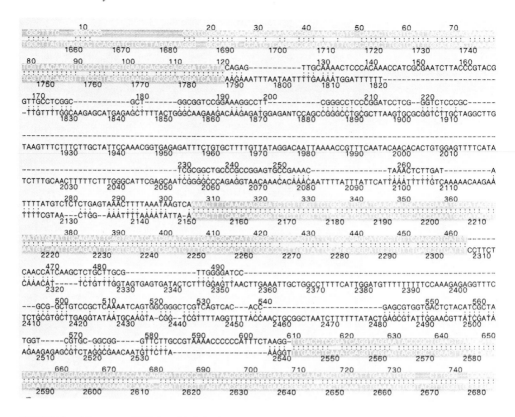

FIGURE 6.   Alignment of ITS-1 and ITS-2 of *N. crassa* and *S. carlsbergensis*. Using the NUCALN program of Wilbur and Lipman[61] *N. crassa* and *S. carlsbergensis* ITS-1 and ITS-2 were aligned. Parameters for the computer program were K-tuple = 3; Window = 20; and GAP = 2. Shaded letters are in mature rRNA sequences. The 3′ end of 17S rRNA extends from position 1 to 119 of *N. crassa* (top sequence) and 1650 to 1786 of *S. carlsbergensis* (lower sequence). 5.8S rRNA extends from position 305 to 462 for *N. crassa* and from 2150 to 2305 for *S. carlsbergensis*. The 5′ end of 27S rRNA beginning at 609 for *N. crassa, S. carlsbergenesis*. The 3′ end of 7S (site 4.5) is at position 2439 for *S. carlsbergenesis*. "N" means unknown nucleotide. "-" means deletion or insertion. ":" means identity.

2.   The presence of intervening sequences interrupts the coding region of 28S rRNA in some copies of the rRNA cistron repeat unit. For purposes of rRNA processing sites, we can ignore some of the introns of this kind in *Drosophila* since transcription of the interrupted genes is minimal at best.[14] However, Lecanidou et al.[65] have shown that approximately 12% of the rRNA cistrons of *B. mori* contain introns in the 28S rRNA coding region. In contrast to the introns extensively described for *D. melanogaster*, the introns of *B. mori* are located at positions similar to those found in *Tetrahymena*. In *Tetrahymena*, as described in a previous section, such introns are removed by an apparent selfsplicing reaction. In *B. mori* there is no evidence concerning the transcription or processing of rRNA of its interrupted genes.

3.   The large ribosomal RNA is split into two pieces of approximately equal sizes (28S A and 28S B). Processing of the 28S rRNA into these two fragments results from the removal of about 75 nucleotides from the pre-28S rRNA.[66]

B. Processing

The most extensive studies on rRNA processing in insects are those performed with *D. melanogaster*. A particularly important finding reported by Long and Dawid[8] is that multiple pathways of rRNA processing seem to operate in the same organism, a result that was later

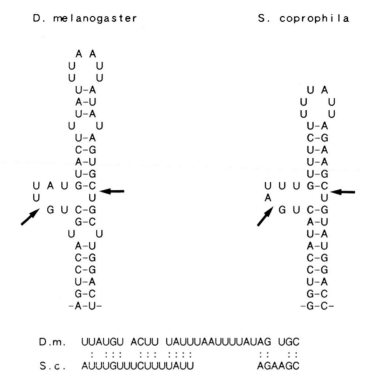

FIGURE 7.   Sequences cleaved to form "5.8S" and "2S" *(hypothetical structures).* Arrows mark termini of mature rRNA. D.m. = *Drosophila melanogaster;*
S.c. = *Sciara coprophila.* The lower portion of the figure aligns the sequences
separating 5.8S and 2S rRNAs of the two species.

extended to include the mouse. Alternate pathways of processing suggest that certain combinations or juxtapositions of sequences are not needed to form an enzyme recognition site.

Mandal and Dawid[67] have reported that the 3′ terminus of the complete pre-rRNA is
identical to that for 28S rRNA. Such a result implies that transcription termination is responsible for generating the 3′ end of 28S rRNA, not a processing (cleavage) event. A
similar result has been reported for *Xenopus laevis.*[68] However, mouse,[69] yeast,[57] and
Tetrahymena[70] have a few additional nucleotides on the 3′-end of pre-rRNA that are not
found in mature 28S rRNA (Figure 8). The discrepancy among these different organisms'
mechanism to produce the 3′ terminus of 28S rRNA may reflect species differences or
differential stability of the 3′ terminus of pre-rRNA. For example, removal of pre-rRNA
sequences to produce the mature 28S rRNA may be a relatively early event for *D. melanogaster* and a late event for the mouse. Recent results suggest that transcripts which extend
further than 28 nucleotides from the 3′ terminus of mature 28S rRNA can be found in the
mouse.[71] Such a result indicates the formation of the 3′ end indicated in Figure 8 results
from a processing event, not termination.

C. Sequences at Processing Sites

A limited number of rRNA processing sites of insects is available. Figure 9 illustrates
site 1 for *D. melanogaster,*[72] a comparison of site 2 for *B. mori,*[73] and *D. melanogaster,*[74]
the sequences interrupting the 5.8S rRNA of *D. melanogaster* and *S. coprophila* and a
comparison of sites 3 and 4 from the latter two species.[62,63] In most instances there is no
outstanding homology other than that for the mature rRNA sequences. However, a 12
nucleotide tract downstream of 5.8S ("2S") is somewhat related to a similarly "conserved"

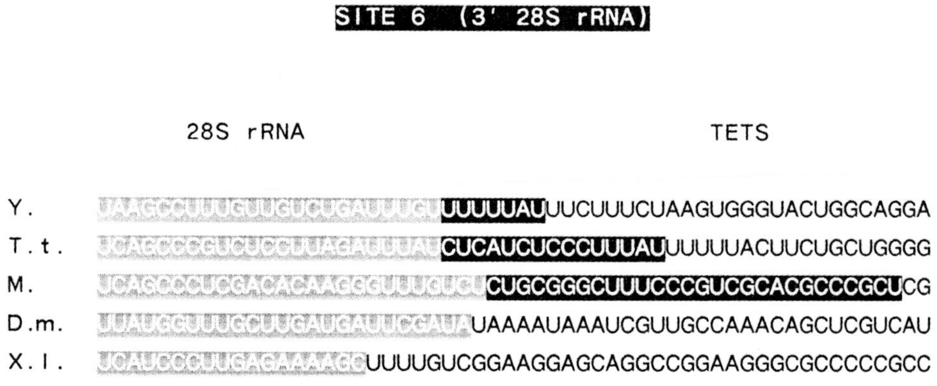

SITE 6 (3' 28S rRNA)

28S rRNA   TETS

Y.   UAAGCCUUUGUUGUCUGAUUUGU**UUUUUAU**UUCUUUCUAAGUGGGUACUGGCAGGA

T.t.   UCAGCCCGUCUCCUUAGAUUUAU**CUCAUCUCCCUUUAU**UUUUUACUUCUGCUGGGG

M.   UCAGCCCUCGACACAAGGGUUUGUCU**CUGCGGGCUUUCCCGUCGCACGCCCGCU**CG

D.m.   UUAUGGUUUGCUUGAUGAUUCGAUA**UAAAAUAAAUCGUUGCCAAACAGCUCGUCAU**

X.l.   UCAUCCCUUGAGAAAAGC**UUUUGUCGGAAGGAGCAGGCCGGAAGGGCGCCCCCGCC**

FIGURE 8.   Terminal external transcribed spacer. Y. = *Saccharomyces carlsbergenesis*,[57] T.t. = *Tetrahymena thermophilia*, M. = mouse,[69] D.m. = *Drosophila melanogaster*,[67] X.l. = *Xenopus laevis*.[68] Shaded letters are sequences of 28S rRNA (mature); solid letters are sequences reported in pre-rRNA *not* found in mature rRNA. Normal letters are sequences adjacent to TETS. See text for comments concerning the mouse.

tract in vertebrates (see Section VI). Figure 7 includes a secondary structure model for the cleavage sites to produce "5.8S" and "2S". The two structures are similar and could form a recognition site for a processing enzyme. Another explanation for the cleavage in the 5.8S rRNA is that in the pre-rRNP or ribosome, these nucleotides are exposed to nucleolytic degradation because they are unprotected. The latter explanation could also be made for the cleavage of 28S rRNA into two approximately equal size fragments. However, a potential structure for the sequence that is cleaved from the pre-28S rRNA might be a processing structure. The region removed from *D. melanogaster* 28S rRNA is one that is variable both in sequence and structure among 28S rRNAs of various species. Cleavage at or near the base of the stem (Figure 10) could, then, remove the 75 nucleotides.[66,75]

VI. PLANT AND MAMMALIAN NUCLEAR rRNA PROCESSING

A. Organization of rDNA
1. Repeat Unit

As opposed to yeast or *Dictyostelium discoideum*, 5S rRNA genes and pre-rRNA genes in higher eukaryotes are not linked but represent two separate families of tandemly repeated units. The large differences in the length of the pre-rRNA repeat unit among higher eukaryotes are mostly accounted for by variations in the size of the nontranscribed spacer regions (NTS). While these NTS may exhibit some degree of length heterogeneity in some species, like in *Drosophila*[76] or in *Xenopus laevis*[77] (size range, 4 to 7 kb), they generally retain similar sizes within rather large phylogenic groups. For example in all mammals they are particularly large (i.e., 20 to 30 kb as observed in mouse,[78] rat,[79,80] calf,[81] human,[82,83] and apes[84]), whereas in higher plants they appear to be relatively short, i.e., 2 to 3 kb.

2. Transcription Unit

With the exception of insects, all metazoan animals studied so far share a common basic pattern of rRNA transcriptional organization, with the covalent continuity of each of the mature 28S and 5.8S rRNAs not disrupted by internal "hidden" breaks, generated by subsequent processing events. This pattern (Figure 11) is also common to higher plants. The variations in the size of the transcription unit among these eukaryotes are mainly restricted within the transcribed spacer regions. It seems noteworthy that the strong constraint observed against the size variation of mature large rRNAs is even more severe for 18S than for 28S rRNA. Transcribed spacers generally have similar lengths in closely related species while

SITE 1 (5′ 18S rRNA)

```
                                          LETS                                      18S rRNA
D.m.  AATATTTATATTGCTTATTTCAATTCAAAAAATATGAATGAAATATGAAAAGAAAACATTATTCTGGTTGATCCTGCCAGTAGTTATATGCTTGTCTCA
```

SITE 2 (3′ 18S rRNA)

```
             18S rRNA                ITS 1
B.m.  GUAGGUAACCUGCGGAAGGAUCAUUAACGGGUGAUGGGAAGAAA
D.m.  UAGGUGAACCUGCGGAAGGAUCAUUAUUGUAUAAUAUCCUUACCGUUAAU
```

SITE 3 (5′ 5.8S rRNA)

```
             ITS 1                          5.8S rRNA
D.m.  CAUAAUUGACAUUAUAUAAAAAUGAAUUAUAAAACUCUAAGCGGUGGAUCAC
S.c.  UUAACAAUAAAUUUGUAAAAAAAAUUUAAUUCAACCCUAAGCGGGGGAUCAC
```

SITE 4 (3′ 5.8S rRNA "2S")

```
         5.8S rRNA ("2S")              ITS 2
D.m.  CUACAUAUGGUUGAGGGUUGUAAGACUAUGCUAAUUAAGUGCUUAUAAAUUUU
S.c.  CCACACAUGGUUGAGGGUCCGUUAGAUUACUUAAUAAAAAUUGCAUUUUUAUUA
```

FIGURE 9. Processing sites for insects. See Figure 1 for definition of sites 1 through 6, LETS, and ITS. D.m. = *Drosophila melanogaster*; B.m. = *Bombyx mori*; S.c. = *Sciara coprophila*. Shaded letters are found in mature rRNA sequences. For sequences from other species, see Figure 13 and Reference 7.

D. MELANOGASTER 28S SPLIT REGION

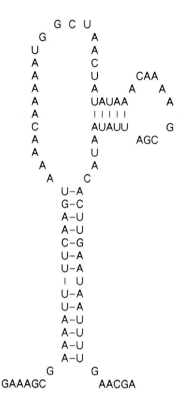

FIGURE 10.   Fragment removed from 28S rRNA of *D. melanogaster* (Hypothetical structure).

distinctive size patterns can be recognized among different phylogenic branches (Figure 11). The three higher plants analyzed so far, i.e., wheat,[85] barley, and flax,[86,87] possess particularly short ITS regions (total size, about 0.3 kb) as compared to vertebrates. Among vertebrates,[88] cold blooded vertebrates are distinguished by their shorter ITS (total size, about 0.8 kb). While birds and mammals have both large ITS (total approximate size, 2 kb), they differ in that mammals contain particularly long ETS (about 4 kb).

B. Evolution of the Structure of Mature and Spacer Domains in Pre-rRNA
1. Mature rRNAs

The complete nucleotide sequences of 18S rRNA have been determined in four vertebrates: *Xenopus*,[89] rat,[90,91] mouse,[92] and rabbit.[93] The degree of conservation is extremely high: mouse and rat sequences are 99% homologous and there is 94% homology between the amphibian and rodent sequences. All these vertebrate sequences exhibit about 80% homology with yeast. Two complete plant 17S rRNA sequences have been recently determined for rice[94] and maize,[95] which are 97% homologous to each other. They are also about 80% homologous to the vertebrates and the yeast sequences. It is remarkable that this strong conservation is far from being uniform since base changes and size differences are clearly restricted to a few definite areas of the molecule, which have retained the same relative location throughout the evolution. These rapidly evolving regions are mostly responsible for two trends in the evolution of 18S rRNA from lower to higher eukaryotes, i.e., size increase and higher G + C content. These two trends and a same pattern of strongly conserved tracts

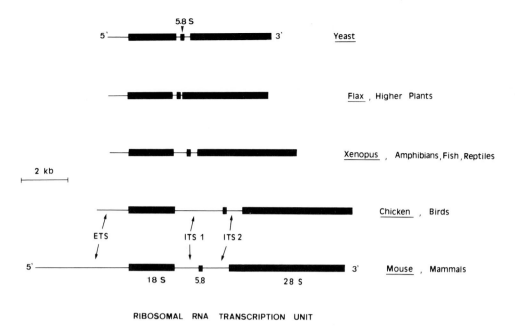

FIGURE 11.   Organization of the rRNA transcriptional unit in vertebrates and higher plants as compared to yeast. The different primary transcripts, which are represented with a common scale, have been aligned by reference to the 5' end of mature 18S rRNA. Precursor-specific RNA spacers are denoted by a thin line and mature rRNAs by black boxes.

interspersed with more rapidly evolving regions also apply to 28S rRNA, as observed when the complete sequences of mouse[96] and *Xenopus laevis*[97] are compared to yeast[98] and *Physarum polycephalum*.[99] The constraint for sequence and size conservation is clearly much lower for 28S than for 18S rRNA, and a number of extensively divergent sequence tracts can be detected within 28S rRNA between two closely related species, such as mouse[96] and rat.[100]

2. Spacer RNA Domains
a. ITS

Both ITS regions have been completely sequenced in *X. laevis*[101] and *X. borealis*,[102] mouse,[103] and rat,[104] while an ITS-2 sequence has been determined in the chick. All these vertebrate ITS clearly depart from mature rRNA sequences considered as a whole in their having a particularly high G + C content (70 to 88%). The ITS regions of another metazoan animal, sea urchin, share similar overall features.[105]

No complete sequence data have been reported so far for transcribed ribosomal spacers in plants. It may be worthwhile mentioning that partial determinations (about 0.2 to 0.3 kb flanking both ends of mature 17S rRNA) for maize[95] also indicate a very high GC content of these regions (72%) similar to what was observed in metazoan animals. This high G + C content clearly distinguishes ITS in metazoa and higher plants from their counterparts in lower eukaryotes (yeast, *Dictyostelium*,[106] slime mold *Physarum polycephalum*[99], and insects *Drosophila melanogaster*[62] and *Sciara coprophila*[63]). It is remarkable that for the four complete ITS-1 sequences available so far in vertebrates, in each case the G and C content are almost identical (Table 1), a feature clearly related to their high potential for internal secondary structure folding, as discussed below.

Unlike mature rRNAs, none of the vertebrate ITS sequences shows any significant homology when compared to its yeast counterpart, even in the immediate vicinity of the

Table 1

SIZE AND BASE CONTENT OF INTERNAL TRANSCRIBED SPACERS IN EUKARYOTIC PRE-rRNAS

	ITS-1				ITS-2			
	Size (Nucl.)	G (%)	C (%)	G + C (%)	Size (Nucl.)	G (%)	C (%)	G + C (%)
Dictyostelium	331	14	13	27	575	24.5	17.5	42
Physarum	—	—	—	—	492	26	23	49
Yeast	355	20	16	36	234	23	15	38
Sea urchin	367	33.5	37.5	71	338	35	39.5	74.5
Xenopus laevis	557	41.3	42.9	84.2	262	34.3	53.6	88
X. borealis	554	38.8	39.7	78.5	336	37.5	44.3	81.8
Chick	—	—	—	—	665	41.3	41.3	82.7
Mouse	999	35.0	35.0	70	1089	37.1	37.1	74.3
Rat	1067	37.1	37.4	74.5	765	36.8	43.2	80.0

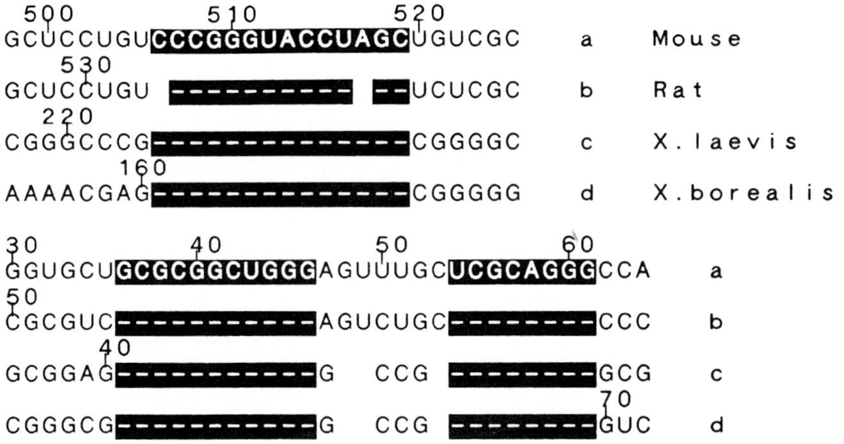

```
 500          510           520
GCUCCUGUCCCGGGUACCUAGCUGUCGC      a      Mouse
   530
GCUCCUGU ============ ==UCUCGC     b      Rat
   220
CGGGCCCG===============CGGGGC      c      X. laevis
       160
AAAACGAG===============CGGGGG      d      X. borealis


30        40        50        60
GGUGCUGCGCGGCUGGGAGUUUGCUCGCAGGGCCA      a
50
CGCGUC============AGUCUGC========CCC     b
   40
GCGGAG==========G  CCG  ========GCG      c
                                 70
CGGGCG==========G  CCG  ========GUC      d
```

FIGURE 12. Conserved sequence tracts in vertebrate ITS. The conserved tracts are in black boxes (identical nucleotides are denoted by a line, with deletions indicated by a space for ITS-1, top, and for ITS-2, bottom. They correspond to the "tract 1" regions identified within each ITS of the two *Xenopus* species.[102]

processing sites. No significant homology can be detected either among vertebrates when distant species (such as an amphibian and a rodent) are compared, except for three short tracts: one is located immediately downstream 5.8S rRNA (see following section), the others are detected within ITS-1 and ITS-2, respectively (Figure 12). Taking into account the very rapid rate of divergence of the remaining parts of the vertebrate ITS, it seems possible that these common tracts have not been conserved by chance but through a selective constraint which could be indicative of some functional role. However, more extensive comparative data are needed to further test this possibility.

Much more significant homologies in ITS sequences are observed when comparing closely related species, such as the mouse[103] and the rat.[104] The conservation is definitely far higher for ITS-1 than for ITS-2. It is noteworthy that homologous tracts are frequently interspersed with regions that have both extensively diverged and varied in size between both rodents, consistent with the occurrence of repeated insertion and deletion events during evolution. The same holds true when comparing a pair of amphibian sequences.[102]

Unlike what is found in nontranscribed ribosomal spacers, no definite pattern of internal repetition can be detected in ITS sequences which could be correlated to the processes responsible for the size variation of these regions during evolution.

Finally, considering the sequence data available so far, the high divergence rate of ITS might suggest that most of ITS regions are devoid of sequence-specific functions.

However, the conservation in vertebrates of a strong potential for the formation of long helical stems could be indicative of some form of structure-specific functions(s). The spacers are obvious candidates for interfacing with transiently associated preribosomal proteins during the nucleolar stages of ribosome biogenesis.[108]

b. ETS

Among higher eukaryotes, *Xenopus laevis* is the only species for which a complete ETS sequence is available. Similar to *Xenopus* ITS, its G + C content is particularly high (83%). Large portions (0.7 to 1 kb) of the 5' terminal domain of the rRNA primary transcript have been sequenced in three mammals: human,[108] rat,[109,110] and mouse.[111-113] Unlike human ETS, both rodent ETS share some conserved sequence features in this region, including the presence of very long stretches of poly(U) which map between 0.3 and 0.5 kb downstream from the initiation point. Two tracts (about 20 nucleotides) common to the three mammals[108] have been identified at the 5' terminus of the primary transcript and around an early RNA processing site about 0.4 to 0.6 kb downstream.

C. Pre-rRNA Structures at Processing Sites

1. Identification of RNA Processing Points

Recent results[114] seem to indicate that most of the mature rRNA termini in higher eukaryotes are directly generated by primary and endonucleolytic cuts, without any further trimming of rRNA ends, in contrast to the prokaryotic mode of rRNA processing.

In addition to the minimal number of processing sites which are required for generating mature rRNA termini, an early cleavage event has been identified in mouse[112,113] (and possibly also in rat[109] and human[108]) within the ETS, about 650 nucelotides downstream from the initiation site. It remains to be seen whether this early excision of a long leader segment from mouse precursor, which takes place rapidly after the synthesis of the primary transcript has been completed, is a general process in higher eukaryotes and how it relates to the control of rRNA biosynthesis. It is worthwhile mentioning that this cleavage site in human[108] and mouse appears to map in the immediate vicinity of a common tract, GAUCGAUGUG-GUG, which is also perfectly conserved in rat.[109] While the existence of a processing site within the ETS has not been reported so far in *Xenopus*,[68] it is remarkable that the same sequence-tract has also been strongly maintained in *X. laevis*[107] and *X. clivii*[115]: it is located about 100 nucleotides downstream of the inititation site.

2. Sequences at Cleavage Sites

No clear consensus features emerge when the sequences around the different processing sites are compared within the same species, for any of the higher eukaryotes analyzed in Figure 13. When homologous processing sites are compared among different species, it can be seen that the rapid sequence divergence of the spacer regions also applies to their very distal positions: all terminal nucleotides are different between rodents and amphibians immediately upstream 5.8S rRNA, while only a common terminal triplet can be found immediately downstream of the 3' end of mature 18S rRNA and immediately upstream of the 5' end of mature 28S rRNA, respectively (Figure 13). However, the conservation clearly appears more extensive immediately downstream 5.8S rRNA sequence.

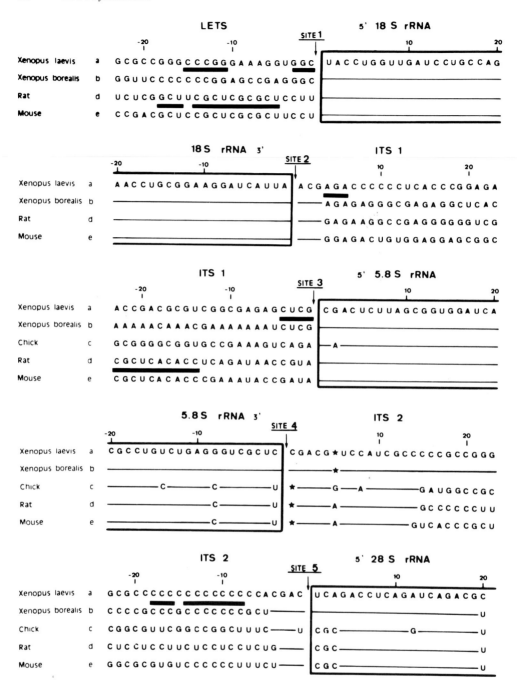

FIGURE 13.   Sequence of vertebrates' ribosomal RNA transcripts encoded by nuclear genes around processing sites. Mature rRNA sequences are boxed. Extensive sequence tracts which have been at least partially conserved among all (or most of) these species are denoted by a line. More partial conservation, restricted to pair(s) of species which are listed next to each other, is indicated by a thick horizontal bar between the two corresponding sequences. Deletions are denoted by a star.

3. Secondary Structure Folding Around Cleavage Sites

Unlike the internal domains of ITS sequences, distal segments of spacers do not exhibit a high potential for extensive base pairings with adjacent mature rRNA sequences and could remain available for a variety of RNA-RNA interactions. Due to the lack of experimental evidence, the potential role of bound preribosomal proteins in inducing RNA conformation

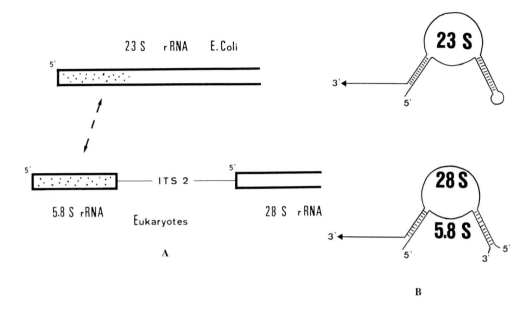

FIGURE 14.   ITS-2 as an eukaryote-specific insertion within the ancestor large rRNA gene. (A) Significant sequence homology can be detected between eukaryotic 5.8S rRNA sequences and the 5′ terminal region of *E. coli* 23S rRNA. (B) The secondary structure of the binary complex between 5.8S and 28S rRNAs in yeast and mouse is perfectly conserved and closely reminiscent of the folding pattern of the 5′ terminal region of prokaryotic 23S rRNA.

features recognized by processing enzymes will not be considered here. It is important to note, however, that a complete set of bound proteins does not seem to be required for the proper maturation of pre-rRNA to take place as shown recently in a relaxed mutant of yeast.[19]

D. Potential Processing Signals

1. Excision of ITS-2 as an Eukaryote-Specific Reaction

Convincing experimental evidence has accumulated recently showing that ITS-2 corresponds to an eukaryote-specific insertion within the ancestral gene coding for the large subunit rRNA as depicted in Figure 14. The eukaryotic large rRNA thus appears to be made up of two separate molecules which are held together by two long base paired interactions as first proposed for yeast.[98] Low, but significant sequence homologies do exist between eukaryotic 5.8S rRNAs and the 5′ terminal domain of prokaroytic 23S rRNA.[116,117] Homologies can also be detected between the immediately adjacent (downstream) region of 23S-rRNA and the 5′ end of eukaryotic 28S rRNAs. The binary complex between 5.8S and 28S rRNA can be folded in a secondary structure which is closely homologous to the model proposed for the 5′ terminal domain of prokaryotic 23S rRNA. The validity of this model has been confirmed first by analysis of mouse sequence data.[118] The secondary structure of the interaction complex has been strongly conserved throughout the evolution of eukaryotes.[119] When considering both the gene arrangement and the precise structure of the mature rRNA binary complex, the removal of ITS-2 from the immediate precursor of eukaryotic large rRNA (32S in vertebrates) then appears related to the excision of an intron, except that mature rRNA junctions are not covalently religated, but held together by strong base pairings. A detailed model involving nucleolar-specific U3 RNA in the recognition process for this maturation reaction through the formation of two duplex structures with both distal segments of ITS-2 has been proposed,[120] as shown in Figure 15. Such an involvement of U3 RNA, which is a component of the nucleolar preribosomes, is further substantiated by its binding to rRNA precursors surviving the removal of bound proteins and by its absence

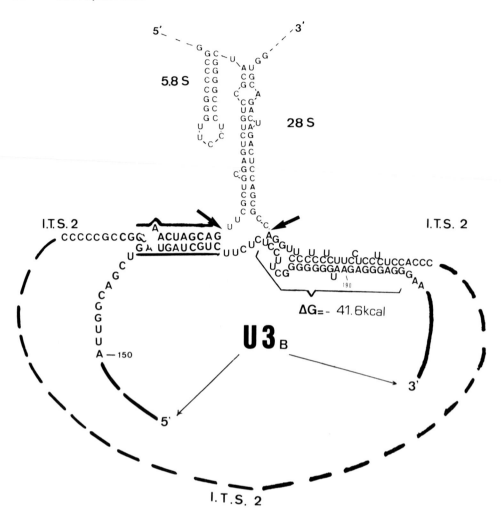

FIGURE 15.   Potential involvement of U3 RNA in the excision of ITS-2 from pre-rRNA. This model has been constructed using rat sequences for U3B RNA[121] and ITS-2.[104]

from mature ribosomal subunits. It is interesting to note that one of the proposed interactions involves the segment of ITS-2 immediately downstream 5.8S rRNA, which appears to have been more conserved during the evolution than the other termini of ITS. Partial sequence data[192] on mouse U3 RNA have recently confirmed that homologous base pairings with distal parts of ITS-2 can also exist in mouse with a perfect conservation of the left-hand stem (Figure 15). Ongoing sequence determinations for U3 RNA in other species should help to test the validity of such a model.

It is noteworthy that the secondary structure folding of ITS-2 regions in mouse[103] may represent a major contribution to the build-up of the mature 5.8S-28S rRNA binary complex since both distal regions of ITS-2 are brought in close vicinity from each other through a series of long hairpin structures, some of them clipping large domains of ITS-2. It must be stressed that homologous folding pattern of ITS-2 can be proposed for rat which could play a role in a conformational switch (summarized in Figure 16) involving both termini of 5.8S rRNA in the elongating transcript; until 28S rRNA is transcribed the 5′ and 3′ ends of 5.8S rRNA are likely to remain bonded by stable base pairings,[122] and the extensive refolding of ITS-2 sequences should favor the switch toward the alternate, more stable pairing with the corresponding 28S rRNA sequences.

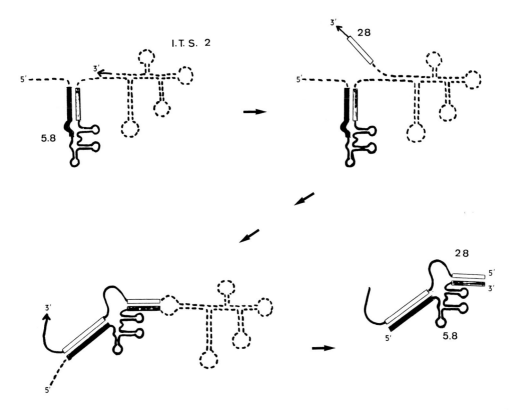

FIGURE 16. Role of ITS-2 sequences in a conformational switch involved in the formation of the mature 5.8S/28S rRNA complex.

An alternative model in which U3 RNA aids in the rearrangement of pre-rRNA has been suggested. In this model the 3′ and 5′ portions of U3 RNA interact with regions near 5.8S and 28S rRNAs in the precursor to melt-out stable intermediates formed in the course of transcription. (Details of this model can be found in References 7 and 123.)

2. Processing of Small Subunit rRNA

In prokaryotes, the RNA spacer sequences which immediately flank mature 16S rRNA within the pre-rRNA are able to form a long basepaired stem which loops out the entire 16S molecule.[24] Such long stable stems cannot be formed in yeast,[58] *Xenopus laevis*,[123a] mouse,[123b] or maize.[95] Only stems of much lower stability can be proposed, which are not supported by comparative analysis of these eukaryotic sequences. An absence of physical interaction between the 5′ end and 3′ cleavage sites (sites 1 and 2) is also consistent with the observation that the two cleavages are temporally separate events in higher eukaryotes: cleavage at site 1 largely precedes cleavage at site 2. More direct evidence that interactions between the regions surrounding the 5′ and 3′ ends of 18S rRNA is not required in specifying cleavage at site 1, has recently been reported by Bowman[123c] using cell transfection and deletion mapping techniques for analyzing the fate of transcripts from truncated mouse rRNA genes.

In the immediate vicinity of sites 1 and 2, RNA spacer sequences have extensively diverged (Figure 13) and no firm indication has been obtained that these spacer regions have retained outstanding secondary structure features. Accordingly, one has to consider the possibility that the distal domains of 18S rRNA, which are strongly conserved in primary and secondary structure, play a major role in signaling cleavage at sites 1 and 2.

3. Processing of Large Subunit rRNA

As discussed previously, the large subunit rRNA is fragmented in eukaryotes with the 5.8S rRNA species being equivalent to the 5′ end of the prokaryotic 23S rRNA molecule. The processing mechanisms involved in generating the ends of the composite "5.8 + 28S" molecule also appear to be distinct from the model proposed for the prokaryotic system.[25] In several eukaryotes, no 3′ RNA spacer has been detected (see Section V.C) which could basepair with sequences immediately upstream of the "5.8S = 28S" molecule, i.e., with the 3′ domain of ITS-1. However, the signals involved in the RNase III-catalyzed excision of prokaryotic 23S rRNA are not unambiguously identified.[29]

4. General Remarks

Finally, it is clear that further comparative evolutionary analyses are needed together with direct biochemical studies in order to pinpoint the molecular features specifying the processing reactions for nuclear pre-rRNAs. Rapid progress may now be expected at these two levels, due to the recent development of adequate methodologies. On one hand, the powerful approach of phylogenetic comparative analysis can now be easily and systematically applied to the study of definite areas of the pre-rRNA (especially the transcribed spacer boundaries): RNA sequence determinations can be rapidly obtained from a wide range of species by using whole unfractionated cellular RNA, without cloning or RNA end-labeling, through reverse transcription of DNA primers derived from highly conserved domains of the molecule.[119] On the other hand, the availability of efficient and selective in vitro transcription systems for truncated genes[123d] should favor the development of an in vitro rRNA processing system necessary for the ultimate identification of the molecular signals involved in each cleavage reaction.

VII. MITOCHONDRIAL rRNAs

A. mit rRNAs in Higher Animals

mit DNAs of all vertebrates share a common pattern of gene organization (see References 124 and 125 for reviews), as clearly indicated by sequence determinations on several mammals[126-129] and by physical mappings on a bird[130] or an amphibian.[131] Within this circular molecule of 15 to 16 kb the small and large rRNA genes are located on the H-strand, very close to each other, in the vicinity of the DNA replication origin of the H-strand. These two RNAs, usually referred to as 12S and 16S, have approximate sizes of 1.0 and 1.6 kb, respectively; they are the smallest known rRNA species, except for the trypanosomatid mit rRNAs.[132] Animal mit rRNA molecules possess a set of primary and secondary structure features in common with both pro- and eukaryotes in likely connection with the conservation of some basic aspects of ribosomal functions. However, comparison of prokaryotic with eukaryotic rRNAs reveals a higher homology than between animal mit rRNA sequences and any of the two major groups of organisms. While this observation could be taken to indicate a more ancient origin than the common ancestor or present-day pro- and eukaryotes,[133] it seems rather to correspond to a particularly high rate of sequence divergence in animal mit DNA.[134] In terms of rRNA gene organization, animal mitochondria also clearly depart from both prokaryotic and eukaryotic nuclear genes in two other aspects: first, no 5S rRNA is present (however, some marginal sequence homologies suggest that a drastically truncated 5S rRNA gene could have become part of the 3′ terminal region of the large (16S) mit rRNA gene[135]). Secondly, the two rRNA genes are "framed" by tRNA genes which are located immediately adjacent to the rRNA termini.[136,137] It is noteworthy that this striking interspersion pattern (Figure 17) also applies to most of the structural genes of the H-strand. Another unique feature of animal mitochondria (possibly restricted to mammals) is the complete symmetry of the transcription.[138] While the rate of transcription of the L-strand is

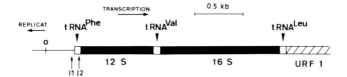

FIGURE 17. rRNA gene organization in human mitochondrial DNA.[124] The left arrow indicates the direction of H-strand synthesis from its origin (marked O). URF 1 stands for unidentified readings frame. Location of both initiation sites for transcription of H-strand are shown by vertical arrows (I_1 and I_2).

several-fold higher than that of the H-strand, most of the steady state mit RNA is transcribed from the H-strand[139] due to a particular instability of the L-strand transcript.[140] The potential involvement of the L-strand transcripts in the processing of H-strand transcripts remains to be considered as mentioned by Attardi.[124]

The organization of the two rRNA genes is suggestive of the formation of a single transcript cleaved by endonucleases at the boundaries of the interspersed tRNAs, which could provide, through their higher order structure, the necessary recognition signals to processing enzyme(s) related to *E. coli* RNase P.[141] Such a putative enzyme should act onto the elongating transcript since no full-size true rRNA precursor can be detected.[124] mit mRNAs encoded by the H-strand appear to be generated through a tRNA punctuated processing of a single polycistronic transcript encompassing most of mit DNA, including both rRNA genes.[142] However, this process cannot account for the production of the bulk of mit rRNA, considering the much higher rate of transcription of rRNA genes.[124,143] The existence of an alternative pathway for the formation of most rRNA has been proposed, involving as a frequent event the premature termination of the polycistronic transcription at the 3' end of the 16S rRNA gene.[144] Sequence features of this region of mammalian mit DNA[145] are indeed reminiscent of transcription termination signals in prokaryotes.[146] The detection of two separate initiation sites in human mit DNA,[147] both located in the region upstream of the 12S rRNA gene, provides an interesting basis for this "alternative pathway" model, which could account for the differential regulation of the production of rRNAs and mRNAs. Each initiation site seems to be specific and determinant of one of the routes for the production of H-strand transcripts, i.e., premature termination after rRNA genes or run-through transcription and synthesis of a long polycistronic molecule encompassing almost the entire H-strand and which is subsequently processed into mit mRNAs as recently reported by Montoya et al.[148] The initiation site responsible for the synthesis of the bulk of rRNA is located about 25 bp upstream of the tRNA^{phe} gene, while the second one has been mapped near the 5' end of the 12S rRNA gene.[148] Obviously the 5' terminal leader region of rRNA precursor (segment I_1I_2, Figure 17) could be involved in determining the transcription termination some 2.5 kb downstream, particularly through mechanisms related to the bacterial attenuation process[149] despite the much longer range effects. However, any hypothetical mechanism must be consistent with the observation that the processing of rRNA precursor does take place before the termination of rRNA gene transcription, which means that the 5' leader RNA segment is no longer covalently bound to the elongating transcript when the RNA polymerase proceeds through the potential termination region. While several mechanisms involving RNA conformational switches can account for this long-range transmission of the signal, one can envision that the RNA leader segment is in some way required for the proper excision of the 12S-16S intergenic tRNA^{val}. This RNA processing event could in turn be a prerequisite for the transcription termination to take place at the 3' end of the 16S rRNA gene. To test this hypothesis would imply a compared analysis of the fate of both populations of rRNA transcripts (i.e., synthesized from each initiation site) using wild type or in vitro mutated

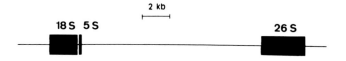

FIGURE 18.   Unusual arrangement of rRNA genes in maize mitochondria.[157]

mit DNA in appropriate in vitro transcription-processing systems. Both the intrinsic characteristics of the mammalian mit genome and the impressive amount of information obtained so far makes it most amenable to the development of such systems in the very near future which should provide unique insights into the mechanistic aspects of rRNA processing at the molecular level.

B. mit RNAs in Higher Plants

Information about the organization of the rRNA genes, their transcription, and the potential processing of rRNA in higher plants (see Reference 150 for review) is much less extensive than in animal mitochondria, mainly due to the much larger size and the heterogeneity of the mitochondrial DNA population, even within a same plant family; the sequence complexities of mit DNA from four species of the cucurbit family range from 350 to 2600 kb.[151] While reiterated sequences in the cucurbitacaes represent only 5 to 10% of mit DNA, these large size variations are likely to reflect extensive sequence rearrangements during the evolution of this family of plants.[151] Like all other nonmitochondrial ribosomes but unique in the mitochondrial class, plant mitochondrial ribosomes contain a separate 5S RNA species,[152] structurally distinct from its nucleus-encoded homologue,[153] but unambiguously related to other 5S rRNAs.[154] As in prokaryotes, the 5S rRNA gene in plant mitochondria is physically linked to the genes coding for the small and large rRNAs, termed 18S and 26S, respectively. The relative arrangement of these genes is strikingly distinct from the prokaryotic pattern since the 5S rRNA gene is located within an unusually large 18S-26S rRNA intergenic spacer, in the immediate vicinity of the 18S rRNA gene[155-157] (see Figure 18). Nevertheless, sequence analyses performed for wheat[158,159] and maize[160] 18S rRNA clearly indicate that mature rRNAs in higher plant mitochondria are structurally much more closely related to prokaryotic rRNAs than their counterparts of animal mitochondria. Comparison of wheat and maize 18S rRNA sequences supports the view of a lower rate of sequence divergence in plant mit DNA compared to animal mit DNA, and provides additional evidence of a relatively recent endosymbiotic origin of plant mitochondria from eubacterial species,[152] structurally distinct from its nucleus-encoded homologue,[153] but unambiguously related to other 5S rRNAs.[154] As in prokaryotes, the 5S rRNA gene in plant mitochondria is only a single base pair separates the 3′ end of tRNA$_f^{met}$ gene from the 5′ end of the 18S rRNA gene in wheat[161] and it may be envisioned that this tRNA structure can act as a processing signal for generating the 5′ end of 18S rRNA from a longer transcript, as proposed for mammalian mit rRNA processing. Although the 18S-5S intergene region has been sequenced in maize mitochondria,[162] the identification of the 3′ end of the mature small subunit rRNA still remains tentative.

VIII. BIOGENESIS OF CHLOROPLAST rRNA IN HIGHER PLANTS

A. Organization of Chloroplast DNA

Recent reviews on the subject have been written by Bedbrook and Kolodner[163] and by Rochaix.[164] Chloroplast DNA from higher plants consists of an homogeneous population of covalently closed circular molecules,[165,166] which according to the species, range in size from 120 to 190 kb.[163] In most cases studied so far, the arrangement of chloroplast DNA

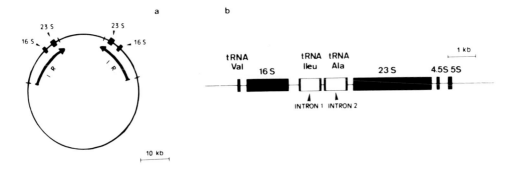

FIGURE 19.   Chloroplast rRNA transcription unit in a higher plant, tobacco. (a) Location of the rRNA gene in chloroplast DNA.[189] The position of the inverted repeat is denoted by arrows. (b) Organization of the rRNA transcription unit.[178]

is characterized by the presence of two chloroplast rRNA transcription units located in a long inverted repeat (22 to 25 kb) which separates two large single copy regions (averaging, respectively, 20 and 80 kb) as depicted in Figure 19a. The loss of one segment of the inverted repeat as observed in two legume species,[167,168] has been correlated to extensive rearrangements of chloroplast DNA during phylogeny,[169] as opposed to high conservation when both segments are present. Several experimental evidences are suggestive of the existence of a correction mechanism operating between the two segments of the inverted repeat, which should play a major role in the evolution of chloroplast genome.[170,171]

B. Structure of rRNA Transcription Unit

Chloroplast ribosomal RNA transcription units have been completely sequenced in two higher plants, maize[172-175] and tobacco,[176-179] and their arrangement analyzed in a variety of other species.[167,169,180-183] Ribosomal RNA genes in chloroplasts are very similar to prokaryotic ones, in terms of organization, size, and sequence. Moreover, recent sequence analyses[184] have shown that 23S chloroplast rRNA gene is more closely related to a blue green alga than to *E. coli,* an observation consistent with the endosymbiotic theory that chloroplasts have derived from an ancestral photosynthetic prokaryote. The large subunit of chloroplast ribosomes contains a 4.5S RNA species unique to higher plants which, on the basis of sequence and secondary structure homologies, appears to represent the counterpart of the 3' terminal domain of prokaryotic 23S rRNA.[185] The four chloroplast rRNA genes are cotranscribed from the same transcription unit in the form a single large precursor,[178,186] in the order 16S-23S-4.5S-5S (Figure 19b). As observed in prokaryotes,[187,188] two tRNA genes are present in the internal spacer regions of chloroplast rRNA operons.[174,178] However, contrary to that observed so far in lower plants chloroplasts, each of these two tRNA genes in higher plants is interrupted by a large intron (0.7 to 1 kb). These two introns which appear largely homologous account for most of the transcribed spacer length. The difference with the prokaryotic transcription unit seems also to extend to the presence of a tRNA^{val} gene in the leader region,[173,176,190] but this must await unequivocal demonstration of the co-transcription of this gene with rRNA since different in vitro transcription experiments have thus far proven inconclusive.[176,190]

As opposed to the mature rRNA regions which are highly homologous with their prokaryotic counterparts, precursor-specific sequences in chloroplast rRNA transcription units have extensively diverged from prokaryotes, whereas they are strongly conserved among different plants like spinach,[190] tobacco, and maize.

C. Processing of rRNA

Analysis of the in vivo transcription products in tobacco[178] indicates that rRNA processing

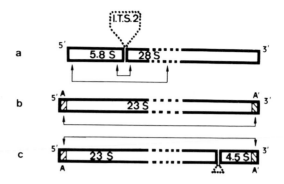

FIGURE 20.   Compared organization of the distal regions
of large subunit rRNA in prokaryotes (b) and in eukaryotes
— either encoded by nuclear genes (a) or by chloroplast genes
of higher plants (c). Dashed terminal regions denoted by A
and A' are able to form an 8 bp long duplex in *E. coli* and
are also complementary in chloroplast 23S and 4.5S rRNAs.
However, region A (12 5' terminal nucleotides in *E. coli*) has
no equivalent in eukaryotic 5.8S rRNA.[117] Transcribed spacer
regions are depicted by looped-out dotted lines. Arrows joined
by a thin line denote potential base pairings.

does not take place until the transcription of the entire unit has been completed. The precise
temporal order and the site for most of the RNA cleavage reactions still remain to be
identified. Results of sequence analyses are consistent with mechanisms related to the currently accepted model of rRNA processing in prokaryotes. Despite extensive divergence in
primary structure, it is remarkable that the sequences flanking mature rRNAs are still able
to form base paired structures[177,179] reminiscent of those proposed to be involved in the
RNase III catalyzed cleavages in *E. coli.*[24,25] However, it must be stressed that the size and
the stability of these double helical stems in chloroplast pre-rRNA are considerably lower
than their prokaryotic counterparts.

Considering the size of its immediate precursor,[185] 4.5S rRNA must remain covalently
linked to the 23S rRNA moiety until the very late stages of processing, i.e., after disruption
of the 16S-23S linkage. It seems noteworthy that the organization of the 3' terminal region
of the large subunit rRNA in chloroplasts of higher plants bears some analogy with what is
observed in the 5' terminal region of the molecule encoded by nuclear genes. In both cases,
the structural equivalent of the terminal region has been separated from the main part of the
gene by an insertion which is specific to eukaryotes (Figure 20). After removal of the spacer
region from the RNA precursor, both mature parts can be held together by base paired
interactions. However, in the case of 4.5S rRNA the only intermolecular duplex proposed
so far involves its seven 3' terminal nucleotides[191] (complementary to the 5' terminal nucleotides of 23S), whereas no other interaction has been identified which could restore the
physical continuity of this molecule with the 3' end of 23S rRNA. Accordingly the potential
analogy of the 23S-4.5S intergenic spacer with an intron sequence is much less clear than
in the case of the 5.8S-28S intergenic spacer in nuclear genes (as discussed in Section
VI.D.1). In tobacco chloroplast,[179] this spacer is 101 bp long. Its primary structure is highly
homologous in maize and shows some resemblance to bacterial insertion elements.[175] The
folding structure model shown in Figure 21 could provide recognition signals for the processing enzyme(s), as suggested by the homology in the secondary structure around the
cleavage sites in both species.

As for 5S chloroplast rRNA, its excision from pre-rRNA does not seem to involve the
cleavage of this sequence from a late intermediate of rRNA processing, but rather corresponds
to an early maturational event.[191]

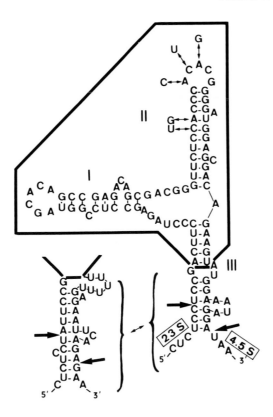

FIGURE 21.   Potential folding of the 23S-4.5S rRNA spacer and adjacent nucleotides in maize chloroplasts. Stems I and II are as proposed by Edwards and Kossel,[175] but the folding of the distal regions of the spacer (stem III) has been modified so as to increase the stability of the structure. The 3′ terminus of mature 23S rRNA and 5′ terminus of mature 4.5S rRNA are denoted by large arrows. The boxed domain can be folded in an identical structure in tobacco chloroplast[179] (the few base changes are denoted by two-headed arrows). The tobacco counterpart of the more variable distal regions (brackets) is shown as an inset.

On the contrary, the splicing of intergenic tRNAs seems to represent the last step in the processing of the transcript. It is noteworthy that these introns share with other intron families (nucleus encoded rRNAs and mit rRNAs of fungi) some distinctive sequence tracts and a number of potential secondary structure features which could be operative in the splicing reaction by bringing intron-exon junctions into vicinity.[43]

IX. CONCLUDING REMARKS

The elaboration of a precursor rRNA to produce equimolar quantities of mature rRNA species seems to have withstood the pressures of evolution with only one example of unlinked rRNA genes reported.[194] However, a variety of rRNA subunits make up the ribosomal particle. The most consistent of the changes from pro- to eukaryotic ribosomes is the fragmentation of the large rRNA (28S) into two pieces (5.8S + 28S). Taking all classes of ribosomal RNAs (i.e., organelle and cytoplasmic) examples of interruptions in all mature rRNA species can be found.

Proceeding from the 5' to the 3' direction on the pre-rRNA various types of discontinuities can be found. To date there is but a single example of an interruption in the rRNA of the small ribosomal subunit. In *Paramecium*[192a] there is a segment near the 3' terminus of 18S rRNA that is cleaved from the precursor resulting in two fragments of 18S rRNA which remain together via hydrogen bonds. In certain insects, 5.8S rRNA is found to be comprised of two fragments ("5.8S" + "2S"). In *D. melanogaster*[62] and *S. coprophila*[63] 26 and 22 nucleotides, respectively are removed from the pre-rRNA to produce "5.8S" + 2S.

The most ubiquitous interruption is that which produces 5.8S and 28S rRNA in eukaryotes. A few years ago it was found that 5.8S rRNA has homologies with the 5' terminal portion of prokaryotic 23S rRNA[116,117] and the 5' terminus of 28S rRNA is homologous to an internal region of 23S rRNA.[98,118] Without exception, nuclear rRNA from yeast to human contains an ITS-2. The size of ITS-2 ranges from about 100 nucleotides to more than 1000 bases (Table 1). Substantial data support the suggestion that 5.8S and 28S rRNAs are hydrogen bonded using 5' and 3' portions of 5.8S rRNA and the 5' portion of 28S rRNA.[118,119] Fragmentation of 28S rRNA near the site of ITS-2 insertion has been found in rRNAs from chloroplast and mitochondria. Two small rRNAs, 7S and 3S, are found in chloroplasts of *Chlamydomonas reinhardii*, which are 282 and 47 nucleotides in length.[195] It appears from some of our analysis (not included) that these interruptions occur in sequences which are found in mature 28S rRNA. *Paramecium* mitochondria[192b] contain a 283 nucleotide RNA that also appears to be the result of a cleavage in the "28S" rRNA sequence. 28S rRNA of insects is comprised of 28S A and 28S B. The two halves of 28S rRNA are produced in *D. melanogaster* as a result of removing about 80 nucleotides from the pre-rRNA sequence.[66] The final example of a fragmented "28S" rRNA is that found in chloroplasts of higher plants. A fragment 103 nucleotides in length is cleaved from the pre-rRNA to produce a "28S" rRNA composed of a large fragment, basepaired with a small (4.55) RNA piece.

Each of the interruptions described above is characterized by "fragmented" mature rRNA species which serve well in ribosomal function. In most cases, the fragments remain together through hydrogen bonds. Intervening sequences are also found in rRNA genes, generating as second class of interruptions. The simple case to be described is that of the intervening sequences found in *D. melanogaster* 28S rRNA genes. These genes are generally not transcribed and therefore, are not processed. However, in several instances the introns are transcribed and subsequently removed and the nucleotides flanking the intron are ligated. The most widely known case is that for the large rRNA of *Tetrahymena*. The most noteworthy property of the excision and ligation is that the reaction requires no protein to catalyze the reaction. Other examples of introns in rRNA genes similar to that in *Tetrahymena* are known, but there is little evidence to date that describes transcription and processing of this type of rRNA genes.

Studies have been done that compare the primary sequences of prokaryotic 16S rRNAs and 235 rRNAs. A consensus secondary structure model has been constructed[197] based in part on evolutionary conservation of primary sequence. Compensatory changes in primary sequence which conserve proposed secondary structure interactions have been taken as pieces of evidence to support the validity of the proposed structure. Similarly, secondary structure models of 18S and 28S rRNAs[75,78] have been formulated. What becomes clear after examining the models is the presence of a "skeleton" structure found in all examples. Additional nucleotides found in some rRNAs can be placed into the basic rRNA structure without perturbing the "skeleton" structure. Thus, it has been possible to directly compare pro- and eukaryotic rRNAs in spite of the substantial size differences.

If we assume this logic holds, we can suggest the fate of a random insertion of new sequences into mature rRNA sequences. Three different categories of insertions can be expected. The first category would be those insertions that are silent; i.e., the overall size of the rRNA would increase but the function of the ribosome would be unaffected. Such extra sequences might evolve into sites that are useful (e.g., they could make a ribosome function better in a hydrophobic or hydrophilic atmosphere). A second category would be related to the first but the bulk or sequence content of the insert would interfere with ribosome function. Such detrimental defects could be removed by clipping off the extraneous material. A third category would be comprised of those insertions that interrupt a critical nucleotide sequence or disrupt an important secondary structure. The third type would be lethal to the cell if it were present in all copies of the rRNA genes, unless the detrimental sequences are removed and the primary sequence (and therefore the secondary structure) is restored. Introns found in *D. melanogaster* and *Tetrahymena* rRNA genes would both be in the third category.

What constitutes an rRNA processing site? In *E. coli* there is good evidence that sequence and structure are required for some processing sites. Additionally, certain processing in *E. coli* requires an RNP. This may not always be the case in eukaryotes since rRNA processing appears to be normal in a yeast mutant that does not form normal protein-RNA particles.[19] Recent experiments performed by Bowman[123b] suggest that the processing signal(s) for the 5′ terminus of 18S rRNA (site 1) is contained in a region which includes LETS and a part of 18S rRNA, but more importantly, does *not* include the 3′ end of 18S rRNA. This is, of course, in contrast to *E. coli* in which the regions flanking 16S rRNA form a site for RNase III cleavage. Bowman's result is supported by the fact that no extensive secondary structures can be formed between the regions flanking 18S rRNA. There are some hints of "conserved" sequences at various points in the rRNA transcripts outside the mature rRNAs but there is little evidence to suggest that they are processing signals and not transcription signals. There remains much to be done.

ACKNOWLEDGMENTS

We wish to thank Drs. Enberg, Cashel, Bowman, Klootwijk, Schlessinger, Boncicelli for communicating results prior to publication. We also acknowledge the advance copies of reviews given us by Drs. G. Attardi and J. D. Rochaix.

REFERENCES

1. **Perry, R. P.,** Processing of RNA, *Ann. Rev. Biochem.,* 45, 605, 1976.
2. **Long, E. O. and Dawid, I. B.,** Repeated genes in eukaryotes, *Ann. Rev. Biochem.,* 49, 727, 1980.
3. **Planta, R. J. and Meyerink, J. H.,** Organization of the ribosomal RNA genes in eukaryotes, in *Ribosomes,* Chambers, G., Craven, G. R., Davies, J., Davis, K., Kahan, L., and Nomura, M., Eds., University Park Press, Baltimore, 1980, 871.
4. **Apirion, D. and Gegenheimer, P.,** Molecular biology of RNA processing in prokaryotic cells, in *Processing of RNA,* Apirion, D., Ed., CRC Press, Boca Raton, Fla., 1984, 35.
5. **Attardi, G.,** RNA synthesis and processing in mitochondria, in *Processing of RNA,* Apirion, D., Ed., CRC Press, Boca Raton, Fla., 1984, 227.
6. **Grant, D. M. and Lambowitz, A. M.,** Mitochondrial rRNA genes, in *The Cell Nucleus,* Vol. 10, Busch, H. and Rothblum, L., Eds., Academic Press, New York, 1982, 387.
7. **Crouch, R. J.,** Ribosomal RNA processing in eukaryotes, in *Processing of RNA,* Apirion, D., Ed., CRC Press, Boca Raton, Fla., 1984, 213.
8. **Long, E. O. and Dawid, I. B.,** Alternative pathways in the processing of ribosomal RNA precursor in *Drosophila melanogaster, J. Mol. Biol.,* 138, 873, 1980.
9. **Bowman, L. H., Rabin, B., and Schlessinger, D.,** Multiple ribosomal RNA cleavage pathways in mammalian cells, *Nucleic Acids Res.,* 9, 4951, 1981.
10. **Glover, D. M. and Hogness, D. S.,** A novel arrangement of the 18S and 28S sequences in a repeating unit of *Drosophila melanogaster* rDNA, *Cell,* 10, 167, 1977.
11. **White, R. L. and Hogness, D. S.,** R loop mapping of the 18S and 28S sequences in the long and short repeating units of *Drosophila melanogaster* rDNA, *Cell,* 10, 177, 1977.

12. **Wellauer, P. K. and Dawid, I. B.,** The structural organization of ribosomal DNA in *Drosophila melanogaster, Cell,* 10, 193, 1977.

13. **Pellegrini, M., Manning, J., and Davidson, N.,** Sequence arrangement of the rDNA of *Drosophila melanogaster, Cell,* 10, 213, 1977.

14. **Long, E. O. and Dawid, I. B.,** Expression of ribosomal DNA insertions in *Drosophila melanogaster, Cell,* 18, 1185, 1979.

15. **Wild, M. A. and Gall, J. G.,** An intervening sequence in the gene coding for 25S rRNA of *Tetrahymena pigmentosa, Cell,* 16, 565, 1979.

16. **Grabowski, P. J., Zaug, A. J., and Cech, T. R.,** The intervening sequence of the ribosomal RNA precursor is converted to a circular RNA in isolated nuclei of *Tetrahymena, Cell,* 23, 467, 1981.

17. **Bertrand, H., Bridge, P., Collins, R. A., Garriga, G., and Lambowitz, A. M.,** RNA splicing in *Neurospora* mitochondria. Characterization of new nuclear mutants with defects in splicing the mitochondrial large rRNA, *Cell,* 29, 517, 1982.

18. **Garriga, G. and Lambowitz, A. M.,** RNA splicing in *Neurospora* mitochondria. The large rRNA intron contains a noncoded, 5′-terminal guanosine residue, *J. Biol. Chem.,* 258, 14745, 1983.

19. **Waltschewa, L., Georgiev, O., and Venkov, P.,** Relaxed mutant of *Saccharomyces cerevisiae:* proper maturation of rRNA in absence of protein synthesis, *Cell,* 33, 221, 1983.

20. **Miller, O. M., Hamkalo, B. A., and Thomas, C. A.,** Visualization of bacterial genes in action, *Science,* 169, 392, 1970.

21. **Dunn, J. J. and Studier, F. W.,** T7 early RNAs and *Escherichia coli* ribosomal RNAs are cut from large precursor RNAs *in vivo* by ribonuclease III, *Proc. Natl. Acad. Sci. U.S.A.,* 70, 3296, 1973.

22. **Nikolaev, N., Silengo, L., and Schlessinger, D.,** A role of ribonuclease III in processing of ribosomal ribonucleic acid and messenger ribonucleic acid precursors in *Escherichia coli, J. Biol. Chem.,* 248, 7967, 1973.

23. **Lund, E., Dahlberg, J. E., and Guthrie, C.,** Processing of spacer tRNAs of *Escherichia coli,* in *Transfer RNA: Biological Aspects,* Söll, D., Abelson, J., and Schimmel, P. R., Eds., Cold Spring Harbor Laboratory, Cold Spring Harbor, N.Y., 1980, 123.

23a. **Apirion, D., Ghora, B. K., Plautz, G., Misra, T. K., and Gegenheimer, P.,** Processing of rRNA and tRNA in *Escherichia coli:* cooperation between processing enzymes, in *Transfer RNA: Biological Aspects,* Söll, D., Abelson, J., and Schimmel, P. R., Eds., Cold Spring Harbor Laboratory, Cold Spring Harbor, N.Y., 1980, 139.

24. **Young, R. A. and Steitz, J. A.,** Complementary sequences 1700 nucleotides apart form a ribonuclease III cleavage site in *Escherichia coli* ribosomal precursor RNA, *Proc. Natl. Acad. Sci. U.S.A.,* 75, 3593, 1978.

25. **Bram, R. J., Young, R. A., and Steitz, J. A.,** The ribonuclease III site flanking 23S sequences in the 30S ribosomal precursor RNA of *E. coli, Cell,* 19, 393, 1980.

26. **Dahlberg, A. E., Dahlberg, J. E., Lund, E., Tokimatsu, H., Rabson, A. B., Calvert, P. C., Reynolds, F., and Zahalak, M.,** Processing of the 5′ end of *Escherichia coli* 16S ribosomal RNA, *Proc. Natl. Acad. Sci. U.S.A.,* 75, 3598, 1978.

27. **Pace, N. R.,** Protein-polynucleotide recognition and the RNA processing nucleases in prokaryotes, in *Processing of RNA,* Apirion, D., Ed., CRC Press, Boca Raton, Fla., 1984, 1.

28. **Glaser, G. and Cashel, C. M.,** *In vitro* transcripts from *rrn*B ribosomal RNA cistrons originate from two tandem promoters, *Cell,* 16, 111, 1979.

29. **King, T. C. and Schlessinger, D.,** S1 nuclease mapping analysis of ribosomal RNA processing in wild type and processing deficient *Escherichia coli, J. Biol. Chem.,* 258, 12034, 1983.

30. **Seiki, M., Ogasawara, N., and Yoshikawa, H.,** Structure and function of the region of the replication origin of the *Bacillus subtilis* chromosome. I. Isolation and characterization of plasmids containing the origin region, *Mol. Gen. Genet.,* 183, 220, 1981.

31. **Seiki, M., Ogasawara, N., and Yoshikawa, H.,** Structure and function of the region of the replication origin of the *Bacillus subtilis* chromsome. II. Identification of the essential regions for inhibitory functions shown by the DNA segment containing the replication origin, *Mol. Gen. Genet.,* 183, 227, 1981.

32. **Seiki, M., Ogasawara, N., and Yoshikawa, H.,** Identification of a suppressor sequence for DNA replication in the replication origin region of the *Bacillus subtilis* chromosome, *Proc. Natl. Acad. Sci. U.S.A.,* 70, 4285, 1983.

33. **Wawrousek, E. F. and Hansen, J. N.,** Structure and organization of a cluster of six tRNA genes in the space between tandem ribosomal RNA gene sets in *Bacillus subtilis, J. Biol. Chem.,* 258, 291, 1983.

34. **Loughney, K., Lund, E., and Dahlberg, J. E.,** tRNA genes are found between the 16S and 23S rRNA genes in *Bacillus subtilis, Nucleic Acids Res.,* 10, 1607, 1982.

35. **Loughney, K., Lund, E., and Dahlberg, J. E.,** Ribosomal RNA precursors of *Bacillus subtilis, Nucleic Acids Res.,* 11, 6709, 1983.

36. **Friedman, D. I. and Gottesmam, M.,** Lytic mode of Lambda development, in *Lambda II*, Hendrix, R. W., Roberts, J. W., Stahl, F. W., and Weisberg, R. A., Cold Spring Harbor Laboratory, Cold Spring Harbor, N.Y., 1983, 21.

37. **Cashel, M.,** personal communication.

38. **Robertson, H. D., Webster, R. D., and Zinder, N. D.,** Purification and properties of ribonuclease III from *Escherichia coli, J. Biol. Chem.,* 243, 82, 1968.

39. **Crouch, R. J.,** Ribonuclease III does not degrade deoxyribonucleic acid-ribonucleic acid hybrids, *J. Biol. Chem.,* 249, 1314, 1974.

40. **Robertson, H. D. and Dunn, J. J.,** Ribonucleic acid processing activity of *Escherichia coli* ribonuclease III, *J. Biol. Chem.,* 250, 3050, 1975.

41. **Cech, T. R., Zaug, A. J., Grabowski, P. J., and Brehm, S. L.,** Transcription and splicing of the ribosomal RNA precursor of *Tetrahymena,* in *The Cell Nucleus, rDNA,* Vol. 10, Busch, H. and Rothblum, L., Eds., Academic Press, New York, 1982, 171.

42. **Engberg, J.,** personal communication.

43. **Michel, F. and Dujon, B.,** Conservation of RNA secondary structures in two intron families including mitochondrial-, chloroplast- and nuclear-encoded members, *EMBO J.,* 2, 33, 1983.

44. **Cech, T. R., Tanner, N. K., Tinoco, I., Jr., Weir, B. R., Zuker, M., and Perlman, P. S.,** Secondary structure of the *Tetrahymena* ribosomal RNA intervening sequence: structural homology with fungal mitochondrial intervening sequences, *Proc. Natl. Acad. Sci. U.S.A.,* 80, 3903, 1983.

45. **Nomiyama, H., Sakaki, Y., and Takagi, Y.,** Nucleotide sequence of a ribosomal RNA gene intron from slime mold *Physarum polycephalum, Proc. Natl. Acad. Sci. U.S.A.,* 78, 1376, 1981.

46. **Allet, B. and Rochaix, J.-D.,** Structure analysis at the ends of the intervening DNA sequences in the chloroplast of 23S ribosomal genes of *C. reinhardii, Cell,* 18, 55, 1979.

47. **Kramer, R. A., Camerson, J. R., and Davis, R. W.,** Isolation of bacteriophage λ containing yeast ribosomal RNA genes: screening by *in situ* RNA hybridization to plaques, *Cell,* 8, 227, 1976.

48. **Free, S. J., Rice, P. W., and Metzenberg, R. L.,** Arrangement of the genes coding for ribosomal ribonucleic acids in *Neurospora crassa, J. Bacteriol.,* 137, 1219, 1979.

49. **Tobata, S.,** Nucleotide sequences of the 5S ribosomal RNA genes and their adjacent regions in *Schizosaccharomyces pombe, Nucleic Acids Res.,* 9, 6429, 1981.

50. **Mao, J., Appel, B., Schnack, J., Sharp, S., Yamada, H., and Soll, D.,** The 5S RNA genes of *Schizosaccharomyces pombe, Nucleic Acids Res.,* 10, 487, 1982.

51. **Bachmann, B. J.,** Linkage map of *Escherichia coli* K-12, edition 7, *Microbiol. Rev.,* 47, 180, 1983.

52. **Schweizer, E., MacKechnie, C., and Halvorson, H. O.,** The redundancy of ribosomal and transfer RNA genes in *Saccharomyces cerevisiae, J. Mol. Biol.,* 40, 261, 1969.

53. **Chathpadhyay, S. K., Kohne, D. E., and Dutta, S. K.,** Ribosomal RNA genes of *Neurospora:* isolation and characterization, *Proc. Natl. Acad. Sci. U.S.A.,* 69, 3256, 1972.

54. **Nath, K. and Bollon, A. P.,** Organization of the yeast ribosomal gene cluster via cloning and restriction analysis, *J. Biol. Chem.,* 252, 6562, 1977.

55. **Chambers, C., Crouch, R. J., and Dutta, S. K.,** Unpublished results.

56. **Klootwijk, J., deJonge, P., and Planta, R. J.,** The primary transcript of the ribosomal repeating unit in yeast, *Nucleic Acids Res.,* 6, 27, 1979.

57. **Veldman, G. M., Klootwijk, J., deJonge, P., Leer, R. J., and Planta, R. J.,** The transcription termination site of the ribosomal RNA operon in yeast, *Nucleic Acids Res.,* 8, 5179, 1980.

58. **Veldman, G. M., Klootwijk, J., van Heerikhuizen, H., and Planta, R. J.,** The nucleotide sequence of the intergenic region between the 5.8S and 26S rRNA genes of the yeast ribosomal RNA operon. Possible implications for the interaction between 5.8S and 26S rRNA and the processing of the primary transcript, *Nucleic Acids Res.,* 9, 4847, 1981.

59. **Klootwijk, J.,** personal communication.

60. **Verbeet, M. P., Klootwijk, J., van Heerikhuizen, H., Fontijn, R., Vreugdenhil, E., and Planta, R. J.,** Molecular cloning of the rDNA of *Saccharomyces rosei* and comparison of its transcription initiation region with that of *Saccharomyces carlsbergensis, Gene,* 23, 53, 1983.

61. **Wilbur, W. J. and Lipman, D. J.,** Rapid similarity searches of nucleic acid and protein data banks, *Proc. Natl. Acad. Sci. U.S.A.,* 80, 726, 1983.

62. **Pavlakis, G. N., Jordan, B. R., Wurst, R. M., and Vournakis, J. N.,** Sequence and secondary structure of *Drosophila melanogaster* 5.8S and 2S rRNAs and the processing site between them, *Nucleic Acids Res.,* 7, 2213, 1979.

63. **Jordan, B. R., Latil-Damotte, M., and Jourdan, R.,** Coding and spacer sequences in the 5.8S-2S region of *Sciara coprophila* ribosomal DNA, *Nucleic Acids Res.,* 8, 3565, 1980.

64. **Fujiwara, H., Kawata, Y., and Ishikawa, H.,** Primary and secondary structure of 5.8S rRNA from the silk gland of *Bombyx mori, Nucleic Acids Res.,* 10, 2415, 1982.

65. **Lecanidou, R., Eickbush, T. H., and Kafatos, F. C.,** Ribosomal DNA genes of *Bombyx mori:* a minor fraction of the repeating units contain insertions, *Nucleic Acids Res.,* 12, 4703, 1984.

66. **Delanversin, G. and Jacq, B.,** Sequence of the central break region of *Drosophila* 26S precursor ribosomal RNA, *C. R. Acad. Sci. Ser. C:,* 296, 1041, 1983.

67. **Mandal, R. K. and Dawid, I. B.,** The nucleotide sequence of the transcription termination site of ribosomal RNA in *Drosophila melanogaster, Nucleic Acids Res.,* 9, 1801, 1981.

68. **Sollner-Webb, B. and Reeder, R. H.,** The nucleotide sequence of the initiation and termination sites for ribosomal RNA transcription in *X. laevis, Cell,* 18, 485, 1979.

69. **Kominami, R., Mishima, Y., Urano, Y., Sakai, M., and Muramatsu, M.,** Cloning and determination of the transcription termination site of ribosomal RNA gene of the mouse, *Nucleic Acids Res.,* 10, 1963, 1982.

70. **Din, N., Engberg, J., and Gall, J. G.,** The nucleotide sequence at the transcription termination site of the ribosomal RNA gene in *Tetrahymena thermophila, Nucleic Acids Res.,* 10, 1503, 1982.

71. **Grummt, I., Ohrlein, A., Maier, U., Hassouna, N., and Bachellerie, J. P.,** Transcription of mouse ribosomal DNA terminates 565 base-pairs downstream of 28S rRNA and involves the interaction of termination factors with repititious sequences in the non-transcribed spacer, *Cell,* in press, 1985.

72. **Simeone, A. and Boncinelli, E.,** 5' Cleavage site of *D. melanogaster* 18S rRNA, *FEBS Lett.,* 167, 249, 1984.

73. **Samols, D. R., Hagenbüchle, O., and Gage, L. P.,** Homology of the 3' terminal sequences of the 18S rRNA of *Bombyx mori* and the 16S rRNA of *Escherichia coli, Nucleic Acids Res.,* 7, 1109, 1979.

74. **Jordan, B. R., Latil-Damotte, M., and Jourdan, R.,** Sequence of the 3'-terminal portion of *Drosophila melanogaster* 18S rRNA and the adjoining spacer: comparison with the corresponding prokaryotic and eukaryotic sequences, *FEBS Lett.,* 117, 227, 1980.

75. **Michot, B., Hassouna, N., and Bachellerie, J. P.,** Secondary structure of mouse 28S rRNA and general model for the folding of the large rRNA in eukaryotes, *Nucleic Acids Res.,* 12, 4259, 1984.

76. **Dawid, I. B., Wellauer, P. K., and Long, E. O.,** Ribosomal DNA in *D. melanogaster:* isolation and characterization of cloned fragments, *J. Bol. Biol.,* 126, 749, 1978.

77. **Wellauer, P. K., Dawid, I. B., Brown, D. D., and Reeder, R. H.,** The molecular basis for length heterogeneity in ribosomal DNA from *X. laevis, J. Mol. Biol.,* 105, 461, 1976.

78. **Kominami, R., Urano, Y., Mishima, Y., and Muramatsu, M.,** Organization of ribosomal RNA gene repeats of the mouse, *Nucleic Acids Res.,* 9, 3219, 1981.

79. **Stumph, W. E., Wu, J. R., and Bonner, J.,** Determination of the size of rat ribosomal DNA repeating units by electron microscopy, *Biochemistry,* 18, 2864, 1979.

80. **Braga, E. A., Yussifov, T. N., and Nosikov, V. V.,** Structural organization of rat ribosomal genes. Restriction endonuclease analysis of genomic and cloned ribosomal DNAs, *Gene,* 20, 145, 1982.

81. **Meunier-Rotival, M., Cortadas, I., Macaya, G., and Bernardi, G.,** Isolation and organization of calf ribosomal DNA, *Nucleic Acids Res.,* 6, 2109, 1979.

82. **Wellauer, P. K. and Dawid, I. B.,** Isolation and sequence organization of human ribosomal DNA, *J. Mol. Biol.,* 128, 289, 1979.

83. **Higuchi, R., Stang, H. D., Browne, J. K., Martin, M. O., Huot, M., Lipeles, J., and Salser, W.,** Human ribosomal RNA gene spacer sequences are found interspersed elsewhere in the genome, *Gene,* 15, 177, 1981.

84. **Arnheim, N., Krystal, M., Schmickel, R., Wilson, G., Ryder, O. and Zimmer, E.,** Molecular evidence for genetic exchanges among ribosomal genes on nonhomologous chromosomes in man and apes, *Proc. Natl. Acad. Sci. U.S.A.,* 77, 7323, 1980.

85. **Gerlach, W. L. and Bedbrook, J. R.,** Cloning and characterization of rRNA genes from wheat and barley, *Nucleic Acids Res.,* 7, 1869, 1979.

86. **Goldsbrough, P. B. and Cullis, C. A.,** Characterization of the genes for ribosomal RNA in flax, *Nucleic Acids Res.,* 9, 1301, 1981.

87. **Ellis, T. H. N., Goldsbrough, P. B., and Castleton, J. A.,** Transcription and methylation of flax rDNA, *Nucleic Acids Res.,* 11, 3047, 1983.

88. **Schibler, U., Wyler, T., and Hagenbüchle, O.,** Changes in size and secondary structure of the ribosomal transcription unit during vertebrate evolution, *J. Mol. Biol.,* 94, 503, 1975.

89. **Salim, M. and Maden, B. E. H.,** Nucleotide sequence of *Xenopus laevis* 18S ribosomal RNA inferred from gene sequence, *Nature London,* 291, 205, 1981.

90. **Torczynski, R., Bollon, A. P., and Fuke, M.,** The complete nucleotide sequence of the rat 18S ribosomal RNA gene and comparison with the respective yeast and frog genes, *Nucleic Acids Res.,* 11, 4879, 1983.

91. **Chan, Y. L., Gutell, R., Noller, H. F., and Wool, I. G.,** The nucleotide sequence of a rat 18S ribosomal RNA gene and proposal for the secondary structure of 18S ribosomal RNA, *J. Biol. Chem.,* 259, 224, 1984.

92. **Raynal, F., Michot, B., and Bachellerie, J. P.,** Complete nucleotide sequence of mouse 18S rRNA gene: comparison with other available homologs, *FEBS Lett.,* 167, 263, 1984.

93. **Connaughton, J. F., Rairkar, A., Lockard, R. E., and Kumar, A.,** Primary structure of rabbit 18S ribosomal RNA determined by direct RNA sequence analysis, *Nucleic Acids Res.,* 12, 4731, 1984.

94. **Takaiwa, F., Oono, K., and Sugimura, M.,** The complete nucleotide sequence of a rice 17S rRNA gene, *Nucleic Acids Res.,* 12, 5441, 1984.

95. **Messing, J., Carlson, J., Hagen, G., Rubenstein, I., and Oleson, A.,** Cloning and sequencing of the ribosomal RNA genes in maize: the 17S region, *DNA,* 3, 31, 1984.

96. **Hassouna, N., Michot, B., and Bachellerie, J. P.,** The complete nucleotide sequence of mouse 28S rRNA gene. Implications for the process of size increase of the large subunit rRNA in higher eukaryotes, *Nucleic Acids Res.,* 12, 3563, 1984.

97. **Ware, V. C., Tague, B. W., Clark, C. G., Gourse, R. L., Brand, R. C., and Gerbi, S. A.,** Sequence analysis of 28S ribosomal RNA from the amphibian *Xenopus laevis, Nucleic Acids Res.,* 11, 7795, 1983.

98. **Veldman, G. M., Klootwijk, J., deRegt, V. C. H. F., Planta, R. J., Branlant, C., Krol, A., and Ebel, J. P.,** The primary and secondary structure of yeast 26S rRNA, *Nucleic Acids Res.,* 9, 6935, 1981.

99. **Otsuka, T., Nomiyama, H., Yoshida, H., Kukita, T., Kuhara, S., and Sakaki, Y.,** Complete nucleotide sequence of the 26S rRNA gene of *Physarum polycephalum:* its significance in gene evolution, *Proc. Natl. Acad. Sci. U.S.A.,* 80, 3163, 1983.

100. **Chan, Y. L., Olvera, J., and Wool, I. G.,** The structure of rat 28S rRNA inferred from the sequence of nucleotides in a gene, *Nucleic Acids Res.,* 11, 7819, 1983.

101. **Hall, L. M. C. and Maden, B. E. H.,** Nucleotide sequence through the 18S-28S intergene region of a vertebrate ribosomal transcriptional unit, *Nucleic Acids Res.,* 8, 5993, 1980.

102. **Furlong, J. C. and Maden, B. E. H.,** Patterns of major divergence between the internal transcribed spacers of ribosomal DNA in *Xenopus borealis* and *Xenopus laevis,* and of minimal divergence within ribosomal coding regions, *EMBO J.,* 2, 442, 1983.

103. **Michot, B., Bachellerie, J. P., and Raynal, F.,** Structure of mouse rRNA precursors. Complete sequence and potential folding of the spacer regions between 18 and 28S rRNA, *Nucleic Acids Res.,* 11, 3375, 1983.

104. **Subrahmanyam, C. S., Cassidy, B., Busch, H., and Rothblum, L. I.,** Nucleotide sequence of the region between the 18S rRNA sequence and the 28S rRNA sequence of rat ribosomal DNA, *Nucleic Acids Res.,* 10, 3667, 1982.

105. **Hindenach, B. R. and Stafford, D. W.,** Nucleotide sequence of the 18S-28S rRNA intergene region of the sea urchin, *Nucleic Acids Res.,* 12, 1737, 1984.

106. **Ozaki, T., Hoshikawa, Y., Iida, Y., and Iwabuchi, M.,** Sequence analysis of the transcribed and 5' non-transcribed regions of the rRNA gene in *Dictyostelium discoideum, Nucleic Acids Res.,* 12, 4171, 1984.

107. **Kumar, A. and Warner, J. R.,** Characterization of ribosomal precursor particles from HeLa cell nucleoli, *J. Mol., Biol.* 63, 233, 1972.

108. **Financsek, I., Mizumoto, K., Mishima, Y., and Muramatsu, M.,** Human ribosomal RNA gene: nucleotide sequence of the transcription initiation region and comparison of three mammalian genes, *Proc. Natl. Acad. Sci. U.S.A.,* 79, 3092, 1982.

109. **Rothblum, L. I., Reddy, R., and Cassidy, B.,** Transcription initiation site of rat ribosomal DNA, *Nucleic Acids Res.,* 10, 7345, 1982.

110. **Harrington, C. A. and Chikaraishi, D.,** Identification and sequence of the initiation site for rate 45S ribosomal RNA synthesis, *Nucleic Acids Res.,* 11, 3317, 1983.

111. **Bach, R., Grummt, I., and Allet, B.,** The nucleotide sequence of the initiation region of the ribosomal transcription unit from mouse, *Nucleic Acids Res.,* 9, 1559, 1981.

112. **Grummt, I.,** Mapping of a mouse ribosomal DNA promoter by *in vitro* transcription, *Nucleic Acids Res.,* 9, 6093, 1981.

113. **Miller, K. G. and Sollner-Webb, B.,** Transcription of mouse rRNA genes by RNA Polymerase. I. In vitro and in vivo initiation and processing sites, *Cell,* 27, 165, 1981.

114. **Bowman, L. H., Goldman, W. E., Goldberg, G. I., Herbert, B., and Schlessinger, D.,** Location of the initial cleavage sites in mouse pre-rRNA, *Mol. Cell. Biol.,* 3, 1501, 1983.

115. **Bach, R., Allet, B., and Crippa, M.,** Sequence organization of the spacer in the ribosomal genes of *Xenopus clivii* and *Xenopus borealis, Nucleic Acids Res.,* 9, 5311, 1981.

116. **Nazar, R. N.,** A 5.8S rRNA-like sequence in prokaryotic 23S rRNA, *FEBS Lett.,* 119, 212, 1980.

117. **Jacq, B.,** Sequence homologies between eukaryote 5.8S rRNA and the 5' end of prokaryote 23S and RNA: evidences for a common evolutionary origin, *Nucleic Acids Res.,* 9, 2913, 1981.

118. **Michot, B., Bachellerie, J. P., and Raynal, F.,** Sequence and secondary structure of mouse 28S rRNA 5' terminal domain. Organization of the 5.8-28S rRNA complex, *Nucleic Acids Res.,* 10, 5273, 1982.

119. **Qu, L. H., Michot, B., and Bachellerie, J. P.,** Improved methods for structure probing in large RNAs: a rapid ''heterologous'' sequencing approach is coupled to the direct mapping of nuclease accessible sites. Application to the 5' terminal domain of eukaryotic 28S rRNA, *Nucleic Acids Res.,* 11, 5903, 1983.

120. **Bachellerie, J. P., Michot, B., and Raynal, F.,** Recognition signals for mouse pre-rRNA processing. A potential role of U3 nucleolar RNA, *Mol. Biol. Rep.,* 9, 79, 1983.

121. **Reddy, R., Henning, D., and Busch, H.,** Nucleotide sequence of nucleolar U3B RNA, *J. Biol. Chem.,* 254, 11097, 1979.

122. **Kelly, J. M. and Maden, B. E. H.,** Chemical modification studies and the secondary structure of Hela cell 5.8S rRNA, *Nucleic Acids Res.,* 8, 4521, 1980.

123. **Crouch, R. J., Kanaya, S., and Earl, P. L.,** A model for the involvement of the small nucleolar RNA (U3) in processing eukaryotic ribosomal RNA, *Mol. Biol. Rep.,* 9, 75, 1983.

123a. **Maden, B. E. H., Moss, M., and Salim, M.,** Nucleotide sequence of an external transcribed spacer in *Xenopus laevis* rDNA: sequences flanking the 5' and 3' ends of 18S rRNA are non-complementary, *Nucleic Acids Res.,* 10, 2387, 1982.

123b. **Raynal, F., Michot, B., and Bachellerie, J. P.,** Primary sequence of the 5' terminal region of mouse 18S rRNA and adjacent spacer. Implications for rRNA processing, *FEBS Lett.,* 161, 135, 1984.

123c. **Bowman, L. H.,** personal communication.

123d. **Melton, D. A., Kreig, P. A., Rebagliati, M. R., Maniatis, T., Zinn, K., and Green, M. R.,** Efficient in vitro synthesis of biologically active RNA and RNA hybridization probes from plasmids containing a bacteriophage SP6 promoter, *Nucleic Acids Res.,* 12, 7035, 1984.

124. **Attardi, G.,** RNA synthesis and processing in mitochondria, in *Processing of RNA,* Apirion, D., Ed., CRC Press, Boca Raton, Fla., 1984, 227.

125. **Grant, D. M. and Lambowitz, A. M.,** Mitochondrial rRNA genes, in *The Cell Nucleus,* Vol. 10, Busch, H. and Rothblum, L., Eds., Academic Press, New York, 1982, 387.

126. **Anderson, S., Bankier, A. T., Barrell, B. G., de Bruijn, M. H. L., Coulson, A. R., Drouin, J., Eperon, I. C., Nierlich, D. P., Roe, B. A., Sanger, F., Schreier, P. H., Smith, A. J. H., Staden, R., and Young, I. G.,** Sequence and organization of the human mitochondrial genome, *Nature (London),* 290, 457, 1981.

127. **Saccone, C., Cantatore, P., Gadaleta, G., Gallerani, R., Lanave, C., Pepe, G., and Kroon, A. M.,** The nucleotide sequence of the large rRNA gene and the adjacent tRNA genes from rat mitochondria, *Nucleic Acids Res.,* 9, 4139, 1981.

128. **Bibb, M. J., Van Etten, R. A., Wright, C. T., Walberg, M. W., and Clayton, D. A.,** Sequence and gene organization of mouse mitochondrial DNA, *Cell,* 26, 167, 1981.

129. **Anderson, S., de Bruijn, M. H. L., Coulson, A. R., Eperon, I. C., Sanger, F., and Young, I. G.,** The complete sequence of bovine mitochondrial DNA: conserved features of the mammalian mitochondrial genome, *J. Mol. Biol.,* 156, 683, 1982.

130. **Glaus, K. R., Zassenhaus, H. P., Fechheimer, N. S., and Perlman, P. S.,** in *The Organization and Expression of the Mitochondrial Genome,* Kroon, A. M. and Saccone, C., Eds., Elsevier/North Holland, Amsterdam, 1980, 131.

131. **Rastl, E. and Dawid, I. B.,** Expression of the mitochondrial genome in *Xenopus Laevis:* a map of transcripts, *Cell,* 18, 501, 1979.

132. **Simpson, L. and Simpson, A.,** Kinetoplast RNA of *L. tarentolae, Cell,* 14, 169, 1978.

133. **Eperon, I. C., Anderson, S., and Nierlich, D. P.,** Distinctive sequence of human mitochondrial rRNA genes, *Nature (London),* 286, 460, 1980.

134. **Miyata, T., Hayashida, H., Kikuno, R., Hasegawa, M., Kobayashi, M., and Koike, K.,** Molecular clock of silent changes in mitochondrial genes over those in nuclear genes, *J. Mol. Evolution,* 19, 28, 1982.

135. **Nierlich, D. P.,** Fragmentary 5S rRNA gene in the human mitochondrial genome, *Mol. Cell Biol.,* 2, 207, 1982.

136. **Ojala, D., Merkel, C., Gelfand, R., and Attardi, G.,** The tRNA genes punctuate the reading of genetic information in human mitochondrial DNA, *Cell,* 22, 393, 1980.

137. **Crews, S. and Attardi, G.,** The sequences of the small rRNA gene and the phenylalanine tRNA gene are joined end to end in human mitochondrial DNA, *Cell,* 19, 775, 1980.

138. **Murphy, W. I., Attardi, B., Tu, C., and Attardi, G.,** Evidence for complete symmetrical transcription *in vivo* of mitochondrial DNA in HeLa cells, *J. Mol. Biol.,* 99, 809, 1975.

139. **Cantatore, P. and Attardi, G.,** Mapping of nascent light and heavy strand transcripts on the physical map of HeLa cell mitochondrial DNA, *Nucleic Acids Res.,* 8, 2605, 1980.

140. **Aloni, Y. and Attardi, G.,** Symmetric *in vivo* transcription of mitochondrial DNA in HeLa cells, *Proc. Natl. Acad. Sci. U.S.A.,* 68, 1957, 1971.

141. **Bothwell, A. L. M., Stark, B. C., and Altman, S.,** Ribonuclease P substrate specificity: cleavage of a bacteriophage ϕ 80-induced RNA, *Proc. Natl. Acad. Sci. U.S.A.,* 73, 1912, 1976.

142. **Ojala, D., Montoya, J., and Attardi, G.,** The tRNA punctuation model of RNA processing in human mitochondria, *Nature (London),* 290, 470, 1981.

143. **Gelfand, R. and Attardi, G.,** Synthesis and turnover of mitochondrial RNA in HeLa cells: the mature ribosomal and messenger RNA species are metabolically unstable, *Mol. Cell. Biol.,* 1, 497, 1981.

144. **Attardi, G., Cantatore, P., Ching, E., Crews, S., Gelfand, R., Merkel, C., Montoya, J., and Ojala, D.,** The remarkable features of gene organization and expression of human mitochondrial DNA, in *The Organization and Expression of the Mitochondrial Genome,* Kroon, A. M., and Saccone, C., Eds., Elsevier/North Holland, Amsterdam, 1980, 103.

145. **Dubin, D. T., Montoya, J., Timko, K. D., and Attardi, G.,** Sequence analysis and precise mapping of the 3′ ends of HeLa cell mitochondrial rRNAs, *J. Mol. Biol.,* 157, 1, 1982.

146. **Rosenberg, M. and Court, D.,** Regulatory sequences involved in the promotion and termination of RNA transcription, *Ann. Rev. Genet.,* 13, 319, 1979.

147. **Montoya, J., Christianson, T., Levens, D., Rabinowitz, M., and Attardi, G.,** Identification of initiation sites for heavy strand and light strand transcription in human mitochondrial DNA, *Proc. Natl. Acad. Sci. U.S.A.,* 79, 7195, 1982.

148. **Montoya, J., Gaines, G. L., and Attardi, G.,** The pattern of transcription of the human mitochondrial rRNA genes reveals two overlapping transcription units, *Cell,* 34, 151, 1983.

149. **Yanofsky, C.,** Attenuation in the control of expression of bacterial operons, *Nature (London),* 289, 751, 1981.

150. **Leaver, C. J. and Gray, M. W.,** Mitochondrial genome organization and expression in higher plants, *Ann. Rev. Plant Physiol.,* 33, 373, 1982.

151. **Ward, B. L., Anderson, R. S., and Bendich, A. J.,** The mitochondrial genome is large and variable in a family of plants (cucurbitacae), *Cell,* 25, 793, 1981.

152. **Leaver, C. J. and Harmey, M. A.,** Higher plant mitochondrial ribosomes contain a 5S rRNA component, *Biochem. J.,* 157, 275, 1976.

153. **Cunningham, R. S., Bonen, L., Doolittle, W. F., and Gray, M. W.,** Unique species of 5S, 18S and 26S rRNA in wheat mitochondria, *FEBS Lett.,* 69, 116, 1976.

154. **Spencer, D. F., Bonen, L., and Gray, N. W.,** Primary sequence of wheat mitochondrial 5S RNA: functional and evolutionary implications, *Biochemistry,* 20, 4022, 1981.

155. **Bonen, L. and Gray, N. W.,** Organization and expression of the mitochondrial genome of plants. I. The gene for wheat mitochondrial ribosomal and transfer RNA: evidence for an unusual arrangement, *Nucleic Acids Res.,* 8, 319, 1980.

156. **Iams, K. P. and Sinclair, J. H.,** Mapping the mitochondrial DNA of *Zea mays:* ribosomal gene localization, *Proc. Natl. Acad. Sci. U.S.A.,* 79, 5926, 1982.

157. **Stern, D. B., Dyer, T. A., and Lonsdale, D. M.,** Organization of the mitochondrial ribosomal RNA genes of maize, *Nucleic Acids Res.,* 10, 3333, 1982.

158. **Bonen, L., Cunningham, R. S., Gray, M. W., and Doolittle, W. F.,** Wheat embryo mitochondrial 18S rRNA: evidence for its prokaryotic nature, *Nucleic Acids Res.,* 4, 663, 1977.

159. **Spencer, D. F., Schnare, M. N., and Gray, N. W.,** Pronounced structural similarities between the small subunit rRNA genes of wheat mitochrondria and *E. coli, Proc. Natl. Acad. Sci. U.S.A.,* 81, 493, 1984.

160. **Chao, S., Sederoff, R., and Levings, C. S.,** Nucleotide sequence and evolution of the 18S rRNA gene in maize mitochondria, *Nucleic Acids Res.,* 12, 6629, 1984.

161. **Gray, M. W. and Spencer, D. F.,** Wheat mitochrondrial DNA encodes a eubacteria-like initiator methionine tRNA, *FEBS Lett.,* 161, 323, 1983.

162. **Chao, S., Sederoff, R. R., and Levings, C. S.,** Partial sequence analysis of the 5S to 18S rRNA gene region of the maize mitochondrial genome, *Plant Physiol.,* 71, 190, 1983.

163. **Bedbrook, J. R. and Kolodner, R.,** Structure of chloroplast DNA, *Annu. Rev. Plant Physiol.,* 30, 593, 1979.

164. **Rochaix, J. D.,** Genetic organization of the chloroplast, *Int. Rev. Cytol,* 93, 57, 1985.

165. **Kolodner, R. and Tewari, K. K.,** The molecular size and conformation of the chloroplast DNA from higher plants, *Biochim. Biophys. Acta,* 402, 372, 1975.

166. **Kolodner, R. and Tewari, K. K.,** Denaturation mapping studies on the circular chloroplast DNA from pea leaves, *J. Biol. Chem.,* 250, 4888, 1975.

167. **Chu, N. W., Oishi, K. K., and Tewari, K. K.,** Physical mapping of the pea chloroplast DNA and localization of the ribosomal RNA genes, *Plasmid,* 6, 279, 1981.

168. **Koller, B. and Delius, H.,** Vicia faba chloroplast DNA has only one set of ribosomal RNA genes as shown by partial denaturation mapping and R-loop analysis, *Mol. Gen. Genet.,* 178, 261, 1980.

169. **Palmer, J. D. and Thompson, W. F.,** Chloroplast DNA rearrangements are more frequent when a large inverted repeat sequence is lost, *Cell,* 29, 537, 1982.

170. **Herrmann, R. G., Seyer, P., Schedl, R., Gordon, K., Bisanz, C., Winter, P., Hildebrandt, J. W., Wlaschek, M., Alt, J., Driesel, A. J., and Sears, B. B.,** The plastid chromosomes of several dicotyledons, in *Biological Chemistry of Organelle Formation,* Bucher, T., Sebald, W., and Weiss, H., Eds., Springer-Verlag, Berlin, 1980, 97.

171. **Palmer, J. D.,** Chloroplast DNA exists in two orientation, *Nature (London),* 301, 92, 1983.

172. **Schwarz, Z. and Kossel, H.,** The primary structure of 16S rDNA from Zea mays chloroplast is homologous to *E. coli* 16S rRNA, *Nature (London)*, 283, 739, 1980.

173. **Schwarz, Z., Kossel, H., Schwartz, E., and Bogorad, L.,** A gene coding for tRNA^{Val} is located near 5′ terminus of 16S rRNA gene in *Zea mays* chloroplast genome, *Proc. Natl. Acad. Sci. U.S.A.*, 78, 4748, 1981.

174. **Koch, W., Edwards, K., and Kossel, H.,** Sequencing of the 16S-23S spacer in a ribosomal RNA operon of Zea mays chloroplast DNA reveals two split tRNA genes, *Cell*, 25, 203, 1981.

175. **Edwards, K. and Kossel, H.,** The rRNA operon from *Zea mays* chloroplasts: nucleotide sequence of 23S rDNA and its homology with *E. coli* 23S rDNA, *Nucleic Acids Res.*, 9, 2853, 1981.

176. **Tohdoh, N., Shinozaki, K., and Sugiura, M.,** Sequence of a putative promoter region for the rRNA genes of tobacco chloroplast DNA, *Nucleic Acids Res.*, 9, 5399, 1981.

177. **Tohdoh, N. and Sugiura, M.,** The complete nucleotide sequence of a 16S ribosomal RNA gene from tobacco chloroplasts, *Gene*, 17, 213, 1982.

178. **Takaiwa, F. and Sugiura, M.,** Nucleotide sequence of the 16S-23S spacer region in an rRNA gene cluster from tobacco chloroplast DNA, *Nucleic Acids Res.*, 10, 2665, 1982.

179. **Takaiwa, F. and Sugiura, M.,** The complete nucleotide sequence of a 23S rRNA gene from tobacco chloroplasts, *Eur. J. Biochem.*, 124, 13, 1982.

180. **Whitfeld, P. R., Herrmann, R. G., and Bottomley, W.,** Mapping of the ribosomal RNA genes on spinach chloroplast DNA, *Nucleic Acids Res.*, 5, 1741, 1978.

181. **Chu, N. M. and Tewari, K. K.,** Arrangement of the ribosomal RNA genes in chloroplast DNA of Leguminosae, *Mol. Gen. Genet.*, 186, 23, 1982.

182. **Mubumbila, M. Gordon, K. H. J., Crouse, E. J., Burckard, G., and Weil, J. H.,** Construction of the physical map of the chloroplast DNA of *Phaseolus vulgaris* and localization of ribosomal and transfer genes, *Gene*, 21, 257, 1983.

183. **Palmer, J. D.,** Physical and gene mapping of chloroplast DNA from *Atriplex triangularis* and *Cucumis sativa*, *Nucleic Acids Res.*, 10, 1593, 1982.

184. **Kumano, M., Tomioka, N., and Sugiura, M.,** The complete nucleotide sequence of a 23S gene from a blue-green alga, *Anacystis nidulans*, *Gene*, 24, 219, 1983.

185. **Machatt, M. A., Ebel, J. P., and Branlant, C.,** The 3′ terminal region of bacterial 23S ribosomal RNA: structure and homology with the 3′ terminal region of eukaryotic 28S rRNA and with chloroplast 4.5S rRNA, *Nucleic Acids Res.*, 9, 1533, 1981.

186. **Hartley, M. R. and Head, C.,** The synthesis of chloroplast high-molecular weight ribosomal ribonucleic acid in spinach, *Eur. J. Biochem.*, 96, 301, 1979.

187. **Lund, E., Dahlberg, J. E., Lindahl, L., Jaskunas, S. R., Dennis, P. P., and Nomura, M.,** Transfer RNA genes between 16S and 23S rRNA genes in rRNA transcription units of *E. coli*, *Cell*, 7, 165, 1976.

188. **Young, R. A., Macklis, R., and Steitz, J. A.,** Sequence of the 16S-23S spacer region in two rRNA operons of *E. coli*, *J. Biol. Chem.*, 254, 3264, 1979.

189. **Jurgenson, J. E. and Bourque, D. P.,** Mapping of rRNA genes in an inverted repeat in *Nicotiana tabacum* chloroplast DNA, *Nucleic Acids Res.*, 8, 3505, 1980.

190. **Briat, J. F., Dron, M., Loiseaux, S., and Mache, R.,** Structure and transcription of the spinach chloroplast rDNA leader region, *Nucleic Acids Res.*, 10, 6865, 1982.

191. **Hartley, M. R.,** The synthesis and origin of chloroplast low molecular weight rRNA in spinach, *Eur. J. Biochem.*, 96, 311, 1979.

192. **Sri Widada, J. and Bachellerie, J. P.,** Unpublished results.

193a. **Seilhamer, J. J., Olsen, G. J., and Cummings, D. J.,** *Paramecium* mitochondrial genes. I. Small subunit rRNA gene sequence and microevolution, *J. Biol. Chem.*, 259, 5167, 1984.

193b. **Seilhamer, J. J., Gutell, R. R., and Cummings, D. J.,** *Paramecium* mitochondrial genes. II. Large subunit rRNA gene sequence and Microevolution, *J. Biol. Chem.*, 259, 5173, 1984.

194. **Tu, J. and Zillig, W.,** Organization of rRNA structural genes in the archaebacterium *Thermoplasma acidophilum*, *Nucleic Acids Res.*, 10, 7231, 1982.

195. **Rochaix, J.-D. and Darlix, J.-L.,** Composite structure of the chloroplast 23S ribosomal RNA genes of *Chlamydomonas reinhardii* evolutionary and functional implications, *J. Mol. Biol.*, 159, 383, 1982.

196. **Maly, P. and Brimacombe, R.,** Refined secondary structure models for 16S and 23S ribosomal RNA of *E. coli*, *Nucleic Acids Res.*, 11, 7263, 1983.

197. **Stiegler, P., Carbon, P., Ebel, J. P., and Ehresmann, C.** A general secondary structure model for procaryotic and eucaryotic RNAs of the small ribosomal subunits, *Eur. J. Biochem.*, 120, 487, 1981.

Chapter 4

ANALYSIS OF SMALL RNA SPECIES: PHYLOGENETIC TRENDS

Mirko Beljanski and Liliane Le Goff

TABLE OF CONTENTS

I. INTRODUCTION

Genetic information in eukaryotes and prokaryotes is stored in DNA molecules whose specific segments are transcribed by appropriate polymerases into various types of RNAs: messenger RNA (mRNA), ribosomal RNA (rRNA), transfer RNA (tRNA), and many other small RNAs (50 to 360 nucleotides). Information in viruses is stored either in DNA or RNA strands. A large variety of small RNAs is present in the nucleus of eukaryotic cells and/or in the cytoplasm of mammals, plants, viruses, and bacteria. The variety of these RNAs, as well as their "constant" or supposed engagement in various biological processes, raises the question of their possible phylogenetic relationship. Phylogenetic measurement, extensively accomplished for rRNAs from different species, has also been studied for various small RNAs. Thus, sequence data of cytoplasmic 5S rRNAs from metazoan somatic cells have been used to analyze the phylogenetic relationships. Many small RNAs have been characterized by sequence analysis, and by physical means, for some of them by proposed or established biological activity. Thus, purine-rich small RNAs stimulate the translation of mRNA[1] while others induce the transformation of bacteria.[2]

The widespread existence of endo- and exonucleases in all living organisms forces one to consider the participation of these enzymes in producing small RNAs which are more or less rich in purine or pyrimidine nucleotides. Some of these enzymes are involved in the formation of mature "mosaic" messenger RNAs, rRNAs, tRNAs, etc., which leaves intervening sequences "unused".

The over-production of rRNAs before cell division and their disappearance as soon as cell multiplication begins[3,4] raise the question of the function of so many extra copies of rRNAs. They may be the target for specific nucleases which deliver different small RNAs, as in the case of 5.8S rRNA emergence from prokaryotic 23S rRNA. Small RNAs interfere through base pairing with large DNA and/or RNAs during synthesis of cell constituents, cell multiplication, and differentiation.[2,5]

Small RNAs can be synthesized *de novo* from ribonucleoside-5'-triphosphates by *Escherichia coli* Qβ replicase, an enzyme which also replicates Qβ RNA template, as well as satellite RNA of Qβ virion. Polynucleotide phosphorylase (PNPase) from bacteria can also synthesize RNA from ribonucleoside-5'-diphosphates.[6] These facts and the observation that small RNAs from different origins can be transcribed into DNA, contribute to modifying notions about the origin and flow of information in cells. For these and other reasons, scientists have thus studied small RNAs and have visualized their possible participation in the creation of new genes and/or pseudogenes. In this chapter, we shall attempt to give an overview of many different small RNAs by describing their chemical and physical properties as well as their evident, or possible biological role. The phylogenetic trends between these RNAs will be evaluated on the basis of sequence data since at the present time the biological functions of the majority of small RNAs are unknown.

II. TRANSFER RNAs

Present in all living organisms, transfer RNAs are small molecules (#75 nucleotides). They bind activated amino acids, recognize a three-base "codon" (in mRNA) which specifies a particular amino acid, and deliver that amino acid to be incorporated into the polypeptide chain. Alanine transfer RNA (tRNA$_{ala}$) was the first RNA to be sequenced,[7] and its secondary structure has been proposed (cloverleaf diagram) (Figure 1). The importance of the structure, conformation, and interaction of tRNAs with other molecules was recently reviewed.[8] There are about 40 functionally different kinds of tRNAs which seemingly have a complicated molecular geometry and which appear to be of very ancient ancestry.[9] All tRNAs contain a double helical structure as proposed by Holley et al.[7] Three important features characterize

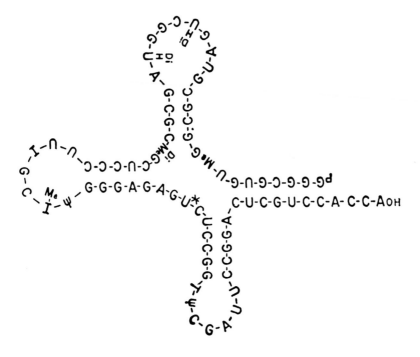

FIGURE 1.   Schematic representation of conformation of the alanine tRNA with short double-stranded regions. (From Holley, R. W. et al., *Science*, 47, 1462, 1965. With permission.)

tRNAs: (1) the presence of unusual bases — pseudouracil, dihydrouracil, hypoxanthine, ribothymine, thiopyrimidines, and various methylated purines; (2) terminal sequences — CCA; and (3) several "invariant" and "semivariant" residues located in the same relative position.[10] The importance and function of modified nucleotides in tRNAs have been reviewed.[11] The sequences of about 120 tRNAs from different biological systems have been determined.[12] There exist tRNAs which initiate protein synthesis; in eukaryotes one tRNA carries methionine and in prokaryotes N-formyl methionine. Since their discovery,[13] the biological activities of tRNAs have been extensively studied in vitro and in vivo.[14]

A. tRNA Genes

Eukaryote and prokaryote tRNA genes are transcribed from DNA as independent units by RNA polymerase III first as precursor transcripts[15-19] (Figure 2). The tRNA precursor cannot be amino-acylated in vitro.[20] The $tRNA_{tyr}$ transcription unit was studied in particular detail because the corresponding gene was synthesized in vitro by chemical and enzymatic means and introduced under in vivo conditions where it functioned correctly.[21,22] Cloned tRNA genes, injected into the large oocyte nucleus of *Xenopus*[23] can also generate correct tRNAs. The number of tRNA genes varies according to species. The haploid yeast genome contains about 300 tRNA genes,[24] *Drosophila*, 600,[25] and *Xenopus*, 8000.[26]

In yeast mitochondria (mt), tRNA genes are clustered,[27] while in HeLa cells and *Xenopus* they appear to be dispersed.[28,29] mt DNA encodes tRNAs which are different from those transcribed in the nucleus (UU in the D stem replaces GU base pairs).[30] Since the degree of homology of mammalian mt tRNAs with non-mt species is between 30 and 50%, the integration of mitochondrial sequences into tRNA evolutionary trees is difficult, although mt tRNA genes are highly conserved.[31] Clustered or dispersed tRNA genes are several times larger than one tRNA length.[32] Several steps in the splicing of the precursor transcripts are required to obtain mature tRNA. The -CCA sequence present in all mature tRNAs is not always encoded in the DNA.[33] In yeast the 3' terminal -CCA residues of $tRNA_{tyr}$ genes are

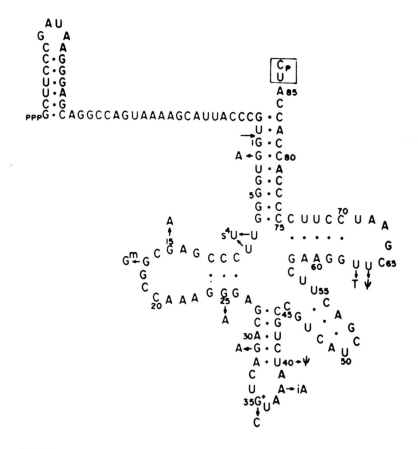

FIGURE 2.   Nucleotide sequence of a precursor to *E. coli* tRNA[tyr]. (From Altman, S. and Smith, J. D., *Nature (London) New Biol.*, 233, 35, 1971. With permission.)

not present but appear to be in *E. coli*.[17,34] They are added later by a nucleotidyl-transferase. The intervening sequences which are transcribed within tRNA genes differ according to the tRNA species and all contain high A + U base composition (A + U/G + C = 2/1).[12]

In prokaryotes tRNA genes are clustered and transcribed in "in vivo" conditions as multicistronic precursor molecules. Transcription starts at the first tRNA gene in a gene cluster and terminates at the last gene. Subsequently, the 5′ and 3′ flanking sequences are removed, liberating mature tRNA.[35]

In retroviruses tRNA molecules are present in large quantities.[36-42] About 200 tRNA molecules are selected from the host cell and included in a virion particle. Most tRNAs bound with 18S or 28S rRNA are unable to accept amino acids unless they are first released by heat treatment. This indicates at least one set of common sequences between tRNAs and rRNAs.

It should be noted that phylogenetically related vertebrates show more similar tRNA-rRNA interaction than do phylogenetically distant vertebrates.[43] tRNA serving as a primer for the reverse transcriptase of genomic RNA of a type C retrovirus binds to 28S rRNA of its host.[43] Nucleotide sequences characteristic of these tRNAs also play a role in the binding of tRNA to 5S rRNA.[44]

Retrovirus tRNA-host rRNA hybridization shows that nucleotide sequence of rRNAs binding the tRNAs are different among various vertebrates. Similar results were obtained with retrovirus genome RNAs, thus indicating a certain degree of nucleotide recognition between retrovirus genome RNAs and rRNAs with tRNA.

III. 5S RIBOSOMAL RNA

5S rRNA was first discovered as a component of a 50S ribosomal subunit of *E. coli*,[45] and thereafter it was further analyzed.[46] The sequence of that 5S rRNA was determined by Sanger et al. in 1968,[47] and since then many prokaryotic and eukaryotic 5S rRNAs have been sequenced. Mammalian 5S rRNAs are considered to be practically identical.[48] This RNA contains about 120 to 121 nucleotides but no methylated or otherwise modified nucleotides. 5S rRNA is transcribed by RNA polymerase III from multiple structural genes. The in vitro synthesized product was identical to 5S rRNA transcribed from plasmid DNA which contains the *X. laevis* oocyte 5S gene.[49] In vitro hybridization of transcribed 5S rRNA with DNA fragments obtained by Eco RI digestion of calf thymus DNA shows that it originates in DNA. Calf thymus chromatin used as an in vitro template does not deliver 5S rRNA.[49]

The number of nucleotide substitutions between 5S rRNAs from vertebrates and invertebrates varies to a significant degree. The 5S rRNA sequences are identical between different genera but are distant from those which belong to the different order. Nucleotide sequences of 5S rRNA from the sponge which is the simplest form of multicellular animals are closely related to invertebrates. Tunicate has affinity to vertebrates.[50] The phylogenetic positions of 5S rRNA from these two species among metazoans were derived from the 5S rRNA sequences by a computer analysis based on the maximum parsimony principle. This method utilizing information contained in the sequence data differs from the conventional taxonomy.[50]

A. Secondary Structure of 5S rRNA

To obtain information about evolutionary changes, it is important not only to compare sequences of 5S rRNAs, but also to establish the secondary structure of this molecule from different species. The archaebacterial 5S rRNA secondary structure resembles typical bacterial 5S rRNA more than the eukaryotic 5S rRNAs do.[51] These observations support the view that the secondary structure of the 5S rRNA remains the "same" in the corresponding site of the stem and has been conserved throughout evolution, although the primary structure of these RNAs has undergone changes.[30]

Recently, with 17 prokaryotic 5S rRNAs used for comparison, a general model for the secondary structure of 5S rRNA has been proposed.[52] In this model eight double helical regions seem to be present and involved in the function of 5S rRNA (Figure 3).

B. 5S rRNA Gene

The genes of 5S rRNA are located in clusters at the telomeres of practically all chromosomes. DNA containing the oocytes 5S rRNA genes has been isolated from *X. laevis* and characterized. One 5S gene is part of a larger repeating unit in the DNA.[53] Sequences of the purified 5S DNA of *X. laevis* have been determined,[54] and the primary structure of the repeating unit in oocyte 5S DNA has been described.[48,55] The spacer sequences differ in base composition and internal repetitiveness. The 5S rRNA genes appear to evolve as a unit. The spacer sequence which is part of the repeating unit in a tandem gene cluster is not part of the gene.[56]

In the 5S DNA, the sequence coding for 5S rRNA alternates with a "spacer" region about 6 times as long.[56] It has been shown that in a cloned oocyte 5S DNA fragment from *X. borealis*, three 5S rRNA genes are separated by about 80 nucleotides.[57] RNA polymerase III binds to a "control region" that differs from promoter regions in prokaryotes. The enzyme correctly transcribes in vitro 5S rRNAs from cloned 5S DNA fragments. Although *Xenopus* and mammalian 5S rRNAs represent each a single homogeneous RNA of 120 nucleotides, they differ by 8 nucleotides.[54] The 5S rRNA sequences of all mammals that have been analyzed, including man, mouse, and kangaroo, are practically identical.[58] The

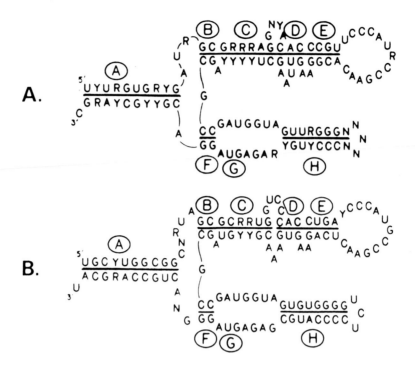

FIGURE 3.   Schematic representation of conformation of 5S rRNAs from Gram positive bacteria (A) and from Gram negative bacteria (B). (From Studnicka, G. M. et al., *Nucleic Acids Res.*, 9, 1885, 1981. With permission.)

5S rRNA sequence in the chicken differs from the mammalian sequence by three base substitutions, two nucleotide additions, and two deletions. The sequence of 5S rRNA from *Bacillus megaterium* is related to that of *B. stearothermophilus* but differs by 23 base replacements and is 3 nucleotides shorter.[59]

The nucleotide sequence of 5S rRNA of vertebrates is apparently conserved throughout evolution. 5S DNA from different species and mouse satellite DNA have undergone frequent unequal homologous or nonhomologous crossing over during evolution.[56] It has been suggested that spacers and particularly repetitive sequences are ''critically important'' to both the stability and the evolutionary flexibility of the multigene family. The introduction of the development of repetitive spacers would tend to enhance the overall duplication/deletion rate in a multigene family.[48] According to these authors, there is a greater divergence of spacers than of gene sequences in homologous 5S gene clusters of different *Xenopus* species. This suggests that the somatic and oocyte genes present in the thousands of copies in *Xenopus* have evolved essentially as separate lines. The difference between 5S rRNA from *Xenopus* somatic and oocyte genes is 6 nucleotides.[48] The *Xenopus* somatic type 5S rRNA does not comprise more than 3% of the genes within 5S DNA. The 5S DNA of chicken embryo fibroblasts delivers an RNA copy of 121 base pairs and spacers of an approximate overall average length of 750 pairs.[60]

C. The Biological Function of 5S rRNA

Reconstitution experiments have demonstrated the importance of 5S rRNA as a structural component in the 50S ribosomal subunit.[45,61] 50S ribosomes that reconstitute without 5S rRNA are void of several proteins and show greatly reduced biological activity. All prokaryote 5S rRNAs tested were active in reconstituting *B. stearothermophilus* 50S ribosomal subunits, while eukaryotic 5S RNAs were not.[62] Secondary structural features in the 5S RNA are important for biological activity.

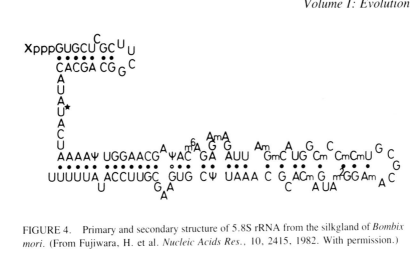

FIGURE 4.   Primary and secondary structure of 5.8S rRNA from the silkgland of *Bombix mori*. (From Fujiwara, H. et al. *Nucleic Acids Res.*, 10, 2415, 1982. With permission.)

Although the participation of 5S rRNA as a structural component of 50S ribosomes has been demonstrated, no particular activity was assigned to this RNA. However, it has been shown that 5S rRNA isolated and purified from rabbit reticulocyte ribosomes exhibits a strong inhibitory effect on the translation of mRNA in a cell-free system. Used in nanogram quantities, 5S rRNA provokes a substantial polysomal breakdown in the presence of ATP and polysomes.[63,64] It was suggested that 5S rRNA may play a role in the peptidyl-transferase activity, where the RNA in question may serve as an intermediate acceptor of the growing peptide chain.[65] It was also assumed that 5S rRNA contributes to the movement of the ribosome relative to mRNA.[66] So far, the biological activity of 5S rRNA has been rather poorly established in comparison to analytical data accumulated for its primary and secondary structure.

D. A Common Ancestral Gene for tRNA and 5S rRNA?

The possible origin of 5S rRNA and tRNA has been considered from the evolutionary point of view.[67] First, precursor tRNA molecules are about the same length as the 5S rRNA.[17] Second, 5S rRNA and tRNA have a similar nucleotide composition[68] and an homology in the secondary structure.[52] Third, many tRNAs exhibit a high degree of sequence similarity (60% homology) with *E. coli* 5S rRNA. The first half of the tRNA sequence exhibits homology with the latter half of the 5S rRNA sequence. An extensive homology exists between several of the tRNAs and the 5S rRNAs from *E. coli*, human KB carcinoma cells, and *Pseudomonas fluorescens*. All these data lead to a hypothetical model for the origin of tRNA and 5S rRNAs in a common ancestral gene.[67]

IV. 5.8S RIBOSOMAL RNA

The 5.8S rRNA has been found as a component of the large ribosomal subunit of pro-karyotes and eukaryotes.[69,70] In prokaryotes, 5.8S rRNA appears to evolve from the 5' end of 23S rRNA.[70] In eukaryotes, 5.8S rRNA (160 nucleotides) has been also found in the large ribosomal subunit where it is base paired with the 28S rRNA molecule.[69,71] The nucleotide sequences of 5.8S rRNA (Figure 4) from over 10 very different species have been reported.[72] The 5.8S rRNA of the silkworm *Bombyx mori* is at the 5' terminal sequence several nucleotides longer than those of other organisms. Since this RNA is base paired with 28S rRNA it would be fascinating to discover how the 5.8S rRNA interact with the 28S rRNA containing the hidden break.[72]

In yeast at all events 5.8S rRNA seems to derive from 7S RNA through enzymatic cleavage.[69] Yeast 5.8S rRNA has been sequenced. It is capable of binding *E. coli* 5S rRNA binding proteins L18 and L25. The binding complexes have ATPase and GTPase activities.[73]

Although it has been suggested that the eukaryotic 5.8S rRNA may play a role in protein synthesis,[73] its biological function remains to be clearly established. This might help to determine to which extent 5.8S rRNA was involved in molecular evolution as well as in the evolution of different species.

V. 7S RNA STRUCTURE

7S RNA was first isolated from eye lens[74] and was thereafter found as a component in normal higher eukaryotic cells: human, rat, murine, chicken,[75] and invertebrates. 7S RNA in RNA viruses[76] is derived from the host. Fingerprints of 7S RNA containing 295 nucleotides (GC content = 60%) from various species and viruses showed an extensive homology. The structure of 7S RNA appears to have been conserved throughout evolution, although there is no consensus about its subcellular localization. It is worthwhile noting that the 5' ends of 7S RNA and that of La 4.5S RNA have extensive (60 to 70%) homologies.[77] About 90 nucleotides of the 5' ends of 7S RNA are homologous with Alu-DNA sequences which are a class of repetitive DNA first found in mammals. These sequences are rich in GC bases (63%). In addition, 45 nucleotides at the 3' end of 7S RNA have been found to be homologous with Alu-DNA sequences. One 300 nucleotide long Alu family sequence contains two binding sites for 7S RNA.[77] The significance of this interaction is not understood. 7S RNA from Novikoff hepatoma has been sequenced.[78] The human 7S L rRNA is heterogeneous within the population of one cell line.[81] This suggests that the 7S L rRNA is encoded by several nonidentical genes.

A. Biological Activity of 7S RNA

Different biological activities of 7S RNA have been suggested. Thus, 7S RNA isolated from eye lens[74] is capable of binding activated amino acids. This RNA, rich in GC pairs (G + C/A + U = 1,9), does not contain pseudouridine and differs from transfer RNA in its size and nucleotide composition. No further investigations have been carried out with this RNA to determine its participation in protein or peptide synthesis. Purified 7S RNA from chicken embryos plays a direct role in heart tissue differentiation and development.[82,83] Since a portion of 7S RNA sequence is homologous to sequences at the origin of DNA replication, it was suggested that 7S RNA may play a role in DNA replication. This may be possible and it should be feasible to demonstrate experimentally whether or not before acting as primer 7S RNA undergoes degradation in order to furnish smaller molecules whose size range is similar to that of RNA primers.

It has been suggested that the 7S L rRNA might be involved in the transport of mRNA from the nucleus into the cytoplasm.[81] Also 7S L sequenced cytoplasmic RNA appears to be involved as a constitutive and indispensable part of the signal recognition particle.[84] Since 7S RNAs do not seem to encode proteins, they probably play a structural role; however, the biological function of 7S RNAs remains to be firmly established in vitro and in vivo.

VI. SMALL RNAS TRANSCRIBED FROM PLASMID

Plasmid DNAs have been shown to code for several low molecular weight RNAs.[85,86] The in vitro transcription by RNA polymerase starts at two sites on the plasmid DNA (Col El DNA).[85] The transcript which starts at one of the sites may be cleaved by RNase H. It is involved as primer in DNA replication. Transcription from the other site on DNA leads to RNA I which inhibits the formation of the RNA primer that exhibits its effect only on homologous DNA template. The RNA I, containing 108 nucleotides, is without effect on initiation of transcription, elongation of RNA chains, or the processing of performed precursors. RNA I is transcribed from DNA and may inhibit the formation of a RNA-DNA hybrid which is a substrate for RNase H.

Plasmid containing mini-cells synthesize the RNA of about 4S to a considerable extent.[87] Multiple drug-resistant plasmid NRl codes for about 10 small RNAs ranging from 60 to 120 nucleotides. They have been characterized by RNase Tl fingerprinting. Some hybridize with RNl DNA digested with restriction endonuclease, indicating that they originate in "the resistance transfer factor region of the plasmid genome." Several of these small RNAs are associated with DNA fragments that contain origins of replication.[86] One stable species of these RNAs migrates at the position of host tRNA. This component contains about 75 nucleotides carrying terminal-CCA residues at 3'. No modified bases have been detected. Thus, this small RNA is not tRNA. Since it has a-CCA residue, one should determine if it is capable of binding amino acids. Two 5S rRNA have been identified. They have no similarity with *E. coli* 5S rRNA.[86] Two RNA species contain 60 to 65 nucleotides possessing pUUAAGp at 5', which indicates that they were not primary transcripts; they possibly resulted through nuclease action on larger RNAs (absence of 5'-PPP). It has been suggested that they might serve as primers for plasmid DNA replication. Curiously enough, a small RNA of about 60 nucleotides has been found solely in the nucleus of plasmid-infected cells. Using a constructed plasmid DNA carrying the 32 nucleotide intervening sequence, it was possible to demonstrate that this small RNA is complementary to an intervening sequence.[88] Sixty nucleotide RNA does not possess a 5'-terminal cap. In contrast to the nucleus, the cytoplasm of uninfected cells contains several small RNA species: 185, 160, 87, 77, and 73 nucleotides. The most abundant is an RNA with 185 nucleotides. These RNAs would seem to act in the translational process. However, direct proof is still unavailable.

VII. SMALL RNAs IN CELLS INFECTED WITH VIRUS

Low-molecular-weight RNAs have been found in cells infected with the SV40 virus, Epstein-Barr virus, vesicular stomatitis virus,[89,90] and defective interfering particles of the vesicular stomatitis virus.[91] In vesicular stomatitis virions, three small RNAs (28, 42, and 70 nucleotides) have been synthesized in vitro in addition to the 47 nucleotide leader sequence.[92] The 47 nucleotide RNA has been sequenced, and three other small RNAs contain (P) ppAA at their 5' terminal. These RNAs are not polyadenylated.

Two species of small RNAs have been found as primary transcripts of the adenovirus 2 Ad-2 genome.[93] Two other species of RNAs (140 bases) appear to be degradation products of viral RNAs. Ad-2 viral small RNAs appear not to be involved in processing adenovirus hnRNA.

A small RNA (50 nucleotides) found in cells co-infected with a standard vesicular stomatitis virus and its interfering particles[94] has sequences complementary to the genome of the defective particles. It does not possess poly A residues and its function is unknown. Apparently its synthesis correlates with the replication of the DNA of defective interfering particles. In contrast, RNA species that have been found in cells infected only with the vesicular stomatitis virus range in size from about 46 bases to 12 kb.[95] Forty-six bases RNA synthesized in vitro has the following sequences:[96]

$$5'(pp)pACGAAGACCACAAAACCAGAU$$

$$AAAAAAUAAAAACCACAAGAGGG(U)C_{OH}3$$

"This sequence is identical to the sequence at the 5' end of the infectious vesicular stomatitis virus RNA and is complementary to the sequence of the 3'OH terminus of this defective interfering particle genome RNA."[96] It should be emphasized that purines are largely in excess over pyrimidines (Pu/Py = 3.0) in this small RNA. Purine-rich RNAs (obtained by degradation of *E. coli* rRNA with pancreatic RNase) and containing 46 nucleotides are actively involved in hematopoiesis, stem cell genesis, and differentiation in animals or humans.[2,5,97]

Small RNAs are encoded by some viruses and expressed during late lytic infection. Small cytoplasmic RNA from cells infected with SV40 virus is 65 nucleotides in length and contains pyrimidine rich sequences (U = 43%; C = 33%; G = 12%; A = 12%). Several clusters contain 5 to 7 pyrimidines, one has 11. This RNA is homologous with the SV40 early mRNAs.[98] It may be involved in the control of late transcription. It should be stressed that pyrimidine rich small RNAs inhibit protein in vitro biosynthesis.[1]

A. Retrovirus Associated Small RNAs

Several small RNAs represent a discrete subclass of the complement of small host RNAs.[99] The 4.5S RNA has been found associated with Moloney leukemia virus and spleen focus-forming virus. These RNAs contain many extra uridylate residues at their 3' termini.[100] The 4.5S RNA from the spleen focus-forming virus contains more than 30 components that vary with the length of the poly U.[100] The 4.5S RNA molecules are provided by the host cells.

B. Viroid RNA Molecules

The viroid of the potato spindle tuber disease (PSTV) is an RNA that represents a covalently closed ring of 359 ribonucleotides,[101] which is approximately the size of the tetrahymena intervening sequence.[102] These latter authors have described "unit-length and linear (+)-strand" RNAs that seem to be intermediates in viroid RNA replication. Does the mature circular viroid involve end-to-end joining of a linear intermediate or does it arise like tetrahymena intervening sequence RNA cyclization via cleavage-ligation at an internal point in a linear molecule?[102] Modified nucleotides have not been detected in any complete or partial PSTV fragment. A stretch of 18 purines, mainly adenosines,[101] exists. At least 60% of PSTV is represented by sequences in the DNA of several *Solanaceous* host species. Linear PSTV RNA molecules may be converted into circular molecules (circle) by an RNA ligase purified from wheat germ,[102] in which circles are indistinguishable from circles extracted from infected plants. Non-*Solanaceous* species contain few or no sequences related to only a small portion of PSTV. It has been suggested that PSTV originates in genes in normal *Solanaceous* plants.[103] The mechanism of infection of plants by PSTV RNA is not completely understood.[104]

VIII. LOW-MOLECULAR-WEIGHT NUCLEAR RNAs

Eukaryotic cells contain, structually and functionally, small RNAs differing from other types of known cellular RNAs.[105-107] In HeLa cells some of the small RNA (Sn RNAs) components are located mainly in the nucleoplasm and one mainly in the nucleoli.[106] All USn RNAs: U1, U2, U3, U4, U5, and U6 exist in RNP particles. The analysis of several of these RNAs revealed the presence of a 5' cap (a base $m^{2,2,7}$-methylguanosine) that differs from the 5' cap of mRNA (m^7-G).[77] This explains why Sn RNAs are not translated in the cytoplasm. Among Sn RNAs, U3 RNA is particularly localized in the nucleus and has not been found in any other site in the cell. It may play a role in transcription of the rRNA from the rDNA templates by maintaining stable open gene complexes.[77] Detailed analysis of Sn RNAs has been described.[80] The U RNAs contain a large number of more or less clustered uridine residues. U5 RNA is very enriched in uridine (35%). RNase Tl fingerprints of the purified U3 RNA, U2, and U1 are practically identical (# 300 nucleotides) in HeLa cells, human normal fibroblasts, and Novikoff hepatoma cells. These uridine rich low molecular weight RNAs appear to have been conserved throughout evolution.[108]

The sequence of 18 nucleotides close to the 5' end of Ul RNA containing 165 nucleotides is complementary to the nucleotides from intron transcript adjacent to the splice point.[109] Some nucleotide sequences of U1 RNA and U2 RNA are expected to form a hybrid with intron-exon borders.[110-112] These RNAs and other U RNAs may play a role in the splicing of premessenger RNAs. However, the results obtained in vitro have not been demonstrated *in situ,* where U RNAs exist as an RNA-protein complex.[113] U2 nuclear RNA hybridizes to heterogeneous nuclear RNAs but not nucleolar RNA. The nucleolus contains the precursors of ribosomal RNAs. Using a human cloned U2 DNA probe, it has been demonstrated that U2 RNA is paired with complementary sequences in heterogeneous nuclear RNA in vivo.[114] Both U2 and U1 RNAs have complementary sequences with intron-exon borders in mRNAs precursor molecules.[110] It is interesting to recall here that pseudogenes, complementary to the small nuclear RNAs U1, U2, and U3, are dispersed and abundant in the human genome. Three pseudogenes — U1.101, U2.13, and U3.5 sequences[115] — are flanked by short direct repeats of 16 to 19 bp.[116]

Although USn RNAs have been purified and sequenced, their precise biological function has not been assigned.

A. Phylogenetic Trends of USn RNAs

The genes of U1, U2, and U3 RNAs appear to have been highly conserved throughout evolution. The genes of these RNAs have been isolated from human genome. This genome appears to contain more pseudogenes for small RNAs than real genes. Notable similarities have been found between sequences of the U1 RNA of *Drosophila* and U1 RNA sequences from other origins.[117] There are similarities between the USn RNAs of vertebrates and invertebrates, as far as primary structure is concerned. Dinoflagellates considered as eukaryotes contain U1-U6 Sn RNA which have a differing molecule center.[80]

IX. OTHER SMALL RNAs IN PROKARYOTES AND EUKARYOTES

Several small RNAs isolated from *Alcaligenes faecalis* bacteria, sediment on the sucrose gradient at 3.9S, 4.3S, 5.5S, and 6.3S.[118] 5.5S rRNA particularly studied differs by its base ratio (G + C/A + U = 2.0) and hyperchromicity (38%) from transfer RNAs (G + C/A + U = 1.35 and hyperchromicity 28%) and from rRNAs. The 5.5S RNA possesses the capacity to accept amino acids in the presence of each of the four ribonucleoside-5'-triphosphates and bacterial enzymes. It may play a role in peptide formation.[119] Showdomycin-resistant *E. coli* cells excrete into the culture medium a 6S RNA (G + A/C + U = 1.82) that acts as a transforming agent and can be in vitro and in vivo transcribed into DNA.[120]

A small stable 10S RNA has been identified and characterized from *E. coli.*[121] Ribonuclease P, unique among all the RNA processing enzymes, contains an RNA moiety that is required for its function.[122,123] Two RNAs are present in this enzyme. One of them, termed M2, is identical or similar to 10S RNA from *E. coli.* Also, ribonuclease P from yeast and *Bacillus subtilis* contains both protein and RNA components.[124] Using a cloned segment of DNA for complementation it has been shown that it codes for an RNA species of 340 bases. Sequence analysis of RNA indicates that this RNA is highly G + C rich.[125] It originates from DNA. It would be interesting to determine if GC rich 5.5S RNA from *Alcaligenes faecalis* originates from 10S RNA bound to RNAse.

An RNA termed 2RNA has been isolated and purified from whole cell soluble RNA from *E. coli.* The molecular weight of this RNA is significantly higher than that of 5S RNA used as a control.[126] The 2RNA may be similar to 5.5S of RNA found in *A. faecalis.*[118]

La 4.5S RNA (80 to 100 bases) is small RNA that emerges in *E. coli* from the progenitor of 23S rRNA through fragmentation at the 3' end.[127] This finding is in agreement with previously reported data on the presence of small RNAs in *E. coli*[128] and *Agrobacterium tumefaciens.*[129,130] La 4.5S RNA has been shown to be a group of hydrogen RNAs bounded to poly A containing nuclear of cytoplasmic RNA present in cultured Chinese hamster ovary cells.[131] This RNA from different cells has been sequenced[80,132] (Figures 5 and 6). It is less conserved than USn RNAs. La 4.5S RNA (96 residues) terminates with pppGp at its 5' end

	10	20	30	40	50
	pppGCCGGUAGUG	GUGGCGCACG	CCGGUAGGAU	UUGCUGAAGG	AGGCAGAGGC
	U				

	60	70	80	90	
	AGAGGGAUCA	CGAGUUCGAG	GCCAGCCUGG	GCUACACAUU	UUUU$_{OH}$

FIGURE 5.   Nucleotide sequence of La 4.5 RNA of mouse and hamster. (From Harada, F. and Kato, N., *Nucleic Acids Res.*, 8, 1273, 1980. With permission.)

	10	20	30	40	U 50
	pppGGCUGGAGAG	AUGGC(UC)AGC	CGUUAAAGGC	UAGGCUCACA	ACCAAAAAUA

	60	70	80	90	98
	UAAGAGUUCG	GUUCCCAGCA	CCCACGGCUG	UCUCUCCAGC	CACCUUUU(U)$_{OH}$

FIGURE 6.   Nucleotide sequence of La 4.5 IRNA of Novikoff hepatoma (modified). (From Busch, H. et al. *Annu. Rev. Biochem.*, 51, 617, 1982. With permission).

and with a short oligo (U) sequence of variable length at its 3' end. La 4.5S RNA contains some regions of sequence that have been found in other small RNAs, all transcribed by RNA polymerase III.

A new class of small RNA molecules, tcRNAs (translational control RNA), associated with the function of mRNAs has been found in differentiating muscle.[133] Among them two species have been purified and sequenced. A 102 nucleotide species (tcRNA$_{102}$) has the poly A binding properties. It contains 40% uracil and 5% cytosine. Another RNA (tcRNA$_{89}$) is less rich in uracil than tcRNA$_{102}$, but contains 8% of cytosine residues. Both RNAs are bound to mRNAs and are transcribed from DNA. The authors have suggested the role of mRNA-associated small RNAs in translational control. One might postulate that tcRNA$_{102}$ containing 40% of uracil would be involved in inhibiting the translation process as it has been shown for oligoribonucleotides which are particularly rich in uracil (47%).[1]

A small RNA, termed unique membrane RNA, has been purified from rat liver microsomes and characterized.[134] This RNA (3.14S) differs from tRNA and 5S rRNA and is not the result of degradation of high molecular weight RNA. It accepts amino acids rather poorly and is not present in ribosomes. Its biological function has not been defined.

Small RNAs have also been found in Ehrlich ascites tumor cells. Electrophoresis on polyacrylamide gel has shown that five of them migrate more slowly than 5S rRNA and three components migrate between 5S and 4S RNA. All these RNAs are localized in the nucleus and account for 0.7 to 2.9% of total nuclear RNAs.[135] These small RNAs have a stability comparable to rRNAs. Their synthesis is more or less efficiently inhibited with the actinomycin D that inhibits the transcription of rRNAs and 5S rRNA from DNA. It remains to be seen whether RNAs are generally the primary transcript and what their biological significance may be.

X. SMALL RNA MOLECULES AND OLIGORIBONUCLEOTIDES AS PRIMERS FOR DNA REPLICATION, TRANSCRIPTION, AND TRANSLATION

Neither prokaryotic nor eukaryotic polymerases have been reported to catalyze *de novo* and in vitro DNA synthesis.[2,136-140] Several lines of evidence demonstrate that oligoribonucleotides and small RNAs are both necessary to initiate DNA replication.[1,2,138,140-143] This variety of RNAs does modify the rhythm of DNA replication and transcription.

What is the origin of such active, small RNAs? They may appear during transcription of

DNA but may also be generated by ribonucleases which degrade giant molecules of various RNAs. It is difficult to collect large quantities of such naturally occurring small RNAs. To get around this difficulty, purified ribosomal RNAs (23S + 16S) was degraded from show-domycin resistant *E. coli* and wild type with different ribonucleases including pancreatic RNase. All RNases used contained phosphatase activity which removes the 2' or 3' phosphate from terminal nucleotide. A family of RNA fragments of 25 to 55 nucleotides is to be found in an RNase digest. They have been selected by their size and characterized by base ratio (purines/pyrimidines ranging from 5.8 to 0.75). Purine-rich RNA fragments interact in vitro with DNA from different origins but do not promote the in vitro synthesis of DNAs from various sources. RNA fragments bind to DNA by hydrogen bonds. This hybrid can be characterized.[2,137,144] Only those RNA fragments that are rich in purine nucleotides are excellent in vitro primers for the conversion of φX174 and λ DNA single strands to replicative form.[137] These data have been confirmed by isolation of the primer:

$$pppAGGGCGAAAAACCGUCU_{A_UA}GGGCGAUGG$$

which is required for conversion of phage fd single strand DNA to replicative form.[145] In cited cases, RNA primers contain approximately twice as many more purine bases (G + A/C + U = 2.2 to 2.4) complementary to pyrimidines present at the origin of the phage DNA replication site: TGCTCCCCCAACTTGOH(3' end).[146,147] The fact that purine-rich RNA fragments isolated from bacterial ribosomal RNAs act as excellent primers for phage DNA replication[137] is explained by the observation that these fragments bind to the complementary bases of phage DNA.[2]

An RNA primer for DNA replication by mitochondrial DNA polymerase is generated by RNase H on a precursor transcript A transcript of 700 nucleotides synthesized from a promoter 508 bases upstream of the origin.[148] A second smaller transcript is copied from the opposite strand of the template and its length (100 nucleotides) is complementary with the promoter proximal region of the primer.[149] There is an RNA of 200 nucleotides which binds to the nascent transcript in order to prevent hybrid formation (RNA primer DNA), but not transcription.[148]

A. Small RNAs in the Translation Process

Participation of small RNAs and/or oligoribonucleotides in the translation of mRNAs into proteins has been postulated[141,150] and thereafter confirmed by experimental data.[143,151,152] Oligoribonucleotides have been found attached to initiation factors involved in protein biosynthesis. Such an RNA molecule called "i-RNA"[143] is rich in adenine (46%) and poor in guanine (7%).[153] This "i-RNA" found in reticulocytes is present in ribosomes from various sources. Although it participates in translation of mRNAs, it lacks specificity for an mRNA from a given species. A small "translating control RNA" rich in uridine, inhibits in vitro the translation of mRNA.[152] Small RNAs, isolated from newborn rats, alter the translation of mRNA in a system of wheat germ.[143] Such an RNA, containing poly U stretches, inhibits the translation of homologous or hererologous mRNA.[154] RNA without poly U appears to activate the translation.

Salt-washed ribosomes of dormant and developing *Artemia salina* embryos contain two distinct factors involved in the translation of mRNAs into proteins.[1] Both compounds are oligoribonucleotides. The inhibitor (6000 mol wt) is rich in pyrimidines (47% U, 11% A, 26% C, 16% G), sensitive to RNase A, and resistant to RNase Tl. The activator (9000 mol wt) is rich in guanine (33% U, 10% A, 6% C, 51% G), sensitive to RNase Tl, and resistant to RNase A.

Bases	Inhibitor (%)	Activator (%)
U	47	33
A	11	10
C	26	6
G	16	51
Purine/pyrimidine	0.37	1.6

The activator of translation counteracts the effect of the inhibitor. It has been found in the same fractions of developing embryos. There is little activator in undeveloped cysts. The amount of these oligoribonucleotides appears to depend on hydration of RNases.[1] Whether such RNases degrade some ribosomal or nuclear RNAs present in excess in the studied material or whether they appear during degradation of mRNAs and splicing of giant precursors RNAs has yet to be determined.

B. Free Intervening Sequences from Genes and Pseudogenes as Active Biological Molecules?

In eukaryotes, intervening sequences (introns) which interupt the coding regions of many genes specifying various RNAs are transcribed as part of precursor RNAs and then removed by the splicing process. This process consists of the excision of introns and thereafter ligation of the functional units, exons. It was reported that the intervening sequence of the tetrahymena ribosomal RNA precursor is excised as a linear molecule which cyclizes itself in the absence of the enzyme.[155] During cyclization, the AAAUAG fragment is lost. Such a purine rich fragment may be involved in the regulation of DNA replication[2,144] or in protein biosynthesis.[1]

Two classes of genomic sequence have recently been implicated in the formation of processed genes. These are multigene families that correspond to the middle repetitive Alu-genes[156] and the multiple Sn RNA genes that encode the ubiquitous small nuclear RNAs prevalent in human and other genes.[116] Members of both gene families are transcribed into small RNAs of generally unknown function.[157] When treated with the restriction endonuclease Alu 1 gene provides two fragments, one of # 170 bp and one of # 120 bp. At least half of the 300 nucleotides and the 300 nucleotide irDNA (inverted repeated DNA) belong to a single sequence family termed the Alu family.[158]

In Chinese hamster cells each poly(A)-terminated nuclear RNA has one molecule of low molecular weight RNA (100 nucleotides) hydrogen bounded to it.[131] This small RNA may interact with sequences transcribed from the Alu family of interspersed repeated DNA.[159] These authors have suggested that the Alu family of interspersed repeated sequence may function as origins of DNA replication in mammalian cells. This remains to be confirmed by direct experimental data. Alu-like elements dispersed throughout the mammalian genome are substrates for the Alu restriction endonuclease.[160] They range in length from 135 bp in the mouse to 300 in man. Alu-like elements display a high degree of sequence homology with the small cytoplasmic 7S and 4.5S RNAs.[156] The 4.5S RNA, which has been found in mouse and hamster but not in human cells, binds through hydrogen bonds to nuclear and cytoplasmic poly(A)$^+$ RNA.[131] The sequence homology is observed for 120 bases starting from the 5' end of the Alu monomer.[161]

Pseudogenes are partial duplicates of structural genes. In most cases, they appear not to be transcribed into RNAs, although there is no definite certainty about this. Pseudogene sequences have been characterized for gene encoding small nuclear RNAs U1, U2, U3.[162] They also have been produced through an RNA intermediate.[116] During the processing of this intermediate, some RNA sequences are lost. Their possible function and that of small nuclear RNAs encoded by pseudogene sequences have to be determined. "Processed genes and Alu-like elements are heritable and inserted into DNA in cells. 1% of mammalian DNA contains sequences expressed as proteins, the remaining 99% may be derived and maintained by a continuous flux of such elements."[163]

C. Small RNAs: Molecular Selfreplication

The Qβ virus which infects *E. coli* has been used as a model for studying the mechanism of molecular selfreplication. The isolated Qβ replicating enzyme has the capacity to reproduce viral RNA in a cell-free system.[164] This single-stranded RNA of 4500 nucleotides possesses infectious potential. A noninfectious "satellite" RNA containing 220 nucleotides purified from *E. coli* cells has also been used as template and in vitro is replicated by the Qβ replicase. The nucleotide sequence of the relevant parts of Qβ RNA has been analyzed.[165] In this small RNA, the sequence-CCC has been identified[166] as the recognition site for interaction with Qβ replicase. A remarkable property of Qβ replicase is that it is capable of synthesizing RNA from four ribonucleoside-5'-triphosphates, even in the absence of RNA template. This enzyme accepts only few natural RNAs and some synthetic ribopolymers as template for RNA in vitro synthesis.[167,168]

RNA synthesis always starts with the incorporation of GTP. The 3' terminating nucleotide of the template does not need to be cytidylic acid. However, cytidylic acid polymers are accepted as template only when they contain at their 3' end a cytidylic acid sequence or more than 5 nucleotides.[168] Randomly mixed ribopolymers containing cytidylic acid poly(C) are accepted as active template with Qβ replicase.[169]

Transforming 6S RNA excreted by showdomycin-resistant *E. coli* differs from other RNAs species found in the wild type strain. This small RNA is purine rich (G + A/C + U = 1.82) in comparison to RNAs from the wild type of *E. coli* (G + A/C + U = 1.0). In Cs_2SO_4 gradient centrifugation, 6S RNA sediments exclusively in the region of RNA and does not contain DNA nor does it hybridize with DNA from the same strain. It was unexpected to observe that 6S RNA can interfere in the in vitro activity of polynucleotide phosphorylase (PNPase) in the exclusive presence of all four ribonucleoside-5'-diphosphates (XDPs) in the incubation mixture.[170] PNPase[6] from the wild strain synthesizes, from equal amounts of XDPs, a poly AGUC whose ratio is 1:1:1:1. In the presence of 6S RNA, the rate of synthesis is increased several-fold and the base ratio of the synthesized product G + A/C + U = 1.75 is practically identical to that found for 6S RNA (G + A/C + U = 1.80). RNase A abolishes template activity of 6S RNA. XTPs cannot replace XDPs. Rifampicin is without effect on the activity of PNPase, while it inhibits that of DNA-dependent RNA polymerase. Crude soluble extracts of 250-fold purified enzyme gives the same results. Ribosomal RNAs, tRNAs, polyovirus RNAs, and poly AGUC are not used as template by PNPase. On sucrose gradient, the ^{14}C-synthesized product made in the presence of 6S RNA, sediments at a position differing from that of the ^{14}C polymer-synthesized in the absence of 6S RNA.[171]

PNPase from showdomycin-resistant *E. coli,* incubated in the presence of all four XDPs each used in equal amounts, synthesizes in vitro AGUC polymer in which the amount of purine bases is twice that of pyrimidines. This shows that modified PNPase may synthesize an RNA that does not hybridize with DNA. PNPase from *E. coli* wild strain is capable of synthesizing small purine rich RNA (G + A/C + U = 1.5 to 1.8) from XDPs at 70°C, acting as primers in in vitro replication of DNAs from bacteria and mammals, but it has a low effect on phage DNAs under the same experimental conditions.[170] Also, purine rich RNA synthesized by PNPase under normal conditions but in the presence of RNase A acts as primer for replication of some DNAs.[171] Thus, it may be imagined that PNPase and RNase, both capable of behaving as "thermoresistant enzymes," might have played a decisive role in synthesizing different RNA primers or even templates for the formation of cell constituents.

D. Transcription of Small Cellular RNAs into DNA

Interest in the conversion of RNA molecules into DNA has recently increased, since this process of transcription seems to be the only one that may explain the creation of pseudogenes.[163] It has been suggested that small RNA molecules transcribed first on DNA may carry the information back to chromosomal DNA but in different segments of DNA than those from which they originate. This concept implies the existence in eukaryotes and prokaryotes of an enzyme, reverse transcriptase, capable of transcribing small RNAs into DNA.

Reverse transcriptase has been discovered in animal tumor viruses,[172] and several reports have been published on transcription of RNA templates into complementary DNA by enzymes found in chick embryos,[173] in bacteria,[120,170,174,175] in plants,[176] in monkey placenta,[177] in normal human lymphocytes,[178] and in fungi.[3,4] 28S ribosomal RNA from *Drosophila melanogaster* can be transcribed into DNA-like material by a DNA polymerase I purified to about 90%. The size of the ^{3}H-DNA product is about 4S, but it may be longer.[179] 25S ribosomal RNA transcription was found to occur to some extent in the absence of the added oligonucleotide primer.[180]

A free RNA-bound "reverse transcriptase", isolated from *E. coli* extracts, synthesizes in vitro DNA, providing that all four ribonucleoside-5′-triphosphates are present in the incubation mixture. Both template RNA and synthesized ^{3}H-DNA have a size of about 6S. ^{3}H-DNA is complementary to template RNA.[120] It is possible to distinguish the RNA-bound enzyme from known DNA-dependent DNA polymerase on several criteria. A particular purine-rich small RNA (6S) excreted by showdomycin-resistant *E. coli* has a transforming potential.[128,181] When added to the culture medium of a tumorogenic strain of *A. tumefaciens,* it is possible to isolate a series of transformants which have partially or totally lost their tumor potential. Different biochemical changes have been observed in these transformants. Hybridization data showed that the RNA has penetrated into recipient bacteria and one copy at least was found in the DNA of complete transformants. In vitro this RNA can be transcribed into complementary DNA. It possesses selfpairing regions and probably a 3′ OH looped terminus required for transcription. When these different small RNAs are injected into *Datura stramonium* inverted stem section under axenic conditions and in the presence of auxin, they lead to the appearance of characteristic Crown-gall tumors[176,182-184] (Figures 7 and 8).

One can easily isolate the endogeneous RNA-bound RNA-dependent DNA polymerase from the microsomal pellet fraction of the eukaryotic fungus *Neurospora crassa*.[3,4] Here also, free template RNA and the synthesized DNA sedimented on sucrose density gradients at approximately 5 to 6S. ^{3}H-DNA is complementary to template RNA as judged by hybridization data. It is quite probable that the eukaryotic organism may well possess RNA-bound "reverse transcriptase". It will be of importance to determine which small RNAs and/or intervening sequences might be transcribed into complementary DNA in eukaryotes. If such data can be accurately established using different species, this would strongly support the concept of pseudogenes formation, as recently suggested.[163] In this respect it has been recently reported that *Drosophila melanogaster* contains nonvirales particles containing 4S, 4.5S, 5S, and 6S molecules of RNA as the major constituents. In addition, these particles contain particle-bound reverse transcriptase.[185] This observation strongly supports several findings reported many years ago[3,4,120] and discussed here.

XI. CONCLUDING REMARKS

The majority of small RNAs from eukaryotes, prokaryotes, and virions described here have been sequenced and their secondary structure proposed. Other small RNAs not sequenced but characterized by chemical and physical means have been studied mostly for their biological activities. Many small RNAs appear to originate from DNA as direct transcripts, while others emerge from RNA precursor transcripts or even from mature RNAs. One common feature of small RNAs is that they bind with relative efficiency to large RNAs (rRNAs, mRNAs, viral genome RNAs), and DNAs through hydrogen bonding. This process

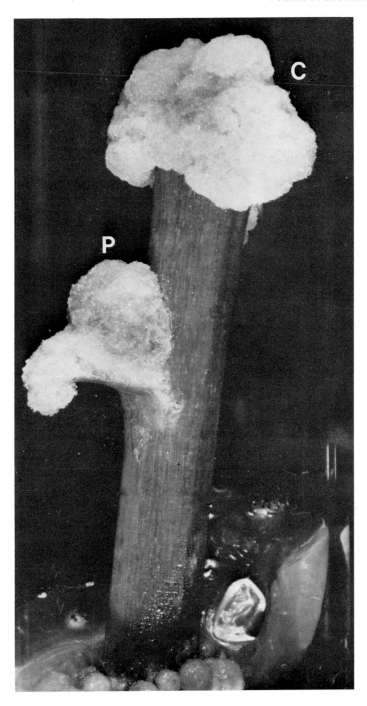

FIGURE 7. Demonstration of the oncogenic capacity of the tumor-inducing small-size RNA from *A. tumefaciens*. Tumor obtained by inoculation of RNA into a fragment of inverted stem of *Datura stramonium* cultured in vitro. C, auxinic callus tissue; P, overgrowth tissue obtained with RNA. (From Le Goff, L. et al., *Can. J. Microbiol.*, 22, 694, 1976. With permission.)

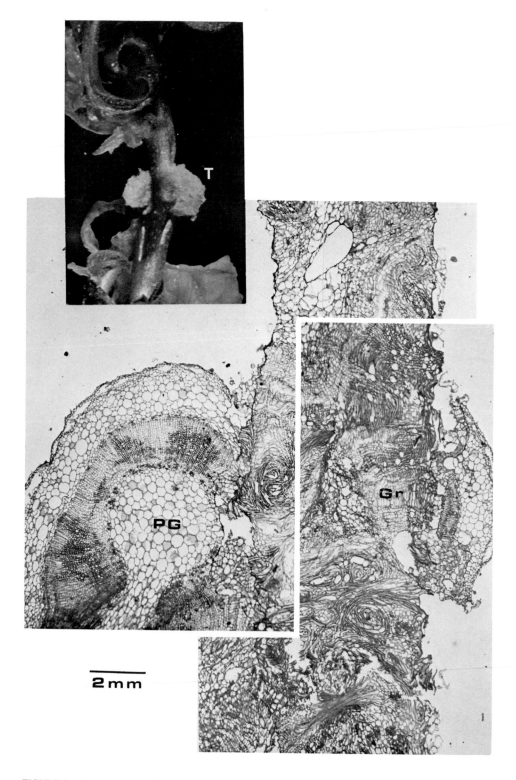

FIGURE 8.   Demonstration of the oncogenic capacity of the tumor-inducing small-size RNA from *A. tumefaciens*. Development of secondary tumor (T) after grafting of the primary overgrowth tissue. Schematic section of a tumor fragment (Gr) engrafted upon stem of healthy plant (PG). (From Aaron-Da Cunha, M. I. et al., *C. R. Soc. Biol.*, 169(3), 755, 1975. With permission.)

necessarily implies sequence homology in certain regions. Comparison of sequences suggests that there exist relative phylogenetic trends between several small and large RNAs, as well as between different species of small RNAs themselves. From this, one may conclude that small RNAs have largely been conserved throughout evolution of various species; understanding of the biological activities of the RNAs should greatly elucidate this conservation process.

Over the last decade, few scientists have been involved in studying the extent to which small RNAs can in vitro and *in situ* be transcribed into DNA. Since the recent discovery of pseudogenes, this problem has come to the fore with particular urgency, partly because some regions of pseudogenes are transcribed into small RNAs. It is at present postulated but not proven as far as we know, that small RNAs, via transcription to DNA, are the only candidates for the creation of new genes or pseudogenes. Also, it should not be forgotten that some small RNAs can be in vitro synthesized *de novo* in the absence of DNA or RNA template. However, it remains difficult to ascertain that this could happen *in situ* in different biological systems. The role of such RNAs may have been of importance during the early stages of the appearance of RNAs or DNAs, i.e., at the creation of life.

As shown here, many small RNAs have been characterized by excellent analytical methods. However, their respective biological roles in the cells of different species need to be demonstrated accurately in order to elucidate their possible phylogenetic relationship and their part in the life of the cell.

REFERENCES

1. **Lee-Huang, S., Sierra, J. M., Naranjo, R., Filipowitz, W., and Ochoa, S.,** Eukaryotic oligonucleotides affecting mRNA translation, *Arch. Biochem. Biophys.,* 180, 276, 1977.
2. **Beljanski, M.,** The regulation of DNA replication and transcription. The role of trigger molecules in normal and malignant gene expression, *Experimental Biology and Medicine,* Vol. 8, Wolsky, A., Ed., S. Karger, Basel, 1983.
3. **Dutta, S. K., Beljanski, M., and Bourgarel, P.,** Endogenous RNA-bound RNA-dependent DNA polymerase activity in *Neurospora crassa, Exp. Mycol.,* 1, 173, 1977.
4. **Dutta, S. K., Mukhopadhyay, D. K., and Bhattacharyya, J.,** RNase-sensitive DNA polymerase activity in cell fractions and mutants of *Neurospora crassa, Biochem. Genet.,* 18, 743, 1980.
5. **Beljanski, M., Plawecki, M., Bourgarel, P., and Beljanski, M.,** Leukocyte recovery with short-chain RNA-fragments in cyclophosphamide-treated rabbits, *Cancer Treat. Rep.,* 67, 611, 1983.
6. **Ochoa, S.,** Enzymic synthesis of polynucleotides. III. Phosphorolysis of natural and synthetic ribopolynucleotides, *Arch. Biochem. Biophys.,* 69, 119, 1957.
7. **Holley, R. W., Apgar, J., Everett, G. A., Madison, J. T., Marquisee, M., Merrill, S. H., Penswick, J. R., and Zamir, A.,** Structure of a ribonucleic acid, *Science,* 147, 1462, 1965.
8. **Sundaralingam, M.,** Nucleic acid principles and transfer RNA, in *Biomolecular Structure Conformation and Function,* Evol. Proc. Int. Symp., Vol. 1, Srinivasan, R., Subramanian, E., and Yathindra, N., Eds., Pergamon Press, Oxford, 1981, 259.
9. **Woese, C. R.,** *The Genetic Code: The Molecular Basis for Genetic Expression,* Harper & Row, New York, 1967.
10. **Nishimura, S.,** Modified nucleosides in tRNA, in *Transfer RNA: Structure, Properties and Recognition,* Schimmell, P. R., Söll, D., and Abelson, J. N., Eds., Cold Spring Harbor Laboratory, Cold Spring Harbor, N.Y., 1979, 59.
11. **McCloskey, J. A. and Nishimura, S.,** Modified nucleosides in transfer RNA, *Acc. Chem. Res.,* 10, 403, 1977.
12. **Sprinzl, M., Grüter, F., and Gauss, D. H.,** Compilation of tRNA sequences, *Nucleic Acids Res.,* 5(Suppl.), r15, 1978.
13. **Hoagland, M. B., Zamecnik, P. C., and Stephenson, M. L.,** Intermediate reactions in protein biosynthesis, *Biochim. Biophys. Acta,* 24, 215, 1957.
14. **Zachau, H. G.,** Transfer ribonucleic acids, *Angew. Chem. Int. Ed.,* 8, 711, 1969.

15. **Burdon, R. H., Martin, B. T., and Lal, B. M.,** Synthesis of low molecular weight ribonucleic acid in tumor cells, *J. Mol. Biol.,* 28, 357, 1967.

16. **Darnell, J. E.,** Ribonucleic acids from animal cells, *Bacteriol. Rev.,* 32, 262, 1968.

17. **Altman, S. and Smith, J. D.,** Tyrosine tRNA precursor molecule polynucleotide sequence, *Nature (London) New Biol.,* 233, 35, 1971.

18. **Abelson, J.,** The organization of tRNA genes, in *Transfer RNA: Biological Aspects,* Söll, D., Abelson, J. N., and Schimmel, P. R., Eds., Cold Spring Harbor Laboratory, Cold Spring Harbor, N.Y., 1980, 211.

19. **Mazzara, G. P. and McClain, W. H.,** tRNA synthesis, in *Transfer RNA: Biological Aspects,* Söll, D., Abelson, J. N., and Schimmel, P. R., Eds., Cold Spring Harbor Laboratory, Cold Spring Harbor, N.Y., 1980, 3.

20. **O'Farrell, P. Z., Cordell, B., Valenzuela, P., Rutter, W. J., and Goodman, H. M.,** Structure and processing of yeast precursor tRNA containing intervening sequences, *Nature (London),* 274, 438, 1978.

21. **Ryan, M. J., Brown, E. L., Belagaje, R., Khorona, H. G., and Fritz, H. J.,** Cloning of two chemically synthesized genes for a precursor to the Su^{3+} suppressor $tRNA^{Tyr}$, in *Transfer RNA: Biological Aspects,* Söll, D., Abelson, J. N., and Schimmel, P. R., Eds., Cold Spring Harbor Laboratory, Cold Spring Harbor, N.Y., 1980, 245.

22. **Khorana, H. G.,** Total synthesis of a gene, *Science,* 203, 614, 1979.

23. **Cortese, R., Melton, D. A., Tranquilla, T., and Smith, J. D.,** Cloning of nematode tRNA genes and their expression in the frog oocyte, *Nucleic Acids Res.,* 5, 4593, 1978.

24. **Schweizer, E., Mackechnie, C., and Halvorson, H. O.,** Redundancy of ribosomal and transfer RNA genes in *Saccharomyces cerevisiae, J. Mol. Biol.,* 40, 261, 1969.

25. **Weber, L. and Berger, E.,** Base sequence complexity of the stable RNA species of *Drosophila melanogaster, Biochemistry,* 15, 5511, 1976.

26. **Clarkson, S. G. and Birnstiel, M. L.,** Clustered arrangement of tRNA genes of *Xenopus laevis, Cold Spring Harbor Symp. Quant. Biol.,* 38, 451, 1974.

27. **Martin, M. C., Rabinowitz, M., and Fukuhara, H.,** Yeast mitochondrial DNA specifies tRNA for 19 amino acids. Deletion mapping of the tRNA genes, *Biochemistry,* 16, 4672, 1977.

28. **Angerer, L., Davidson, N., Murphy, W., Lynch, P., and Attardi, G.,** An electron microscope study of the relative positions of the 4S and ribosomal RNA genes in HeLa cell mitochondrial DNA, *Cell,* 9, 81, 1976.

29. **Dawid, I. B., Klukas, C. K., Ohi, S., Ramirez, J. L., and Upholt, W. B.,** Structure and evolution of animal mitonchondrial DNA, in *The Genetic Function of Mitochondrial DNA,* Saccone, C. and Kroon, A. M., Eds., North Holland, Amsterdam, 1976, 3.

30. **Wittmann, H. G.,** Components of bacterial ribosomes, *Ann. Rev. Biochem.,* 51, 155, 1982.

31. **Cantatore, P., De Benedetto, C., Gadaleta, G., Gallerani, R., Kroon, A. M., Holtrop, M., Lanave, C., Pepe, G., Quagliariello, C., Saccone, C., and Sbisa, E.,** The nucleotide sequences of several tRNA genes from rat mitochondria: common features and relatedness to homologous species, *Nucleic Acids Res.,* 10, 3279, 1982.

32. **Clarkson, S. G., Birnsteil, M. L., and Purdon, I. F.,** Clustering of transfer RNA genes of *Xenopus laevis, J. Mol. Biol.,* 79, 411, 1973.

33. **Deutscher, M. P.,** Synthesis and functions of the -C-C-A terminus of transfer RNA, *Prog. Nucleic Acid Res. Mol. Biol.,* 13, 51, 1973.

34. **Daniel, V., Sarid, S., and Littauer, U. Z.,** Bacteriophage induced transfer RNA in *Escherichia coli, Science,* 167, 1682, 1970.

35. **Gruissem, W. and Seifart, K. H.,** Transcription of 5SRNA genes *in vitro* is feedback-inhibited by HeLa 5SRNA, *J. Biol. Chem.,* 257, 1468, 1982.

36. **Beaudreau, G. S., Sverak, L., Zischka, R., and Beard, J. W.,** Attachment of ^{14}C amino acids to BA1 strain A (myeloblastosis) avian tumor virus RNA, *Natl. Cancer Inst. Monogr.,* 17, 791, 1964.

37. **Bonar, R. A., Sverak, L., Bolognesi, D. P., Langlois, A. J., Beard, D., and Beard, J. W.,** Ribonucleic acid components of BA 1 strain A (myeloblastosis) avian tumor virus, *Cancer Res.,* 27, 1138, 1967.

38. **Travnicek, M.,** RNA with amino acid-acceptor activity isolated from an oncogenic virus, *Biochim. Biophys. Acta,* 166, 757, 1968.

39. **Erikson, E. and Erikson, R. L.,** Isolation of amino acid-acceptor RNA from purified avian myeloblastosis virus, *J. Mol. Biol.,* 52, 387, 1970.

40. **Erikson, E. and Erikson, R. L.,** Association of 4S ribonucleic acid with oncornavirus ribonucleic acids, *J. Virol.,* 8, 254, 1971.

41. **Sawyer, R. C. and Dahlberg, J. E.,** Small RNAs of Rous sarcoma virus: characterization of two dimensional polyacrylamide gel electrophoresis and fingerprint analysis, *J. Virol.,* 12, 1126, 1973.

42. **Waters, L. C. and Mullin, B. C.,** Transfer RNA in RNA tumor viruses, *Prog. Nucleic Acid Res. Mol. Biol.,* 20, 131, 1977.

43. **Yang, W. K. and Hwang, D. L. R.**, *In vitro* selective binding of tRNAs to rRNAs of vertebrates, in *Transfer RNA: Biological Aspects*, Söll, D., Abelson, J. N., and Schimmel, P. R., Eds., Cold Spring Harbor Laboratory, Cold Spring Harbor, N.Y., 1980, 517.

44. **Richter, D., Erdman, V. A., and Sprinzl, M.**, Specific recognition of GTψ C loop (loop IV) of tRNA by 50S ribosomal subunits from *E. coli, Nature (London) New Biol.*, 246, 132, 1973.

45. **Elson, D.**, A ribonucleic acid particle released from ribosomes by salt, *Biochim. Biophys. Acta*, 53, 232, 1961.

46. **Rosset, R. and Monier, R.**, A propos de la présence d'acide ribonucléique de faible poids moléculaire dans les ribosomes d' *E. coli, Biochim. Biophy. Acta*, 68, 653, 1963.

47. **Sanger, F., Brownlee, G. G., and Barrell, B. G.**, Sequence of 5S ribosomal RNA, *Structure, Function, and Transfer RNA 5SRNA*, Proc. Meet. Fed. Euro. Biochem. Soc., 4th ed., Froeholm, L. O., Ed., Academic Press, New York, 1968, 1.

48. **Fedoroff, N. V. and Brown, D. D.**, The nculeotide sequence of oocyte 5S DNA in *Xenopus laevis*. I. The AT-rich spacer, *Cell*, 13, 701, 1978.

49. **Furth, J. J., Ormsby, J., Su, C. Y., and Averyhart-Fullard, V.**, Transcription of calf thymus 5S ribosomal DNA and calf thymus satelite DNA by RNA polymerase III, *Fed Proc. Fed. Am. Soc. Exp. Biol.*, 41, 1294, 1982.

50. **Komiya, H., Hasegawa, M., and Takemura, S.**, Nucleotide sequences of 5SrRNA from sponge *Halichondria japonica* and tunicate *Halocynthia roretzi* and their phylogenetic positions, *Nucleic Acids Res.*, 11, 1964, 1983.

51. **Woese, C. R., Magrum, L. J., and Fox, G. E.**, Archaebacteria, *J. Mol. Evol.*, 11, 245, 1978.

52. **Studnicka, G. M., Eiserling, F. A., and Lake, J. A.**, A unique secondary folding pattern for 5SRNA corresponds to the lowest energy homologous secondary structure in 17 different prokaryotes, *Nucleic Acids Res.*, 9, 1885, 1981.

53. **Brown, D. D., Wensink, P. C., and Jordan, E.**, Purification and some characteristics of 5SDNA from *Xenopus laevis*, *Proc. Natl. Acad. Sci. U.S.A.*, 68, 3175, 1971.

54. **Brownlee, G. G., Cartwright, E. M., and Brown, D. D.**, Sequence studies of the 5SDNA of *Xenopus laevis*, *J. Mol. Biol.*, 89, 703, 1974.

55. **Miller, J. R., Cartwright, E. M., Brownlee, G. G., Fedoroff, N. V., and Brown, D. D.**, The nucleotide sequence of oocyte 5SDNA in *Xenopus laevis*. II. The GC-rich region, *Cell*, 13, 717, 1978.

56. **Brown, D. D. and Sugimoto, K.**, 5SDNAs of *Xenopus laevis* and *Xenopus mulleri*: evolution of a gene family, *J. Mol. Biol.*, 78, 397, 1973.

57. **Korn, L. J. and Brown, D. D.**, Nucleotide sequence of *Xenopus borealis* oocyte 5SDNA: comparison of sequences that flank related eukaryotic genes, *Cell*, 15, 1145, 1978.

58. **Ford, P. J. and Brown, R. D.**, Sequences of 5S ribosomal RNA from *Xenopus mulleri* and the evolution of 5S gene-coding sequences, *Cell*, 8, 485, 1976.

59. **Pribula, C. D., Fox, G. E., Woese, C. R., Sogin, M., and Pace, N.**, Nucleotide sequence of *Bacillus megaterium* 5SRNA, *FEBS Lett.*, 44, 322, 1974.

60. **Brownlee, G. G. and Cartwright, E. M.**, The nucleotide sequence of the 5SRNA of chicken embryo fibroblasts, *Nucleic Acids Res.*, 2, 2279, 1975.

61. **Nomura, M. and Erdmann, V. A.**, Reconstitution of 50S ribosomal subunits from dissociated molecular components, *Nature (London)*, 228, 744, 1970.

62. **Wrede, P. and Erdmann, V. A.**, *Escherichia coli* 5SRNA binding proteins L18 and L25 interact with 5.8SRNA but not with 5SRNA from yeast ribosomes, *Proc. Natl. Acad. Sci. U.S.A.*, 74, 2706, 1977.

63. **Wreschner, D. H.**, The role of ribosomal RNA in protein synthesis. Inhibition of translation by reticulocyte 5S ribosomal RNA, *FEBS Lett.*, 94, 139, 1978.

64. **Wreschner, D. H.**, Reticulocyte 5S ribosomal RNA inhibition of cell-free protein synthesis. Novel responses in ribosomal behaviour, *FEBS Lett.*, 94, 145, 1978.

65. **Raacke, I. D.**, A model for protein synthesis involving the intermediate formation of peptidyl-5SRNA, *Proc. Natl. Acad. Sci. U.S.A.*, 68, 2357, 1971.

66. **Kao, T. H. and Crothers, D. M.**, A proton-coupled conformational switch of *Escherichia coli* 5S ribosomal RNA, *Proc. Natl. Acad. Sci. U.S.A.*, 77, 3360, 1980.

67. **Mullins, D. W., Jr., Lacey, J. C., Jr., and Hearn, R. A.**, 5SrRNA and tRNA: evidence of a common evolutionary origin, *Nature (London) New Biol.*, 242, 80, 1973.

68. **Spirin, A. S. and Gavrilova, L. P.**, *The Ribosome*, Springer-Verlag, New York, 1969.

69. **Veldman, G. M., Brand, R. C., Klootwijk, J., and Planta, R. J.**, Some characteristics of processing sites in ribosomal procursor RNA of yeast, *Nucleic Acids Res.*, 8, 2907, 1980.

70. **Nazar, R. N.**, A 5.8SrRNA-like sequence in prokaryotic 23SrRNA, *FEBS Lett.*, 119, 212, 1980.

71. **Sankoff, D., Cedergren, R. J., and McKay, W.**, A strategy for sequence phylogeny research, *Nucleic Acids Res.*, 10, 421, 1982.

72. **Fujiwara, H., Kawata, Y., and Ishikawa, H.**, Primary and secondary structure of 5.8S rRNA from the silk gland of *Bombix mori*, *Nucleic Acids Res.*, 10, 2415, 1982.

73. **Lee, J. C. and Henry, B.,** Binding of rat ribosomal proteins to yeast 5.8S ribosomal ribonucleic acid, *Nucleic Acids Res.,* 10, 2199, 1982.

74. **Virmaux, N., Mandel, P., and Urban, P. F.,** Evidence of two soluble RNA types in eye lens, *Biochem. Biophys. Res. Commun.,* 16, 308, 1964.

75. **Erikson, E., Erikson, R. L., Henry, B., and Pace, N. R.,** Comparison of oligonucleotides produced by RNase Tl digestion of 7SRNA from avian and murine oncornaviruses and from uninfected cells, *Virology,* 53, 40, 1973.

76. **Bishop, J. M., Levinson, W., Quintrell, N., Sullivan, D., Fanshier, L., and Jackson, J.,** Low molecular weight RNAs of Rous sarcoma virus. II. 7SRNA, *Virology,* 42, 927, 1970.

77. **Busch, H.,** The function of the 5' cap of mRNA and nuclear species, *Perspect. Biol. Med.,* 19, 549, 1976.

78. **Li, W. Y., Reddy, R., Henning, D. H., Epstein, P. M., and Busch, H.,** Nucleotide sequence of 7SRNA. Homology to Alu DNA and La 4.5SRNA, *J. Biol. Chem.,* 257, 5136, 1982.

79. **Reddy, R., Henning, D., and Busch, H.,** The primary nucleotide sequence of U4RNA, *J. Biol. Chem.,* 256, 3532, 1981.

80. **Busch, H., Reddy, R., Rothblum, L., and Choi, Y. C.,** SnRNAs, SnRNPs and RNA processing, *Annu. Rev. Biochem.,* 51, 617, 1982.

81. **Ullu, E., Murphy, S., and Melli, M.,** Human 7SL RNA consists of 140 nucleotide middle-repetitive sequence inserted in an Alu sequence, *Cell,* 29, 195, 1982.

82. **Desphande, A. K., Niu, L. C., and Niu, M. C.,** Requirement of informational molecules in heart formation, in *The Role of RNA in Reproduction and Development,* Niu, M. C. and Segal, S. J., Eds., North-Holland/American Elsevier, New York, 1973, 229.

83. **Desphande, A. K. and Siddiqui, M. A. Q.,** Acetylcholinesterase differentiation during myogenesis in early chick embryonic cells caused by inducer RNA, *Differentiation,* 10, 133, 1978.

84. **Walter, P. and Blobel, G.,** Signal recognition particles contains a 7SRNA essential for protein translocation across the endoplasmic reticulum, *Nature (London),* 229, 691, 1982.

85. **Tomizawa, J., Tateo, I., Selzer, G., and Som, T.,** Inhibition of ColEl RNA primer formation by a plasmid-specified small RNA, *Proc. Natl. Acad. Sci. U.S.A.,* 78, 1421, 1981.

86. **De Wilde, M., Davies, J. E., and Schmidt, F. J.,** Low molecular weight RNA species encoded by a multiple drug resistance plasmid, *Proc. Natl. Acad. Sci. U.S.A.,* 75, 3673, 1978.

87. **Roozen, K. J., Fenwick, R. G., Jr., and Curtiss, R.,** Synthesis of ribonucleic acid and protein in plasmid-containing minicells of *Escherichia coli* K12, *J. Bacteriol.,* 107, 21, 1971.

88. **Campos, R., Javanovich, S., and Villarreal, L. P.,** A small RNA complementarity to an intervening sequence produced late in SV40 infection, *Nature (London),* 291, 344, 1981.

89. **Lerner, M. R., Andrews, N. C., Miller, G., and Steitz, J. A.,** Two small RNAs encoded by Epstein-Barr virus and complexed with protein are precipitated by antibodies from patients with systemic lupus erythematosus, *Proc. Natl. Acad. Sci. U.S.A.,* 78, 805, 1981.

90. **Colonno, R. J. and Banerjee, A. K.,** A unique RNA species involved in initiation of vesicular stomatitis virus RNA transcription *in vitro, Cell,* 8, 197, 1976.

91. **Emerson, S. V., Dierks, P. M., and Parson, J. T.,** *In vitro* synthesis of a unique RNA species by a T particle of vesicular stomatitis virus, *J. Virol.,* 23, 708, 1977.

92. **Testa, D., Chanda, P. K., and Banerjee, A. K.,** Unique mode of transcription *in vitro* by vesicular stomatitis virus, *Cell,* 21, 267, 1980.

93. **Mathews, M. B.,** Genes for VA-RNA in adenovirus 2, *Cell,* 6, 223, 1975.

94. **Rao, D. D. and Huang, A. S.,** Synthesis of a small RNA in cells coinfected by standard and defective interfering particles of vesicular stomatitis virus, *Proc. Natl. Acad. Sci. U.S.A.,* 76, 3742, 1979.

95. **Colonno, R. J. and Banerjee, A. K.,** Complete nucleotide sequence of the leader RNA synthesized *in vitro* by vesicular stomatitis virus, *Cell,* 15, 93, 1978.

96. **Semler, B. L., Perrault, J., Abelson, J., and Holland, J. J.,** Sequence of a RNA templated by the 3' OH RNA terminus of defective interfering particles of vesicular stomatitis virus, *Proc. Natl. Acad. Sci. U.S.A.,* 75, 4704, 1978.

97. **Plawecki, M. and Beljanski, M.,** Comparative study of *Escherichia* endotoxin, hydrocortisone, and Beljanski-Leukocyte-Restorer activity in cyclophosphamide-treated rabbits, *Proc. Soc. Exp. Biol. Med.,* 168, 408, 1981.

98. **Alwine, J. C. and Khoury, G.,** Simian virus 40-associated small RNA: mapping on the simian virus 40 genome and characterization of its synthesis, *J. Virol.,* 36, 701, 1980.

99. **Sawyer, R. C., Harada, F., and Dahlberg, J. E.,** Virion-associated RNA primer for Rous sarcoma virus DNA synthesis: isolation from uninfected cells, *J. Virol.,* 13, 1302, 1974.

100. **Harada, F., Kato, N., and Noshino, H. O.,** Series of 4.5S RNAs associated with poly(A)-containing RNAs of rodent cells, *Nucleic Acids Res.,* 7, 909, 1979.

101. **Gross, H. J., Domdey, H., Lossow, C., Jank, P., Raba, M., Alberty, H., and Saenger, H. L.,** Nucleotide sequence and secondary structure of potato spindle tuber viroid, *Nature (London),* 273, 203, 1978.

102. **Branch, A. D., Robertson, H. D., Greer, C., Gegenheimer, P., Peebles, C., and Abelson, J.,** Cell-free circularization of viriod progeny RNA by an RNA ligase from wheat germ, *Science,* 217, 1147, 1982.

103. **Hadid, A., Jones, D. M., Gillespie, D. H., Wong-Stall, F., and Diener, T. O.,** Hybridization of potato spindle tuber viroid to cellular DNA of normal plants, *Proc. Natl. Acad. Sci. U.S.A.,* 73, 2453, 1976.

104. **Semancik, J. S.,** Small pathogenic RNA in plants. The viroids, *Annu. Rev. Phytopathol.,* 17, 461, 1979.

105. **Weinberg, R. A. and Penman, S.,** Small molecular weight monodisperse nuclear RNA, *J. Mol. Biol.,* 38, 289, 1968.

106. **Ro-Choi, T. S. and Busch, H.,** Low-molecular-weight nuclear RNAs, in *The Cell Nucleus,* Vol. 3, Busch, H., Ed., Academic Press, New York, 1974, 151.

107. **Zieve, G. and Penman, S.,** Small RNA species on the HeLa cell: metabolism and subcellular localization, *Cell,* 8, 19, 1976.

108. **Nohga, K., Reddy, R., and Busch, H.,** Comparison of RNase Tl fingerprints of U1, U2, and U3 small nuclear RNAs of HeLa cells, human normal fibroblasts and Novikoff Hepatoma cells, *Cancer Res.,* 41, 2215, 1981.

109. **Rogers, J. and Wall, R.,** A mechanism for RNA splicing, *Proc. Natl. Acad. Sci. U.S.A.,* 77, 1877, 1980.

110. **Lerner, M. R., Boyle, J. A., Mount, S. M., Wolin, S. L., and Steitz, J. A.,** Are SnRNPs involved in splicing?, *Nature (London),* 283, 220, 1980.

111. **Lerner, M. R. and Steitz, J. A.,** Snurps and scyrps, *Cell,* 25, 298, 1981.

112. **Lazar, E., Jacob, M., Krol, A., and Branlant, C.,** Accessibility of U1RNA to base pairing with a single-stranded DNA fragment mimicking the intron extremities at the splice junction, *Nucleic Acids Res.,* 10, 1193, 1982.

113. **Yang, V. W., Lerner, M. R., Steitz, J. A., and Flint, S. J.,** A small nuclear ribonucleoprotein is required for splicing of adenoviral early RNA sequences, *Proc. Natl. Acad. Sci. U.S.A.,* 78, 1371, 1981.

114. **Calvet, J. P., Meyer, L. M., and Pederson, T.,** Small nuclear RNA U2 is base-paired to heterogeneous nuclear RNA, *Science,* 217, 456, 1982.

115. **Denison, R. A., Van Arsdell, S. W., Bernstein, L. B., and Weiner, A. M.,** Abundant pseudogenes for small nuclear RNAs are dispersed in the human genome, *Proc. Natl. Acad. Sci. U.S.A.,* 78, 810, 1981.

116. **Van Arsdell, S. W., Denison, R. A., Bernstein, L. B., Weiner, A. M., Manser, T., and Gesteland, R. F.,** Direct repeats flank three small nuclear RNA pseudogenes in the human genome, *Cell,* 26, 11, 1981.

117. **Mount, S. M. and Steitz, J. A.,** Sequence of U1 RNA from *Drosophila melanogaster:* implications for U1 secondary structure and possible involvement in splicing, *Nucleic Acids Res.,* 9, 6351, 1981.

118. **Beljanski, M. and Bourgarel, P.,** Isolement et caracterisation d'un ARN matriciel d' *Alcaligenes faecalis, C. R. Acad. Sci. Paris Ser. D.,* 266, 845, 1968.

119. **Beljanski, M. and Beljanski, M.,** ''Acide aminé-acide ribonucléiqüe'' intermédiaire dans la synthèse des liaisons peptidiques, *Biochim. Biophys. Acta,* 72, 585, 1963.

120. **Beljanski, M. and Beljanski, M.,** RNA-bound reverse transcriptase in *Escherichia coli* and in vitro synthesis of a complementary DNA, *Biochem. Genet.,* 12, 163, 1974.

121. **Lee, S. Y., Bailey, S. C., and Apirion, D.,** Small stable RNAs from *Escherichia coli:* evidence for the existence of new molecules and for a new ribonucleoprotein particle containing 6SRNA, *J. Bacteriol.,* 133, 1015, 1978.

122. **Kole, R. and Altman, S.,** Reconstitution of RNase P activity from inactive RNA and protein, *Proc. Natl. Acad. Sci. U.S.A.,* 76, 3795, 1979.

123. **Kole, R. and Altman, S.,** Properties of purified ribonuclease P from *Escherichia coli, Biochemistry,* 20, 1902, 1981.

124. **Stark, B. C., Kole, R., Bowman, E. J., and Altman, S.,** Ribonuclease P: an enzyme with an essential RNA component, *Proc. Natl. Acad. Sci. U.S.A.,* 75, 3717, 1978.

125. **Motamedi, H., Lee, K., Nichols, L. N., and Schmidt, F. J.,** An RNA species involved in *Escherichia coli* ribonuclease P activity, *J. Mol. Biol.,* 162, 535, 1982.

126. **Goldstein, J. and Harewood, K.,** Another species of ribonucleic acid in *Escherichia coli, J. Mol. Biol.,* 39, 383, 1969.

127. **Merten, S., Synenki, R. M., Locker, J., Christianson, T., and Rabinowitz, M.,** Processing of precursors of 21S ribosomal RNA from yeast mitochondria, *Proc. Natl. Acad. Sci. U.S.A.,* 77, 1417, 1980.

128. **Beljanski, M., Beljanski, M., and Bourgarel, P.,** ARN transformant porteur de caractères héréditaires chez *Escherichia coli* showdomycino-résistant, *C. R. Acad. Sci. Paris Ser. D:,* 272, 2107, 1971.

129. **Le Goff, L.,** Contribution à l'étude du Crown-Gall. Participation à la Tumorisation des Cellules Végétales d'Acides Ribonucléiques Particuliers et Action des ces ARN sur l'expression due Pouvoir Oncogéne d' *Agrobacterium tumefaciens,* thesis, Universite du Pierre et Marie Curie, Paris, 1977.

130. **Le Goff, L. and Manigault, P.,** Variations dans l'expression du pouvoir oncogéne d' *Agrobacterium tumefaciens, Physiol. Plant.,* 42, 337, 1978.

131. **Jelinek, W. and Leinwand, L.,** Low molecular weight RNAs hydrogen-bounded to nuclear and cytoplasmic poly(A)-terminated RNA from cultured chinese hamster ovary cells, *Cell,* 15, 205, 1978.

132. **Harada, F. and Kato, N.,** Nucleotide sequences of 4.5S RNA associated with poly(A)-containing RNAs of mouse and hamster cells, *Nucleic Acids Res.,* 8, 1273, 1980.

133. **Mroczkowski, B., McCarthy, T. L., Zezza, D. J., Bragg, P. W., and Heywood, S. M.,** Small RNAs involved in gene expression of muscle-specific proteins, *Exp. Biol. Med.,* 9, 277, 1984.

134. **Gardner, J. A. A. and Hoagland, M. B.,** A unique ribonucleic acid of low molecular weight from rat liver microsomes, *J. Biol. Chem.,* 243, 10, 1968.

135. **Hellung-Larsen, P. and Frederiksen, S.,** Small molecular weight RNA components in Ehrlich ascites tumor cells, *Biochim. Biophys. Acta,* 262, 290, 1972.

136. **Stravianopoulos, J. G., Karkas, J. D., and Chargaff, E.,** Nucleic acid polymerases of the developing chicken embryos: a DNA polymerase preferring a hybride template, *Proc. Natl. Acad. Sci. U.S.A.,* 68, 2207, 1971.

137. **Beljanski, M.,** ARN-amorceurs riches en nucleotides G et A indispensables à la réplication *in vitro* de l'ADN des phages φX174 et lambda, *C. R. Acad. Sci. Paris Ser. D:,* 280, 1189, 1975.

138. **Chargaff, E.,** Initiation of enzymatic synthesis of deoxyribocucleic acid by ribonucleic acid primers, *Prog. Nucleic. Acids Res. Mol. Biol.,* 18, 1, 1977.

139. **Kornberg, A.,** *DNA Replication,* W. H. Freeman & Co., Ed., San Francisco, 1980.

140. **Ogawa, T. and Okazaki, T.,** Discontinuous DNA replication, *Annu. Rev. Biochem.,* 49, 421, 1980.

141. **Furth, J. E. and Natta, C.,** Translational control of β and α globin chains synthesis, *Nature (London) New Biol.,* 240, 274, 1972.

142. **Reichman, M. and Penman, S.,** Stimulation of polypeptide initiation *in vitro* after protein synthesis inhibition in vivo in HeLa cells, *Proc. Natl. Acad. Sci. U.S.A.,* 70, 2678, 1973.

143. **Berns, A., Salden, M., Bogdanowsky, D., Raymondjean, M., Schapira, G., and Blomendel, H.,** Non-specific stimulation of cell-free protein synthesis by a dialysable factor isolated from reticulocyte initiation factors (i-RNA), *Proc. Natl. Acad. Sci. U.S.A.,* 72, 714, 1975.

144. **Beljanski, M. and Plawecki, M.,** Particulars RNA-fragments as promoters of leukocyte and platelet formation in rabbits, *Exp. Cell. Biol.,* 47, 218, 1979.

145. **Geider, K., Berck, E., and Schaller, H.,** An RNA transcribed from DNA at the origin of phage fd single strand to replicative form conversion, *Proc. Natl. Acad. Sci. U.S.A.,* 75, 645, 1978.

146. **Langeveld, S. A., van Mansfeld, A. D. M., Baas, P. D., Jansz, H. S., van Arkel, G. A., and Weisbeek, P. J.,** Nucleotide sequence of the origin of replication in bacteriophage φX174 RF DNA, *Nature (London),* 271, 417, 1978.

147. **Ling, W.,** Pyrimidine sequences from the DNA of bacteriophage fd, fl and φX174, *Proc. Natl. Acad. Sci. U.S.A.,* 69, 742, 1972.

148. **Tomizawa, J. I. and Itoh, T.,** Plasmid ColEl incompatibility determined by interaction of RNA I with primer transcript, *Proc. Natl. Acad. Sci. U.S.A.,* 78, 6096, 1981.

149. **Selzer, G., Som, T., Itoh, T., and Tomizawa, J. I.,** The origin of replication of plasmid p15A and comparative studies on the nucleotide sequences around the origin of related plasmids, *Cell,* 32, 119, 1983.

150. **Goldstein, E. I. and Penman, S.,** Regulation of protein synthesis in mammalian cells. V. Effect of actinomycin D on translation control in HeLa cells, *J. Mol. Biol.,* 80, 243, 1973.

151. **Heywood, S. M., Kennedy, D. S., and Bester, A. J.,** Separation of specific initiation factors involved in the translation of myosin and myoglobin messenger RNAs and the isolation of a new RNA involved in translation, *Proc. Natl. Acad. Sci. U.S.A.,* 71, 2428, 1974.

152. **Bester, A. J., Kennedy, D. S., and Heywood, S. M.,** Two classes of translational control RNA: their role in the regulation of protein synthesis, *Proc. Natl. Acad. Sci. U.S.A.,* 72, 1523, 1975.

153. **Bogdanowsky, D., Hermann, W., and Schapira, G.,** Presence of a new RNA species among the initiation protein factors active in eukaryotes translation, *Biochem. Biophys. Res. Commun.,* 54, 25, 1973.

154. **Zeichman, N. and Brutkrentz, D.,** Isolation of low molecular weight RNAs from connective tissue. Inhibition of mRNA translation, *Arch. Biochem. Biophys.,* 188, 410, 1978.

155. **Zang, A. J., Grabowski, P. J., and Cech, T. R.,** Autocatalytic cyclization of an excised intervening sequence RNA is a cleavage-ligation reaction, *Nature (London),* 301, 578, 1983.

156. **Jagadeerwaran, P., Forget, B. G., and Weissman, S. M.,** Short interspersed repetitive DNA elements in eukaryotes: transposable DNA elements generated by reverse transcription of RNA Pol.III transcripts? *Cell,* 26, 141, 1981.

157. **Zieve, G. W.,** Two groups of small stable RNAs, *Cell,* 25, 296, 1981.

158. **Houck, C. N., Rinehart, F. P., and Schmid, C. W.,** A ubiquitous family of repeated DNA sequences in the human genome, *J. Mol. Biol.,* 132, 289, 1979.

159. **Jelinek, W. R., Toomey, T. P., Leinwand, L., Duncan, C. H., Biro, P. A., Choudary, P. V., Weissman, S. M., Rubin, C. M., Houck, C. M., Deininger, P. L., and Schmid, C. W.,** Ubiquitous interspersed repeated sequences in mammalian genomes, *Proc. Natl. Acad. Sci. U.S.A.,* 77, 1398, 1980.

160. **Schmid, C. W. and Jelinek, W. R.,** The alu family of dispersed repetitive sequences, *Science,* 216, 1065, 1982.

161. **Haynes, S. R. and Jelinek, W. R.,** Low molecular weight RNAs transcribed in vitro by RNA polymerase III from Alu-type dispersed repeats in chinese hamster DNA are also found *in vivo, Proc. Natl. Acad. Sci. U.S.A.,* 78, 6130, 1981.

162. **Hollis, G. F., Hieter, P. A., McBride, O. W., Swan, D., and Leder, P.,** Processed genes: a dispersed human immunoglobulin gene bearing evidence of RNA-type processing, *Nature (London),* 296, 321, 1982.

163. **Sharp, Ph.,** Conversion of RNA to DNA in mammals: Alu-like elements and pseudogenes, *Nature (London),* 301, 471, 1983.

164. **Spiegelman, S., Haruna, I., Pace, N. R., Mills, D. R., Bishop, D. H. L., Claybook, J. R., and Peterson, R.,** Replication of viral RNA, *J. Cell. Physiol. Suppl.,* 70(2), 35, 1967.

165. **Flavell, R. A., Sabo, D. L., Bandle, E. F., and Weissmann, C.,** Site-directed mutagenesis. Effect of an extracistronic mutation on the *in vitro* propagation of bacteriophage QβRNA, *Proc. Natl. Acad. Sci. U.S.A.,* 72, 367, 1975.

166. **Sumper, M. and Luce, R.,** Evidence for de novo production of self-replicating and environmentally adapted RNA structures by bacteriophage Qβ replicase, *Proc. Natl. Acad. Sci. U.S.A.,* 72, 162, 1975.

167. **Weissman, C., Billeter, M. A., Goodman, H. M., Hindley, J., and Weber, H.,** Structure and function of phage RNA, *Annu. Rev. Biochem.,* 42, 303, 1973.

168. **Feix, G. and Sano, H.,** Initiation specificity of the Poly(cytidylic acid)-dependent Qβ replicase activity, *Eur. J. Biochem.,* 58, 59, 1975.

169. **Eikhom, T. S. and Spiegelman, S.,** The dissociation of Qβ-replicase and the relation of one of the components to a Poly-C-dependent Poly-G-polymerase, *Proc. Natl. Acad. Sci. U.S.A.,* 57, 1833, 1967.

170. **Plawecki, M. and Beljanski, M.,** Synthèse in vitro d'un ARN utilisé comme amorceur pour la réplication de l'ADN, *C.R. Acad. Sci. Paris Ser. D:,* 278, 1413, 1974.

171. **Plawecki, M.,** Synthèse et Réplication In Vitro d'ARN Biologiquement Actifs. Analogie avec les ARN d' *Escherichia coli* Showdomycino-Résistants, thesis, Universite du Pierre et Marie, Curie, Paris, 1974.

172. **Temin, H. M.,** Protovirus hypothesis: speculations on the significance of RNA-directed DNA synthesis for normal development and carcinogenesis, *J. Natl. Cancer. Inst.,* 46, iii, 1971.

173. **Kang, C. Y. and Temin, H. M.,** Endogeneous RNA-directed DNA polymerase activity in uninfected chicken embryos, *Proc. Natl. Acad. Sci. U.S.A.,* 69, 1550, 1972.

174. **Loeb, L. A., Tartof, K. D., and Travaglini, E. C.,** Copying natural RNAs with *E. coli* DNA polymerase I., *Nature (London) New Biol.,* 242, 66, 1973.

175. **Gulati, S. C., Kacian, D. L., and Spiegelman, S.,** Conditions for using DNA polymerase I as an RNA-dependent DNA polymerase, *Proc. Natl. Acad. Sci. U.S.A.,* 71, 1035, 1974.

176. **Beljanski, M., Manigault, P., Beljanski, M., and Aaron-da Cunha, M. I.,** Genetic transformation of *Agrobacterium tumefaciens* B₆ by RNA and nature of the tumor inducing principle, *Proc. 1st Int. Congr. IAMS,* 1, 132, 1974.

177. **Mayer, R. J., Smith, R. G., and Gallo, R. C.,** Reverse transcriptase in normal monkey placenta, *Science,* 185, 864, 1974.

178. **Wu, A. M. and Gallo, R. C.,** Reverse transcriptase, *CRC Crit. Rev. Biochem.,* 3, 289, 1975.

179. **Travaglini, E. C. and Loeb, L. A.,** Ribonucleic acid dependent deoxyribonucleic acid synthesis by *Escherichia coli* deoxyribonucleic acid polymerase. I. Characterization of the polymerization reaction, *Biochemistry,* 13, 3010, 1974.

180. **Brizzard, B. L. and De Kloet, S. R.,** Reverse transcription of yeast double-stranded RNA and ribosomal RNA using synthetic oligonucleotide primers, *Biochim. Biophys. Acta,* 739, 122, 1983.

181. **Beljanski, M., Beljanski, M., Manigault, P., and Bourgarel, P.,** Transformation of *Agrobacterium tumefaciens* into a non-oncogenic species by an *Escherichia coli* RNA, *Proc. Natl. Acad. Sci. U.S.A.,* 69, 191, 1972.

182. **Aaron-Da Cunha, M. I., Kurkdjian, A., and Le Goff, L.,** Nature tumorale d'une hyperplasie obtenue expérimentalement, *C.R. Soc. Biol.,* 169(3), 755, 1975.

183. **Le Goff, L., Aaron-Da Cunha, M. I., and Beljanski, M.,** Un ARN extrait d' *Agrobacterium tumefaciens,* souches oncogènes et nononcogènes, élément indispensable à l'induction des tumeurs chez *Datura stramonium, Can. J. Microbiol.,* 22, 694, 1976.

184. **Beljanski, M. and Aaron-Da Cunha, M. I.,** Particular small size RNA and RNA-fragments from different origins as tumor inducing agents in *Datura stramonium, Mol. Biol. Rep.,* 2, 497, 1976.

185. **Shiba, T. and Saigo, K.,** Retrovirus-like particles containing RNA homologous to the transposable element copia in *Drosophila melanogaster, Nature (London),* 302, 119, 1983.

Chapter 5

MITOCHONDRIAL DNAs AND PHYLOGENETIC RELATIONSHIPS

A. J. Birley and J. H. Croft

TABLE OF CONTENTS

I. INTRODUCTION

Microevolutionary processes are currently being investigated by the study of genetic diversity in the mitochondrial genome. In addition to its small size and relative ease of extraction and purification two particular properties of the mitochondrial genome have led to the adoption of mitochondrial DNA (mtDNA) variation in studies concerned with recent phylogenetic comparison. These are the maternal mode of inheritance, and the absence of recombination in sexually reproduced lineages. Theoretical studies representing an extension of neutral gene theory have also been rapid. Not only can an expectation of the rate of gene substitution be formulated for the mitochondrial genome, so too can comparative levels of nuclear and mitochondrial diversity found within populations and with respect to mating behavior, migration rates, and population subdivision. Consequently, measurements of mtDNA diversity are revealing details of evolutionary processes concerned with the recent geographical and ecological differentiation of species. These studies are also likely to be informative in an analysis of rates of gene introgression and convergent evolutionary phenomena.

Studies of mtDNA variation in higher animals, both within and between populations of species and genera, will form the major part of this account. However, studies of the microevolutionary process are beginning in a diversity of life forms including protozoa, yeasts and filamentous fungi, and higher plants. Any principles and assumptions made in comparisons of relatedness between individuals, species, and genera in higher animals will be useful in context of those addressing the evolutionary dynamics of mtDNA evolution in a wider range of organisms. Of course, in some instances, for example, in some fungal species or in isogamous species, it may not be possible to assume either the constraint of maternal inheritance or the absence of genetic recombination. Some informative aspects of mtDNA evolution in fungal species are presented in this account and the authors hope that the theoretical modeling of mtDNA evolution will, in time, be presented on a more general basis which may be able to account for the very large range of diversity found in the mitochondrial genome.

II. THE RANGE OF DIVERSITY OF THE MITOCHONDRIAL GENOME

Although the evidence for the existence of a mitochondrial genetic system had been accumulating since the early descriptions of cytoplasmically inherited respiratory deficient *petite* mutants of yeast,[1,2] the impetus to the study of mitochondrial genetics was not provided until much later when the physical presence of DNA within chick embryo mitochondria was demonstrated.[3,4] Since then much literature on both functional and structural aspects of mtDNA has accumulated. The major function of the organelle in oxidative phosphorylation is fundamentally similar in all eukaryotic systems; however, the most striking feature of the mitochondrial genome is its diversification in size, organization, and structure. The extent and nature of this diversity is great and includes surprising aspects, many of which cannot be explained readily in evolutionary terms at present.

The most striking aspect of this diversity is the variation in size of the mitochondrial genome. In most animals the size of the mtDNA is about 16 kb,[5] whereas in fungi, it can range from about 18 to over 100 kb.[6,7] In higher plants the mitochondrial genome is large, ranging from about 100 to as much as 2500 kb.[8,9] In animals and fungi the mtDNA is present in monomeric form, usually as closed circular molecules, though in *Tetrahymena*[10,11] and *Paramecium*,[12,13] and in the yeast *Hansenula mrakii*,[14] it is present in linear form. In two species of the single-celled alga, *Chlamydomonas*, where the mtDNA consists of only 15 kb, preparations show mainly linear molecules with unique ends, though a small proportion of circular molecules is also present.[15]

In higher plants the structural organization of the mitochondrial genome is less clear and it has been demonstrated that the mitochondrion contains a population of heterogeneous circular molecules of DNA which can be distinguished on the basis of size. In some cases each class is unique and shows no detectable homology with each other, but in other cases sequences may be common to more than one class of circular molecule and multimeric series may be found.[8,16-26] Further variation in the number of classes of molecule is found in plants carrying cytoplasmically inherited male sterility.[27-29] The relationship in plants between the heterogeneous population of circular molecules and the complete mitochondrial genome has not been determined. It is possible that the various classes of mtDNA could represent autonomously replicating regions of the complete genome derived from it by recombination or by amplification.[8,17,30-33] Similar events have been described in several species of fungi giving rise to petite mutations in yeast[34] or senescence-like phenomena in *Podospora*,[35-40] *Aspergillus*,[41-44] and *Neurospora*.[45,46] Alternatively, it is possible that the mitochondrial genome of higher plants is fragmented into several distinct mitochondrial "chromosomes".[16,19,21,25,26] Examples of heterogeneous populations of mtDNA molecules are not frequent in fungal species except in the cases of petite and senescence. However, small mitochondrial plasmid-like molecules have been described in two species of *Neurospora*.[47,48] In these cases there is no detectable homology between the plasmids and the normal mtDNA, and although these plasmids are transcribed, no functional role for them has been detected.

Overall, therefore, the observed variation in structure of the mitochondrial genome indicates a high degree of evolutionary flexibility. Further evidence for this flexibility comes from the observation that during evolution, DNA sequences have been transferred between the mitochondrion, the chloroplast, and the nucleus in plants, fungi, and animals,[49-56] though the mechanism for such transposition is not clear.

When the internal organization and structure of the mitochondrial genome is examined the impression of evolutionary flexibility is reinforced further. Most mitochondrial genomes which have been studied in detail have roughly the same gene content. Thus, each genome can be expected to contain a gene for a large and a small rRNA, three cytochrome oxidase subunits, apocytochrome *b*, ATPase subunit 6, about 22 to 25 tRNAs, and a number (up to about 10) of open reading frames of unidentified function (URFs). Other genes have been identified in some organisms. For example *var1* in yeast,[57] which codes for a protein associated with the small mitochondrial ribosomal subunit, and a gene coding for ATPase subunit 9 (the DCCD-binding protein) in yeast[58,59] and in higher plants[60] are mitochondrially located. The active ATPase subunit 9 gene in *Neurospora, Aspergillus,* and mammals is located in the nucleus, but in both fungal examples a sequence with high homology to this gene has been located in the mitochondrion.[51,52] In addition to the large and small rRNAs, higher plant mitochondria code for a 5S rRNA[61,62] and probably for the α-subunit of the F_1 component of ATPase.[63,64] Though several of the genes listed above have been located in higher plant mtDNA the exact coding content for any plant mitochondrial genome remains to be clarified.[65,66]

The organization of the genes within the mtDNA is extremely variable. In animals evolution has led to a small and extremely economic molecule in which the scope for structural variation is limited. The coding sequences for proteins, tRNAs, and rRNAs in mammals are virtually continuous with no more than a small number of nucleotide pairs between each functional gene; even the overlap of some genes is apparent. The gene order and overall structure of human, mouse, and bovine mtDNA is almost identical.[67-70] There is some variation in size both between and within species, but this is due mainly to variation in the displacement loop (D-loop) region in which the origin of replication is located. In various *Drosophila* species there is considerable variation in size from 15.7 to 19.5 kb.[71] This variation is found almost entirely in a region of the mtDNA which is very rich in AT base pairs.

In fungi, in addition to the considerable size diversity, the internal arrangement of the genes in the mtDNA is extremely variable. Thus, many insertion and deletion events and extensive sequence arrangements have been involved during the evolution of the fungal mitochondrial genome. The extent of this variation in fungi is demonstrated when the mitochondrial genomes in a range of yeast species are examined, though it should be emphasized that these species may represent a wide range of taxonomic relationships. Among the yeasts the size of the mtDNA ranges from about 19 kb in *Torulopsis glabrata*[6] and *Schizosaccharomyces pombe*[72] to about 75 kb in *Saccharomyces cerevisiae*,[73] and over 100 kg in *Brettanomyces custerii*.[7] One yeast, *Hansenula mrakii*, is unusual in that its 55 kb genome is linear and not covalently closed circular molecule as in the other species.[14] The gene order and arrangement within the genome of the yeasts display very few features in common. The size differences can be accounted for by the presence of noncoding sequences between the genes and variation in the extent of the individual genes themselves, much of this being due to the presence or absence of introns; several genes in the larger genomes are of a complex mosaic structure. One yeast, *Koeckera africana*,[74] in common with the water mold, *Achlya bisexualis*,[75] has been found to contain a duplication of the rRNA region. In the case of *Achlya* the repeat is large and contains the functional rRNA genes, whereas in *K. africana* only a part of the large rRNA gene is involved.

In other mycelial fungi, where the gene order has been published, it is again found that there is considerable structural and size variation. The three mycelial species for which most information is available are *Aspergillus nidulans*, *Neurospora crassa*, and *Podospora anserina* where the sizes of the mitochondrial genomes are 32.4, 62, and 91 kb, respectively.[76-78] Differences in the order of the genes between these three species are apparent, but if it is assumed that inversion has been involved in the divergence of these genomes, then only a small number of such events is required to explain those differences. However, it should be emphasized that the mechanism which led to these rearrangements is not understood at the present time.

Detailed nucleotide sequence analysis of mtDNA has also revealed several interesting features, but perhaps the most surprising of these is the observation that the mitochondrion does not utilize the universal genetic code, and moreover that the code used varies somewhat from organism to organism. In the mitochondrial code more groups of four codons can be read by a single tRNA. The tRNAs which correspond to these four-codon families have a U in the first anticodon position which, in the unmodified form, can pair with any of the four bases which may be present in the third position of the codon.[79-82] As a consequence, fewer tRNAs are required in the mitochondrion than for the universal code and 22 different tRNAs were found in mammalian mitochondria,[67-70] 24 in yeast,[81] and 23 in *Neurospora*.[83,84] Mitochondrial tRNAs have many features which distinguish them from their cytoplasmic counterparts. It has been proposed that in mammals, the presence of a modified nucleotide in the position immediately 3' to the anticodon restricts the codon recognition response to U or G "wobble"[85] only, whereas its absence would permit U, C, A, or G wobble, thus resulting in a four-codon family.[82]

The evolutionary relationships of the mitochondrial codes found in various groups of organisms are complex and difficult to determine at the present time because of a relative paucity of data, but some interesting differences have been reported. The UGA termination codon of the universal code is used for tryptophan in fungi and mammals, but possibly not in plants.[86] The codons AGA and AGG code for arginine in the universal code and in the mitochondria of yeast and *Neurospora*, but in mammals they may represent termination codons. In mammals and yeast mitochondrial AUA codes for methionine, but in *Neurospora* it is used for isoleucine where it follows the universal code.[87] Finally, a striking deviation from the universal code is seen in yeast where the four-codon family, CUN (where N = U, C, A, or G) codes for threonine instead of the usual leucine. It seems probable that the investigation of additional groups of organisms will provide evidence of further variation in the mitochondrial genetic codes and that the evolutionary relationships of these various codes will be revealed. It is possible to argue, on the basis of the current data, that the divergence is due either to a relaxation of the constraints imposed on translation in the mitochondrion, or that it represents the persistence of primitive codes.[88]

This brief review of the gross range of diversity in the mitochondrial genome is by no means complete. Mitochondrial variation has been discussed recently in detail in several excellent reviews[5,89,90] and these have also included consideration of the evolutionary aspects of this diversity. In addition several discussions of the origin of mitochondria, probably from an early endosymbiotic relationship, are also available.[91,92] Further discussion of these aspects will not be attempted in this account and in the following sections a consideration of phylogenetic relationships within and between closely related species and populations will be given.

III. STRUCTURAL DIVERSITY OF THE MITOCHONDRIAL GENOME WITHIN AND BETWEEN CLOSELY RELATED SPECIES

A. Extent of Structural Diversity in Related Groups of Animals and Plants

The mitochondrial genome of most animals is very compact and shows little structural diversity between closely related species or strains when compared to that found in fungi or in higher plants. However, several examples of such polymorphisms have been described in animals. In both *Drosophila* and mammals this variation is largely confined to the region of the molecule which carries the origin of replication. In *Drosophila* this is composed of an AT-rich region which can vary in size from about 1 to 5.1 kb. This AT-rich region derived from distantly related species shows little or no homology in molecular hybridization experiments and this is so even between the sibling species *D. melanogaster*, *D. mauritiana*, and *D. simulans*. Moreover, heteroduplex and restriction endonuclease analysis has shown there to be intraspecific variation in this region. In *D. mauritiana* deletions or insertions amounting to as much as 0.7 kb have been observed.[93-97] The variation in the D-loop region of mammals is not as extensive as that in the AT-rich region of *Drosophila*.[98-100] There is little sequence conservation of the D-loop region between unrelated mammals. However, a restriction endonuclease analysis of primate mtDNA showed there to be good restriction site homology between species, but a deletion of 95 bp was located near to the replication origin in the gorilla.[101] Variation in the D-loop region has been found to be extensive between sheep and goats where it is shorter in the goat by about 25% due to a possible deletion of about 150 bp.[102] Structural variation has also been reported in different strains of the protozoan species, *Tetrahymena pyriformis*, where an inverted repeat containing the large rRNA gene is regularly found in the unusual linear genome.[10,103]

In these examples of structural variation in animal mtDNA no obvious single mechanism for the generation of the differences can be proposed. However, one interesting example of variation in the bovine D-loop region has been described.[104] A polymorphism for the length of a poly-C sequence of residues has been found to occur, not only between closely related animals, but possibly also within a single animal and even within a single cell. From 9 to 19 cytosine residues have been found to occur in this homopolymeric region and this variation can also be generated by the serial passaging of the cloned region on a plasmid in *Escherichia coli*. It is proposed that alignment errors during replication are responsible for the generation of this variation, though the general relevance of this mechanism to the length polymorphisms described in other species is not clear.

There is, as yet, little comparative data on polymorphisms within and between closely related species of higher plants. Moreover, the basic structural organization of the plant mitochondrial genome is not yet resolved with the relationships between the various observed mtDNA molecules and plasmid-like DNAs still in need of clarification. Variation in structure has been observed within several species in strains containing male sterile cytoplasms[27-33] and it has also been observed to arise in tissue culture lines in maize.[105,106] In *Nicotiana* structural variation has been detected between the species *N. tabacum* and *N. knightiana*,[107] while in the Cucurbitaceae the mtDNA was found to vary in size from 330 to 2400 kb, in a study of 4 species.[108]

The best demonstration that structural rearrangements are involved during the evolution of higher plants comes from a restriction endonuclease analysis of the mtDNA of maize and of species of teosinte. In the genus *Zea* it has been shown that normal taxonomic groupings correspond very closely to those derived from restriction pattern analysis.[109,110] A detailed analysis of the restriction patterns of maize and of three teosinte species, together with molecular cloning and hybridization experiments, has shown that although there is considerable sequence homology between the mtDNAs derived from these species, there is clear evidence for the structural rearrangement of many of the fragments.[111] The exact nature of the rearrangements and of the mechanism of their generation remains obscure.

B. Intron Variation Within and Between Related Species of Fungi

The mitochondrial genomes of closely related species of fungi show considerable structural and size variation. Such variation has been investigated most fully in the three genera *Saccharomyces*, *Neurospora*, and *Aspergillus*. In all three cases the greater part of this variation is due to a series of insertions/deletions which have been identified as introns located in certain genes, particularly in those coding for cytochrome oxidase subunit 1 and for apocytochrome *b*. Other insertion/deletion events have also been detected elsewhere in the genome. As a consequence of this large amount of structural variation a particularly high degree of restriction pattern polymorphism is found within and/or between closely related species or strains in all three genera. In spite of this, heteroduplex analysis and detailed comparison of restriction site maps show there to be a high degree of sequence homology in those regions outside the various insertions/deletions.

The introns of the fungal mitochondrial genome have been studied in depth because of their role in mRNA processing and possibly also in regulation. In laboratory strains of *S. cerevisiae* in which this problem was first studied, the gene coding for apocytochrome *b* was found to have a complex mosaic structure with 6 exons and 5 introns.[112] Genetic and molecular studies have shown that the processing of the mRNA transcribed from this mosaic gene is a sequential process and that two of the introns, the second and the fourth, contain sequences which code for proteins, the maturases, which are essential for the splicing of subsequent introns.[112,113] Moreover, the maturase postulated for the 4th intron is also required for the processing of one of the introns located in the gene which codes for cytochrome oxidase subunit 1 and which also has a complex mosaic structure. Similar mosaic structures have been described for genes in *Aspergillus nidulans*[114,115] and *Neurospora*[116] and in an increasing number of other species. Detailed sequence analyses of many of these introns have shown them to have a related structure, and a general mechanism for the processing of the mRNA transcripts has been proposed.[117,118]

That there is a degree of sequence or structural homology in these introns suggests that they may have a common and probably ancient origin. The latter theory is further supported by the observation that the intron located in *A. nidulans* apocytochrome *b* gene has exactly the same splice points as the third intron in the *S. cerevisiae* gene,[115,119] and the third intron of *A. nidulans* cytochrome oxidase subunit 1 gene similarly has the same splice points as the second intron in the *Schizosaccharomyces pombe* gene.[120,121] In spite of this, studies of closely related strains or species in *Saccharomyces*, *Neurospora*, and *Aspergillus* show there to be polymorphism in genome structure due to the presence or absence of introns or combinations of introns. This has led to the concept of the "optional intron". In yeast short and long forms of the genes coding for cytochrome oxidase subunit 1 and for apocytochrome

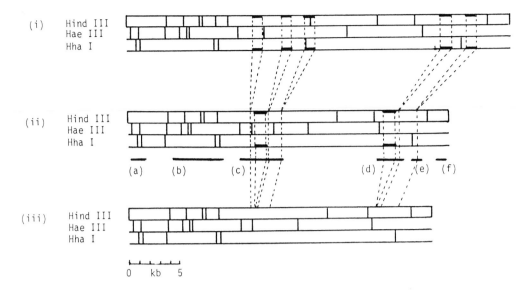

FIGURE 1.   Linearized restriction maps of the circular mitochondrial genomes of (i) *Aspergillus nidulans* var. *echinulatus*, (ii) *A. nidulans*, and (iii) *A. quadrilineatus*. For clarity, the maps for only three restriction enzymes are illustrated. The locations of six genes are indicated: (a) SrRNA, (b) LrRNA, (c) *oxiA* which codes for cytochrome oxidase subunit 1, (d) *cobA* which codes for apocytochrome *b*, (e) the unidentified reading frame URF1, and (f) *oliA* which codes for ATPase subunit 6. The positions of seven optional introns present in these three species are shown and connected by broken lines. Other small inserts/deletions are present but are not shown here. Data compiled from References 114, 120, 134, 138 and 139.

b have been described, with the differences in the lengths of the genes being due to the presence or absence of five optional introns in the cytochrome oxidase gene, and three in the apocytochrome *b* gene.[112,122-126] A similar situation exists between the three *Neurospora* species, *crassa*, *sitophila*, and *intermedia* and also between independent isolates of *crassa*.[116]

In *Aspergillus* mtDNA variation has not yet been found within a single species. For example, isolates of *A. nidulans* derived from England, South Africa, Arizona, and Hungary, and over a time period ranging from about 1939 to 1981, have indistinguishable mitochondrial genomes as determined from restriction site maps, even though there is considerable evidence for nuclear genetic variation within the species.[127,128] Similar observations have been made in other species of *Aspergillus*, though it is still expected that some intraspecific mitochondrial genome variation will be found upon more thorough investigation. However, intron variation and a small amount of restriction site divergence is found between the sibling species, *A. nidulans*, *A. nidulans* var. *echinulatus*, and *A. quadrilineatus* (Figure 1).

There is thus an enigma. On the one hand the evidence suggests that the introns have an ancient origin, yet they may be readily dispensed with. The mechanism by which this variation might be generated is far from clear. Introns can be deleted from the genome in laboratory strains. Thus, the 4th and 5th introns in the *S. cerevisiae* apocytochrome *b* gene have been lost,[113] and the loss of the intron from the large rRNA gene in ω⁻ strains, which is otherwise universally present, is very striking.[129,130]

New combinations of pre-existing introns can be produced by recombination-like processes in the mtDNA. Thus, in yeast in the laboratory, recombination between long and short forms of the apocytochrome *b* gene has resulted in a series of strains in which specific introns were added or subtracted, and even in which all introns were deleted.[131] In *S. cerevisiae* and possibly also in basidiomycetes, as well as in other isogamous organisms such as *Chlamydomonas*, mitochondrial recombination occurs readily during the normal sexual meiotic cycle in the immediate postfusion heteroplasmon. In *Aspergillus*, as in higher plants, the mitochondrial genome is inherited uniparentally and mitochondrial recombination does not

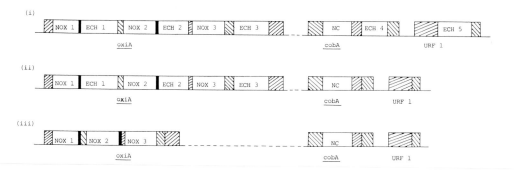

FIGURE 2.   Intron variation in the *oxiA, cobA* and URF1 genes in (i) *Aspergillus nidulans* var. *echinulatus,* (ii) the recombinant mitochondrial genome produced by artificial somatic hybridization between these two species, and (iii) *A. nidulans.* All exons are blocked in or shaded, and all introns are labeled. The symbols NOX and NC refer to introns of *A. nidulans* found in *oxiA* and *cobA,* respectively, and ECH refers to introns found in var. *echinulatus.* Data compiled from References 114, 115, 120, 134, 138, and 139.

occur during the sexual cycle.[132,133] However, in *Aspergillus* and also in higher plants, mitochondrial recombination has been shown to occur with a high frequency in artificially produced somatic hybrids following protoplast fusion.[107,134-136] In the group of species closely related to *A. nidulans* it has been shown that recombinant mitochondrial genomes produced in the laboratory may have a similar intron content to other naturally occurring species. Thus, the recombinant mitochondrial genome produced between *A. nidulans* and *A. nidulans* var. *echinulatus,* for example, has a similar genome to that found in *A. rugulosus* (Figure 2). On the basis of this observation and of a detailed study of the nuclear genetic makeup of artificially induced somatic hybrids it has been proposed that somatic hybridization may play an important role in *Aspergillus* in the evolution of groups of sibling species.[137]

A further explanation for the origin of the structural polymorphisms is that of horizontal transfer. There is little direct evidence in support of this proposal, but the large-scale rearrangements found in fungal and plant mitochondrial genomes and the detection of homologous sequences in the mtDNA, the nucleus, and chloroplast DNA suggest that such processes may have taken place during evolution. The apparent gain or loss of specific introns in the genomes of closely related strains or species is certainly reminiscent of the process of transposition, but there is no clear evidence for the presence of transposons in mtDNA. However, a small transposon-like sequence of 37 bp flanked by a 5 bp direct repeat has been reported within the third intron of the cytochrome oxidase subunit 1 gene of *A. nidulans* which is absent from the otherwise very homologous equivalent intron in *Schizosaccharomyces pombe.* The high degree of homology of these two introns and the much lower homology of the surrounding exons in these two species have also been used as evidence for the horizontal transfer of DNA between in this case, two unrelated species.[120,121] In *A. nidulans,* a small region of about 126 bp is present within the unidentified reading frame, URF 1, which corresponds to a variable region in the mammalian URF 1. This short insert, which is AT-rich, has certain similarities to other *Aspergillus* introns in both the cytochrome oxidase subunit 1 and apocytochrome *b* genes, and it contains GGT sequences which could represent the upstream splice point in these introns.[138] It is then of interest to note that URF 1 in *A. nidulans* var. *echinulatus* contains a large intron which is located in the region occupied by the mini-insert in *A. nidulans.*[139] A final observation, which is of relevance to the possible transposition of intron sequences in fungi, is that in *Podospora anserina* an intron of the cytochrome oxidase subunit 1 gene contains an origin of replication and upon excision from the mtDNA, replicates autonomously as a mitochondrial plasmid.[140]

Extensive variation in the structure of the mitochondrial genomes of plants and fungi, and to a limited extent animals, is present even between what otherwise must be considered to be strains or species which have a very close phylogenetic relationship. The importance of this variation and its mechanism of origin, maintenance, and evolution remain largely obscure at the present time and attempts to quantify the rates of structural change have not been made. Any long-term attempts to understand the evolution of mtDNA clearly must take all these features into account.

IV. DIVERSITY OF THE MITOCHONDRIAL GENOME IN ANIMAL POPULATIONS

Most studies of variation in the animal and particularly the mammalian mitochondrial genome have been carried out by the direct measurement of divergence from nucleotide sequence data of a part, or the whole, of the genome, or as an estimate of genetic diversity obtained from restriction endonuclease digests. Comparison of the sizes of DNA fragments produced by digestion with one or more restriction endonucleases provides the simplest method of assessing variation. The presence or absence of a restriction endonuclease site for the small circular mtDNA molecule provides equivalence to the number of DNA fragments resolved by gel electrophoresis.

The choice and number of restriction endonucleases in relation to the nucleic acid composition of mtDNA are important in the assessment of mtDNA variation and for the estimation of evolutionary divergence.[141] The techniques for the restriction endonuclease mapping of mtDNA are well documented.[142] As already indicated an assessment of mtDNA variation by restriction endonuclease mapping can be carried out with differing degrees of resolution, the highest being provided by the methods of Cann and Wilson.[143] Estimates of nucleotide sequence diversity can be further supplemented by the alignment and ordering of the fragments relative to a known DNA sequence or with respect to the origin and direction of replication,[144,145] so providing information about the distribution of restriction sites within the genome. Where the complete DNA sequence is known for a species, restriction sites may be mapped directly and variation described in relation to functional differentiation of the genome, for example, the rRNA, tRNA, and protein encoding regions. Of course confirmation of the order of restriction fragments can be obtained from double digests with restriction endonucleases as can the presence of insertions and deletions. However, nucleotide substitution at single base pair sites is characteristic of the evolution of mtDNA in higher animals and any length variations are small in size.

The study of restriction site gain or loss is the most frequently used method of assessing mtDNA diversity. The method is subject to a number of limitations which can lead to the underestimation of genome diversity and divergence. For example, small DNA fragments of the order of 160 to 200 bp and less are sometimes poorly resolved. When two identical restriction sites are separated by this minimal level of resolution both site number and site variation may be underestimated. Two further properties of the mitochondrial genome can also lead to an underestimation of genome diversity and divergence. These are the relatively high mutation rate of animal mtDNA and an intrinsic mutational bias in favor of transitions rather than transversions. Such potential biases in estimates of nucleotide sequence divergence will be discussed later in more detail and have been thoroughly investigated by Aquadro and Greenberg[146] for a fully sequenced 900 bp segment of the human mitochondrial genome. These authors compared estimates of genome diversity for restriction sites with those obtained directly from the nucleotide sequence and concluded that studies of restriction site variation may indeed underestimate the extent of mtDNA sequence divergence.

Before outlining the estimation of nucleotide sequence diversity it is appropriate to describe some theoretical expectations of evolution in the mitochondrial genome. The dynamics of mitochondrial genome evolution are, at their simplest, described in terms of the neutral gene theory.[147] Application of the neutral gene theory to the mitochondrial genome has resulted in expectations, based upon population size and mutation rate, of the rate of nucleotide substitution per year or per generation.[148] In this model the average time to fixation of new mutations and the expected level of genetic diversity in populations will be subject only to random genetic drift and recurrent mutation. Such a model does not incorporate any assumptions about natural selection and is therefore, in terms of parameters, the simplest model of variation and evolution. A theoretical description of mtDNA diversity is also heavily dependent upon the biology of mitochondrial genome replication and transmission both between and within generations of the cell or complete life cycle.[148,149]

In most higher organisms the mitochondrial genome is considered to be maternally inherited. Although in some organisms mtDNA in spermatozoa (up to 10^2 molecules) can invade the egg cytoplasm at fertilization,[150,151] the relative abundance of mtDNA in oocytes (up to 10^8 molecules)[152-154] precludes any significant paternal contribution. Indeed, studies of mtDNA transmission in crosses in a variety of organisms such as *Xenopus*,[152] *Neurospora*,[155] *Aspergillus*,[132,133] maize,[156] *Equus*,[157] sheep and goats,[102] and deer-mice[158] support the maternal mode of inheritance. A particularly innovative and critical method for the assessment of the contribution of paternal mtDNA has been described by Lansman et al.[159] Recurrent backcrossing of females of the moth, *Heliothus*, to a common father with a unique mtDNA failed to detect any accumulation of paternal mtDNA in the maternal lineage. The design of the experiment was such as to define a lower limit for the number of paternal mtDNA molecules: 1 in 500. Since, after 91 recurrent backcrosses no paternal mtDNA was detected, the upper limit of paternal leakage (defined by the experimental design) was estimated to be of the order of 1 paternal in 25,000 maternal mtDNA molecules per generation. It should, of course, be recognized that strict maternal inheritance may not be universal. Isogamous species are an obvious example. Other species worthy of careful investigation are those with primitive spermatozoa or where the contribution of the egg-cytoplasm is comparatively small. One example of a completely paternal inheritance has been reported in the fungus *Allomyces*.[160]

Models of mtDNA diversity and evolution which are based on the neutral gene theory are concerned with a representation of gene frequency dynamics in terms of recurrent mutation and the effects of sampling variation (random genetic drift). Formulations of the expectation of genetical diversity within populations, the average time to fixation or loss of neutral mutations, and the rate of substitution of neutral mutations have been presented by Birky et al.[148] The study of microevolution is based essentially upon a consideration of phylogenetic relationships within closely related species and populations. As such both the estimation and behavioral dynamics of mitochondrial gene diversity are central to both the qualitative and quantitative assessment of population diversity and speciation rates. Essential to a model of mtDNA variation dynamics is an understanding of the role of sampling variation or random genetic drift upon the mtDNA genome, both within and between generations. Mitochondrial DNA is propagated in both somatic and germ cell lineages. As discussed earlier strict maternal inheritance allows mtDNA diversity to be evaluated from somatic cell lineages and, although subject to random genetic drift and mutation, the chance substitution of a new base pair in sufficient cell lineages as to be detected by restriction endonuclease digestion is very small, as indicated by the recurrent backcrossing study with the moth, *Heliothus*. Moreover, the dynamics of somatic cell lineages will have no bearing upon species evolution.

A potential problem concerning the dynamics of maternal inheritance has been described by Hauswirth and Laipis[161] in an elegant study of a maternal lineage of Holstein cows. A single base change identified as a transition from adenine to guanine was substituted in the lineage within 4 years. This uniquely identified base substitution could not be attributed to herd mismanagement or paternally inherited mtDNA and represents an exceptionally high rate of evolution. Hauswirth and Laipis emphasized that aspects of mtDNA transmission other than maternal inheritance in higher eukaryotes may be important in any formulation of mtDNA evolution. They suggested that the dynamics of mitochondrial amplification in the maturing oocyte and the dynamics of mitochondrial populations in germ-cell lineages are worthy of more thorough study. Hence, the dynamics of mtDNA replication in germ-cell lineages may be poorly understood. A current estimate of this lineage in cows is 20 to 50 cell generations. At each generation chance segregation will affect the relative fate of mutant and normal alleles, the influence of random genetic drift and other processes in the maternal lineage probably being important in oocyte formation due to cellular death and polar body formation. Of course for organisms in which the germ line is not defined, for example, some fungi, vegetative segregation of mitochondria will be particularly relevant to mtDNA evolution. Hence, in the evolution of mtDNA random genetic drift is operative within a number of levels of biological and evolutionary complexity: within cells of a lineage, between cells in an organism, between individuals of a population, and between populations of a species. Genetic diversity has been defined by Birky et al.[148] for all of these categories and we will follow their nomenclature in the following description of mtDNA dynamics.

A. Theoretical Expectations of mtDNA Diversity

Following neutral gene theory, the long-term average of the rate of substitution of a base pair mutation in cell generations is equal to the rate of neutral base substitution. The influence of random genetic drift upon the rate of fixation, i.e., base pair substitution or loss in a population, together with the expectations of gene diversity within populations, is heavily dependent upon the effective population size, N_{eo}, of organelle genes. This represents the long-term average of the number of organelle genes which contribute to subsequent generations. N_{eo} is defined for the organelle genes as

$$\frac{N_m N_f}{\alpha^2 N_m + \beta^2 N_f}$$

where N_m and N_f are the respective number of male and female organelle genes and α and β are the respective female and male proportional gametic contributions (such that $\alpha + \beta = 1$) to the zygote. For the nuclear genome, in a diploid organism, the effective number of genes, $2N_e$, is defined as

$$\frac{4N_m N_f}{(N_m + N_f)}$$

The parameter N_{eo} is, in fact, an approximation, but a satisfactory one if the loss of genetical diversity within populations has not reached a steady state, and is an underestimate for values of $\beta \leq 0.01$.[148,162] The two parameters N_{eo} and N_e provide a means of comparison of sex-ratio, degree of maternal inheritance, and gene dynamics in the organelle and nuclear genomes. For example, if we let $N_m = N_f$, and assume complete maternal inheritance of the mitochondrial genome, then the ratio $2N_e/N_{eo}$ is 4. Hence, the effective number of organelle genes is $1/4$ of the effective number of nuclear genes when the transmission of the mitochondrial genome is purely through the maternal lineage.

It is seen that the number of male and female organelle and nuclear genomes in the population contribute to this ratio of effective numbers. Variation in β contributes to the degree of maternal inheritance of organelle genomes while variation in the ratio of N_m/N_f reflects the sex ratio or aspects of reproductive biology such as mating behavior and harem formation. Table 1 depicts the behavior of $2N_e/N_{eo}$ when subject to variations in the parameter β and strong variation in N_m/N_f ratio. It is clear that variations from a 1:1 sex ratio will result in considerable deviation from the value of 4, a pattern followed even when there is 10% contribution of the paternal organelle complement.

Table 1
VARIATION IN SEX
RATIO AND EFFECTIVE
NUMBER OF ORGANELLE
GENES

Sex-ratio female:male	$2N_e/N_{eo}$	
	$\beta = 0$	$\beta = 0.1$
1:100	7.92	6.42
1:1	4.00	3.28
100:1	0.08	0.14

Note: Assuming strictly maternal inheritance and when the proportion of paternally transmitted genes (β) is 10%.

Variations of N_e and N_{eo} are reflected in the levels of genetical variation expected in populations when equilibrium is achieved by the balance between recurrent forward mutation to neutral alleles, and fixation or loss due to random genetic drift. When $\beta = 0$, N_{eo} will be equal to N_f and the expected level of gene diversity between individuals within a population, at equilibrium is defined as $H_c = 2N_f\mu/(2N_f\mu + 1)$ for the mitochondrial genome and for the nuclear genome $H = 4N_e\mu/(4N_e\mu + 1)$. The ratio of genetic diversity in the nuclear as opposed to the mitochondrial genome is thus $2N_e/N_f$. This ratio of effective numbers is depicted in Table 1. Hence, besides the study of mtDNA diversity in its own right, a comparative assessment of genetic diversity in the nuclear and the mitochondrial genomes is likely to be of value for an assessment of factors affecting gene diversity within populations or closely related phylogenies. Before any discussion of actual values of $2N_e/N_f$ in natural populations, it is pertinent to discuss the observed levels of mtDNA diversity in natural populations and the procedures for their estimation.

B. Measurement of mtDNA Diversity

The extent of nucleotide variation within populations is generally assessed by the estimation of two quantities, p, the proportion of nucleotides that are polymorphic, and H, the proportion of heterozygotes per nucleotide site. If the sample size (n) is equal to 2, both polymorphism and heterozygosity have the same meaning since these two quantities are, in practice, based upon the haplotypes of different individuals. While H can be readily identified in the context of nonhaploid nuclear genomes, it cannot be directly related to the mitochondrial genome. However, as we shall see later, the estimation of H for mtDNA is still relevant in an evolutionary context. While both p and H can be directly estimated from nucleotide sequence data, they are at present largely available, on a whole genome basis, from mtDNA restriction map data.

For the purposes of within–species phylogenetic comparison, p is estimated from all pairwise combinations among the different individuals that belong to a population sample. A number of estimators are available for p and its variance. The reader is referred to specialist statistical treatments in References 163 to 169 for details of their derivation and application. However, following Engels,[163] for the case of n = 2,

$$\hat{p} = \frac{k}{2j(m - k)}$$

where k is the number of restriction site differences in the sample, m is the total number of restriction sites in the sample, and j is the number of bases which constitute a recognition site. Estimates of p so far derived from all unique pairwise combinations of individuals within a population form the basis of phylogenetic comparisons, networks, and tree construction. For larger sample sizes, that is for the description of nucleotide diversity within a population,[164,166]

$$\hat{p} = \frac{k}{2mj}$$

and following Engels,[163]

$$\hat{H} = \frac{nc - \sum\limits_{i=1}^{m} c_i^2}{jc(n - 1)}$$

where c_i represents the number of individuals in the sample which are cleaved by a restriction endonuclease at the ith site. Since m is the total number of distinct restriction sites in the sample,

$$c = \sum\limits_{i=1}^{m} c_i$$

Estimates of p are at present available mainly for mammalian populations or species. Examples of p in the protozoa, *Paramecium*[170] and *Acanthamoeba*,[171] have also been published and it is expected that estimates will become available for a larger range of phyla- and genera in the near future. Data for mammalian species have been summarized by Avise and Lansman.[172] While values of $\hat{p}$ range from about 0.01 to 0.04 for several ape and rodent species, horses, and cows, those of gorilla and man are about an order of magnitude smaller at 0.006[173] and 0.004,[174] respectively. The low estimated value of p for man can be interpreted to be a reflection of man's recent origin from a small group of individuals.

The parameters $N_e\mu$ and $N_f\mu$ for the nuclear and mitochondrial genome, respectively, have, in a long-term context, an evolutionary interest since they define the expected level of heterozygosity in equilibrium under neutral theory. Their estimation can be achieved by substitution of estimates of H for the population into the appropriate formulas (see above) relating heterozygosity with $N_e\mu$ or $N_f\mu$. Alternatively, following Ewens,[164,166] the quantity $\theta = 4N_e\mu$ for nuclear genes, or $\theta = 2N_f\mu$ for mitochondrial genes, can be estimated as

$$\hat{\theta} = \frac{p}{j \, \log n}$$

The standard errors associated with the estimates of population diversity, $\hat{p}$, $\hat{H}$ and $\hat{\theta}$, are large, and moreover differ according to whether the investigator is concerned with his data in a long-term evolutionary context or in terms of differences between two or a number of populations. This is because questions of an evolutionary interest must allow for a complete repetition of the evolution of the particular genome or DNA segment in question. Consequently, the variance of $\hat{p}$ is much larger than it is for situations in which an investigator is studying a specific DNA segment in a specific population. These aspects of hypothesis testing are lucidly discussed by Ewens.[166]

Estimates of $N_e\mu$ and $N_f\mu$ (Table 2) have been collated for a number of species. In some cases (*) $4N_e\mu$ has been estimated from data on allozyme polymorphism. Standard errors on such estimates are large, but the data have shown some consistencies. The mitochondrial

Table 2
ESTIMATES OF $N_e\mu$ AND $N_f\mu$ FROM NATURAL POPULATIONS

Species	$N_e\mu$ ($\times\ 10^3$)	$N_f\mu$ ($\times\ 10^3$)	$2N_e\mu/$ $N_f\mu$	Ref.
Drosophila melanogaster	1.25	3.0	0.88	93,163,175-177
Geomys pinetum	0.10	4.5	0.080	162,178
Peromyscus polionotus	0.20	2.8	0.154	142,168,178,179
Mus domesticus	0.25	7.8	0.057	180,181
Homo sapiens	0.25	8.7	0.320	163,164,182,183

genome is always more variable than nuclear DNA. Some estimates of nuclear genetic diversity are, of course, biased in that the estimates of $N_e\mu$ are based on allozyme polymorphism and consequently do not include information about introns and noncoding sequences. Indeed, even in cases where $N_e\mu$ has been estimated from restriction endonuclease map data the region in question has not been chosen at random from the whole genome and invariably contains coding sequences even though these may constitute only a proportion of the total map of particular interest. Nonetheless, the fact that $2N_e\mu/N_f\mu$ is consistently very much less than 4 for such phylogenetically distinct species can be reasonably explained on the basis of higher mutation rates and mtDNA replication repair processes in the mitochondrial genome, rather than any variations in reproductive biology or sex-ratio imbalance which would be expected to lead more to variation in $2N_e\mu/N_f\mu$ among species. Hence the mitochondrial genome can be expected to evolve at a much faster rate than the nuclear genome, both in terms of the rate of gene substitution and an increase in the rate of loss of neutral alleles.

The mitochondrial genome of higher eukaryotes therefore has two major properties which make it especially suited for the construction of recent phylogenies or deductive inference of microevolutionary events. These are a simple maternal mode of inheritance and a relatively high level of genetical variation. Variations in mtDNA in maternal lineages are a reflection of maternal ancestry, mutation rates, and past population structure. Although the theoretical treatment of mtDNA diversity within populations and later in the estimation of evolutionary rates has its basis in the neutral gene theory, it should be pointed out that natural selection will also be reflected in present-day molecular diversity. Consequently, if natural selection has been responsible for a proportion of nucleotide substitutions in the mitochondrial genome, phylogenetic relationships may be correspondingly biased.

Mitochondrial DNA diversity is especially suited to the study of phylogeny at a within-species level since it is a sensitive discriminator of genetical differentiation within subdivided populations and closely related species. Hence we can expect mtDNA variation to be of value in the assessment of the degree of genetic isolation for ecologically defined population subdivisions of a species. The investigation of more distant phylogenetic relationships presents some difficulties since the high rate of evolution will cause multiple base pair substitutions to be overlooked, so underestimating the actual number of events between two species or genera.[183,184] The degree of underestimation will be greatest for silent site substitutions and regions of the genome subject to low evolutionary constraint. Additionally, small length mutations relative to two identical restriction sites will cause a number of site differences to apparently vary when the length mutation is within the limits of technique. This feature is a probable explanation for hypervariable restriction sites noted by Lansman et al.[185] in populations of *Peromyscus maniculatus*. Such problems in the estimation of genetical divergence were discussed by Aquadro and Greenberg[146] in a detailed evaluation of genetic divergence from nucleotide sequence data. Base pair substitution in mitochondrial DNA shows a marked excess (32-fold) of transitions over transversions and is the most likely cause of convergent evolution of restriction sites. In their analysis of a 900 bp region of human mtDNA Aquadro and Greenberg identified the convergent loss of a site 3 times in their sample of 7 individuals. In another instance mutation caused the loss of an MnlI restriction site and the formation of a new MnlI site two nucleotides away. Hence, evolutionary divergence, when it is estimated from mtDNA restriction map data, is progressively underestimated with increasing phylogenetic distance.

V. EVOLUTIONARY IMPLICATIONS OF mtDNA DIVERSITY IN ANIMALS

Evolutionary studies of molecular diversity are based upon comparisons among present-day species. Nucleotide variations within populations will, on neutral gene theory, eventually become fixed in or lost from populations, a process which can take a very long time. However, the rate at which new neutral mutants become fixed in populations is equal to the mutation rate. In the following section the evolution of mtDNA sequences is described on a phylogenetic basis. Genetic differentiation of populations of a species as well as more distant relationships among genera are discussed. The measurement of evolutionary rates is outlined and variation in the rate of mtDNA evolution is considered in relation to variation in function or structure of the mitochondrial genome. Finally, within-species phylogeny is discussed in relation to recent observations which strongly implicate between species introgression of mitochondrial genomes.

A. The Rate of mtDNA Evolution
Relative to nuclear DNA, animal mtDNA is characterized by a high rate of evolution. This is manifested largely through base pair substitution. Differences in the size of animal mitochondrial genomes, particularly those of mammals, are very small and confined to length variations usually of the order of a few base pairs. Phylogenetic comparisons and the estimation of evolutionary rates are largely based on restriction map data. As demonstrated by the neutral theory of evolution the rate of gene or base pair substitution is equal to the mutation rate. The high rate of animal mtDNA evolution is likely to be a consequence of a higher mutation rate for the mitochondrial genome. Variation in the mtDNA is proving to be particularly useful for the description of radiation within species and the estimation of species divergence times. In relation to nuclear genes, mtDNA provides the potential as a relational evolutionary clock for the study of times of duplicate gene divergence and for the comparison of evolutionary rates within lineages for noncoding and coding sequences.

The high rate of mtDNA evolution was demonstrated by Brown et al.[183] in 5 pairwise comparisons of primates. The thermostability of H and L strands in hetero- and homoduplex conformation was used to provide an estimate of base mismatches of 22% in between species and within-species comparisons respectively of the guenon, *Cercopithecus aethiops,* and man. This estimate of divergence was similar to p, the number of base substitutions per base pair species divergence, obtained from restriction map data. The percentage of nucleotide sequence difference obtained from heteroduplex thermostability of single-copy nuclear DNA consequently provided a valid comparison of evolutionary rate in the mitochondrial and nuclear genomes. The evolutionary rate of the mitochondrial genome was between 5 and 10 times that of single-copy nuclear DNA. This estimate may be conservative since the upper limit is based on restriction map data but it agrees with that obtained from sequence data analysis.[186] By relating the number of substitutions per base pair to evolutionary divergence in years between pairs of primate species, an estimate of 0.02 substitutions per base pair per million years was obtained for mtDNA.

Comparison of the rate of evolution in nuclear and in mtDNA does not directly imply variation in the mutation rates of these two genomes. While the rate of gene substitution is directly related to the neutral mutation rate, the rate can also be influenced by evolutionary constraint upon the two genomes, implying greater constraint in gene substitution for the nuclear genome. Additionally, the rate of evolution is known to vary not only from protein to protein but also between different regions of nuclear DNA such as introns, exons, leader, and trailing sequences. Hence, the gross comparison of genomes may contain substantial heterogeneity in degree and relative level of evolutionary constraint. A particularly interesting and comparative analysis of evolutionary rates in mitochondrial and nuclear encoded proteins points to an elevation of the mitochondrial mutation rate by a factor of 3 over that of the nuclear genome.[187] These authors examined the rate of evolution for synonymous substitutions in the URF1, cytochrome oxidase subunit 1 and cytochrome *b* mitochondrial genes with that in nuclear encoded amylase and immunoglobulin C_K (I_gC_K). Where K_s^c is defined as the species difference per synonymous nucleotide site, corrected for multiple site substitutions, the data obtained for rat and maize, with species divergence time ($\hat{T}$) of 17×10^6 years (estimated from protein and DNA data) gave K_s^c as about 0.2 for the protein coding nuclear genes and about 1.17 for the mitochondrial genes. The rate of evolution, estimated as K/2T, was found to be about 5.4×10^{-9} for nuclear genes and 35×10^{-9} for mitochondrial genes. The mitochondrial genome can be concluded to evolve about 6.5 times faster than the nuclear genome. The rate of evolution in nuclear pseudogenes is about twice as fast as it is for nuclear synonymous sites and probably approximates to an upper rate of DNA evolution.[188] Hence, the rate of nucleotide substitution in the mitochondrial genome is about three times that in the nucleus and this is strong evidence for a threefold elevation of mutation rate in the mitochondrial genome.

In contrast, Miyata et al.[187] also noted a remarkable similarity between the rates of nucleotide substitution at replacement sites in nuclear and mitochondrial genes. For example, in a human-mouse comparison mtDNA genes gave a rate of 1.25×10^{-9} per site per year and 1.13×10^{-9} for the nuclear encoded β-globin. If we assume a threefold elevation of mutation rate for the mitochondrial genome we must now infer that relative to nuclear coding sequences, replacement site substitution, and hence protein evolution, in the mitochondrial genome is strongly constrained.

Selective constraint upon replacement site substitutions has also been observed from a comparison of the evolution of cytochrome oxidase subunit 2 in *Rattus norvegicus* and *R. rattus*.[184] Only 6% of the nucleotide differences between these two species could lead to an amino acid difference, the other 94% being silent site substitutions. Again, the comparatively low rate of replacement site substitution in the mtDNA contrasts with the average of 30% or more in nuclear genes. The authors drew attention to studies of the kinetics of the mitochondrial enzyme γ-DNA polymerase which is about 5 times more likely than nuclear DNA polymerase α to cause nucleotide infidelity during replication.[189] It is of interest to note in this context that a mutation which specifically increases the rate of mutation in the mitochondrial, but not nuclear, genome has been reported in *Schizosaccharomyces pombe*.[190]

It will be recalled from Table 2 that the observed level of gene diversity in the mitochondrial genome is much higher compared to that in the nuclear genome than expected. This appears to be reflected in the high rate of mtDNA evolution, which can be in excess of tenfold over the nuclear genome. While variation in the mutation rate may account for some of this difference in the rate of evolution it does not appear to account for all. Hauswirth et al.[104] present evidence to suggest the discrepancy may be due to our lack of understanding of mtDNA transmission dynamics. Following an investigation of sequence variation in a maternally related lineage of Holstein cows they discovered intra-individual heterogeneity in mtDNA sequences from bovine tissues and came to the conclusion that germ-line cells may indeed be highly heterogeneous for their mtDNA genomes. The sequence variation, which concerned the D-loop (see earlier discussion), was moreover likely to be generated during replication. Hence, it will be of interest to evaluate transmitted mitochondrial gene numbers and amplification processes in germ-line tissues.

B. Measurement of Evolutionary Rate

Following Kimura[191] the difference in the number of base substitutions per nucleotide site between two DNA sequences can be estimated from the relation

$$K = -\tfrac{1}{4} \log_e[(1 - 2P - 2Q)(1 - 2P - 2R)(1 - 2Q - 2R)]$$

where P, Q, and R are proportions of the total number of sites due to transitions and transversions. Transitional changes in DNA sequence due to A ↔ G and C ↔ T classes of mutation and are defined by the quantity P. Two types of transversion are recognized by the equation, one of which maintains the same hydrogen bonding relationships, defined as Q, and is represented by the changes T ↔ A and C ↔ G. The other class of transversion (R) causes a change in hydrogen bonding, i.e., mutations of the type T ↔ G and C ↔ A. When Q = R the formula reduces to

$$K = -\tfrac{1}{2} \log_e[(1 - 2P - Q) \sqrt{1 - 2Q}]$$

The rate of evolution defined per unit time since two species diverged is thus equal to K/2T.

Constancy in the rate of evolution is well established for particular proteins and nucleic acid sequences[193] and permits calibration of evolutionary divergence from the fossil record. Alternatively pairwise comparisons of the number of base pair differences between species or genera can be used to estimate a common nodal or ancestral origin and an "unrooted" tree. The mechanisms and problems of phylogenetic tree construction will not be attempted here and procedural aspects can be obtained from, for example, Prager and Wilson,[194] Fitch,[195,196] and Felsenstein.[197,198]

The use of an "unrooted" tree by Brown and Simpson[184] for cytochrome oxidase subunit 2 provided a comparison of evolutionary rates with respect to lineage. Comparisons of human, bovine, and rat nucleotide sequences demonstrated reasonable similarity in evolutionary rate in lineages for silent-site substitutions but a relatively faster rate of evolution in the human lineage (from a common species ancestor) for replacement site substitutions. There are, of course, two reasons which could account for such an observation, namely, on the neutral gene theory: a relative relaxation of functional constraint in the human lineage, or in terms of natural selection, an increase in the number of gene substitutions.

More generally, evolutionary trees have been constructed (Figure 3) in relation to divergence time within the sub-genus *Mus*, in particular, the divergence times of mouse species which are commensal with man.[180] Such information has not been obtained from fossil records. Indeed, the time of divergence of commensal mice corresponded with the appearance of *Homo erectus*.

C. Phylogenetic Networks

An alternative demonstration of relatedness is provided by the construction of a phylogenetic "network" and has been used extensively for within-species comparisons of restriction map variation and human nucleotide sequence data. The phylogenetic network is an example of an "unrooted" tree and its construction is illustrated by Lansman et al.[185] for population samples of *Peromyscus maniculatus*. Basically "composite" mitochondrial genotypes are identified from the restriction maps, each derived from a single restriction endonuclease. A composite map is then constructed for every individual from the particular map possessed by that individual for every restriction endonuclease. An example, from the above study, is shown in Figure 4. The interrelationships represent a simple or parsimonius network and also illustrate a particular problem encountered with mtDNA evolution; that of apparent evolutionary convergence in restriction site evolution.

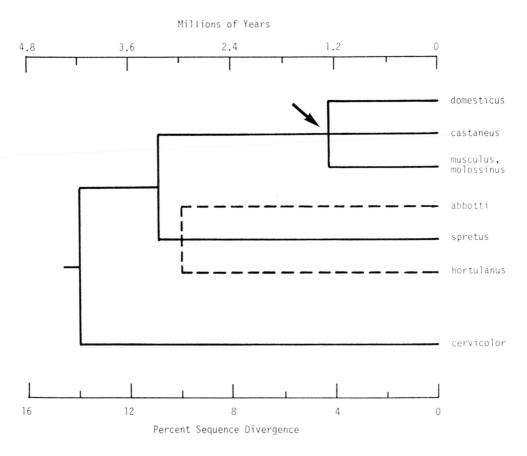

FIGURE 3.   Evolutionary tree for mtDNA, from the subgenus *Mus*, taken from Ferris et al.[180] The tree is based
on restriction map comparisons. Sequence divergence is related to an evolutionary time scale on the basis of a rate
of divergence of 2 to 4%/10[6] years. Broken lines represent the results of preliminary studies. The arrow represents
the origin of commensal mice.

The consequences of a high rate of evolution and the considerable excess of transitions
are potentially manifest as underestimates of evolutionary divergence. Hence, estimations
of evolutionary divergence need either to compensate for problems posed by variation in
the base composition from randomness, or their potential bias quantified. The phylogenetic
consequences of the bias between transitions and transversions were especially noted by
Brown et al.[186] in studies of an 896 bp sequence (containing coding sequences) in primate
evolution. An effect of the excess of transitions and consequently high rate of selected
multiple site substitutions is to produce parallel and even back-mutational change in lineages.
This presents obvious difficulty for the construction of unambiguous phylogenies and determination of branch point order. The magnitude of the problem can be illustrated by a fall
in detected transitions with divergence time and illustrated by Brown et al.[186] for their 896
bp fragment. For primate species which diverged 5 million years ago, an average 91.5% of
transitions were detected. This figure fell progressively with species divergence time until
at 80 million years since divergence only 44.5% of transitions were detected. Such a trend
was observed for both replacement and silent sites and to a lesser extent in tRNA sequences.
The degree of underestimation is illustrated in the following hypothetical lineages derived
from a common ancestor. The phylogenetic change here is one transversion despite seven
transitions having taken place in the two lineages.

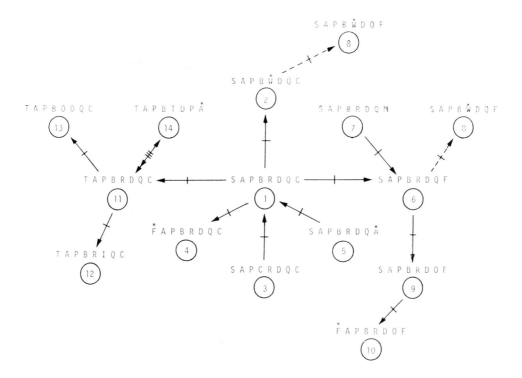

FIGURE 4. An example of a phylogenetic network which was constructed by Lansman et al.[185] for mtDNA clones derived from a population of *Peromyscus maniculatus*. Restriction map variation was studied with 8 enzymes. The different maps associated with each restriction enzyme are assigned a different letter so as to define every mtDNA clone by an 8 letter code or "composite restriction map". The composite maps were then assembled into a sequence network which described their relationship to one another. The network is such that the steps linking the maps are the simplest and in most instances involve a change of one restriction site. Where a nucleotide site is lost, the direction of loss is indicated by an arrow and the bars depict the number of restriction site losses between two mtDNA clones. The exact nature of the change between clones 11 and 12 was not determined. A number of cases of evolutionary convergence of restriction site loss are also shown in this network by asterisks. One apparent case of evolutionary convergence presents a phylogenetic "dilemma"; the ancestral sequence of clone 8 could be either clone 2 or clone 6.

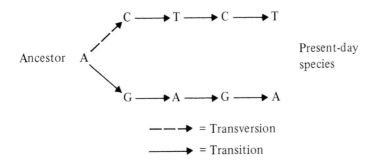

Given such gross underestimation of sequence divergence, the study of mtDNA is obviously most suited to studies of within-species radiation and recent phylogenetic divergence; however, corrections are available for multisite substitution (see for example Brown et al.[186] and Kaplan and Risko,[199] to which the reader is referred for further information).

D. Functional Organization and mtDNA Evolution

Mitochondrial DNA evolution in mammalian species has, on a whole genome basis, been described in relation to function (i.e., in respect of different genes, gene functions, and-

Table 3
RESTRICTION MAP VARIATION AND FUNCTION IN HUMAN AND MOUSE (*MUS DOMESTICUS*) MITOCHONDRIAL GENOMES

Region	Total restriction sites		Variable restriction sites (%)		Inferred muta-tions (%)	
	Human	**Mouse**	**Human**	**Mouse**	**Human**	**Mouse**
Ribosomal RNA	77	30	22	13	70	13
Transfer RNA	50	18	26	17	66	17
Protein encoding	279	141	41	35	88	43
D-Loop	35	10	46	30	255	60
Total genome	441	199	37	37	95	37

Note: The percent inferred mutations concern base pair variations due to small insertions. In this case, following both Cann et al.[200] and Ferris et al.[180] variation at each base pair position is considered to be a single mutation despite the fact that base pair and insertational variation may reflect different mutational processes. The high level of variation in the D-loop through small insertional variations is emphasized in the high resolution studies of man.[200]

physical features such as the D-loop). Complete or partial mtDNA nucleotide sequences, where available, allow the co-alignment of restriction maps and have furthered study in apes and man[101,200] and in the domestic mouse *Mus domesticus*.[180] While complete nucleotide sequence variation in the D-loop region has been studied within the human species,[146] evolution has been detailed in the main by the application of high resolution restriction mapping techniques sufficient to recognize length variation as small as ±2 bp.[143,200] The extent of length variation in the human and possibly other mammalian genomes has been emphasized by the latter studies and will be considered in the present context. Mitochondrial genomes from 112 humans indigenous to 4 continents were mapped with 12 restriction endonucleases to reveal 163 variable and 278 invariant restriction sites. Restriction site variability is, of course, determined as site gain or loss in relation to a standard nucleotide sequence or restriction map. Substantial length variations were initially observed in higher animals, including man, in less refined studies of restriction map variation where they were confined to the D-loop.[68,102,201] In the recent high resolution studies in humans 14 length variants from 112 individuals were identified at sites within the D-loop and in 7 locations at junctions between the coding sequences and within noncoding regions other than the D-loop. Hence, length variations are not only small in size but occur in regions of the genome less critical to coding function. The rate of evolution by length mutation is like that of nucleotide site variation, probably high compared to nuclear DNA. The ratio of length mutation to point mutation, in noncoding DNA, is in the range 0.67 to 1.0 in the mitochondrion and about 0.2 (range 0.03 to 0.5) for noncoding DNA flanking coding regions in the nuclear genome.

Variation in relation to functional or structural differentiation of the mitochondrial genome is depicted in Table 3 for two species, *Mus domesticus*[180] and man.[200] Both studies used the sequence comparison method, i.e., restriction fragments were compared in alignment with the known mitochondrial sequence for each species. Comparisons of the two species are based on the relative ranking within each species of percent variability for the four aspects of structure or function. As expected the D-loop shows the higher variability followed by protein encoding DNA. Ribosomal and tRNA encoding regions are the least variable with some difference in species' rank order.

The rate of evolution for mitochondrially encoded tRNA genes is about 100 times that of nuclear encoded tRNA, being of the order of replacement site evolution. This gross discrepancy is attributed to both the increased mtDNA mutation rate and to a lack of functional constraint in mitochondrial tRNA genes compared to those in the nucleus.[88,186]

Comparative studies of mitochondrial and nuclear gene evolution in relation to lineage has, despite the considerable variations in rate seen above, suggested that evolution in these two genomes is not entirely independent. Studies of mitochondrially encoded subunit 2 of rat cytochrome oxidase and nuclear encoded cytochrome *c* provide examples. Cann et al.[200] noted that the number of amino acid replacements in four lineages of human, bovine, rat, and mouse were, for cytochrome *c* and the mitochondrially encoded cytochrome oxidase, positively correlated. Evolutionary rates are known to vary from protein to protein. As discussed earlier the evolution of the cytochrome oxidase subunit 2 gene is, for replacement site substitutions, greater in human than in bovine and rat lineages. Comparison of the number of replacement site substitutions in cytochrome *c* showed them to have a similar rank relatedness with those of the cytochrome oxidase protein. These two proteins are, of course, functionally related in an electron transport chain. Studies of within species and cross-species reaction kinetics for the two proteins supported the co-evolution of these two proteins, primate vs. non-primate molecular interactions being weakest.[180,202]

E. Species Origin, Hybridization, and Gene Introgression

The strict maternal inheritance of mtDNA enables a new approach to the analysis of species hybrids, barriers, hybridization, and gene introgression. The discovery of parentage and probable age of pathenogenetic lizards of the genus *Cnemidophorus*[203] provided one of the first of such studies. A subspecies of *C. tigris, marmoratus* was identified from studies of relatedness between mtDNA restriction maps. The parthenogenetic species *C. neomexicanus* and *C. tesselatus* were more recently evolved than some races of *C. tigris*, i.e., their restriction maps were not closest to the common root of a phylogenetic network. Hence, mtDNA analysis is in principle likely to be of value in descriptions and dynamics of phylogenetic branching processes in speciation.

More recently gene flow has been strongly implicated between related species, both in *Drosophila*[204] and *Mus*.[205] Mitochondrial DNA was studied in isofemale lines derived from natural populations of *Drosophila pseudoobscura* and *D. persimilis*.[204] Eighteen different restriction maps were identified with 8 restriction endonucleases among 54 lines which had been independently derived from natural populations in which *Drosophila pseudoobscura* and *D. persimilis* were sympatric, and from allopatric populations of *D. pseudoobscura*. Variation in mtDNA was, within technical limitations, attributable to restriction endonuclease site loss or gain. Only one polymorphic insertion of about 0.5 kb was recognized. Comparison of restriction maps in sympatric populations of the two species revealed considerable homology of restriction map (76 to 80%). In contrast, mtDNA maps in allopatric *D. pseudoobscura* were all distinct from those in populations where the two species were sympatric. Powell points out that each species, on the basis of cytogenetic study and protein coding loci, had a characteristic nuclear genome[206-209] and presents evidence strongly supportive of mitochondrial exchange between these species in sympatric populations. Fertile F_1 hybrid females do occur between the species where they are sympatric and the rate of gene exchange has been estimated at 1 in 10^{-4}.

At present any proposed mechanism for the maintenance of genetic homogeneity in the mitochondrial genome in sympatric populations is likely to be speculative. Following Birky et al.,[148] genetic subdivision on the neutral gene theory will be attained for organelle genomes when $Nm < 4$, where N is the subpopulation size, m the migration rate between subpopulations, and both the migration rates and population sizes of males and females are equal. These premises are based upon a two-dimensional "stepping stone" model of subpopulation structure with migration.[210] If the rate of species hybridization is approximated as an upper limit to the migration rate for the mitochondrial genome, N must be of the order of 40,000. Assuming population sizes in nature are very large, it is reasonable to assume that Nm < 4 and the mitochondrial genomes in these two species will at least approach genetic uniformity. Of course, such models, based upon neutral gene theory assume a state of genetic equilibrium. The extent to which such an assumption is realistic cannot be easily answered in terms of the past evolution of the two species. Moreover, if the migration rate of males is substantially larger than it is in females, a substantial increase in subpopulation size for a given migration rate will be needed in order to prevent genetic differentiation of subpopulations. Interestingly, genetical differentiation of subpopulations in respect of the nuclear genome under neutral gene theory will, for a given population size, be effected at a lower migration rate than for the mitochondrial genome. If Nm < 1, local populations will show independent evolution, a result at least in agreement with present observations.

The second study implicating genetic exchange between species for the mitochondrial genome concerns natual populations of *Mus musculus* and *M. domesticus* in Europe.[205] The species hybridize in a narrow North-South zone which geographically demarcates species limits, such that *M. musculus* occupies Scandinavia, Northern Denmark, and Eastern Europe and *M. domesticus* Western Europe. On the basis of genetical variation at eight protein encoding nuclear loci, identified by gel electrophoresis, *M. domesticus* and *M. musculus* are distinct species. Restriction map variation in the mitochondrial genomes of both species provided the basis for parsimonious phylogenetic tree construction. Surprisingly two mtDNA maps of *M. musculus*, originating in Scandinavia (north Denmark and south Sweden) showed greater homology with the lineages of *M. domesticus* evolution than to their own species. Given that the rate of sequence divergence is a constant 2 to 4%/10^6 years, Ferris et al.[205] estimated the age of *M. domesticus* mtDNA in Scandinavian *M. musculus* to be 10^5 years. The divergence time for *M. domesticus* and *M. musculus*, based upon mtDNA evolution is about 10^6 years. Hence, the origin of the Scandinavian mtDNA phenotypes is after species formation and presumably due to gene introgression. Whether or not the introgressive *M. domesticus* mtDNA becomes fixed in Scandinavian mice by chance, or is the result of strong natural selection or its replicative property is at present unclear. Nonetheless, these studies of mtDNA evolution in *Drosophila* and mice provide striking anomalies in our conception of species based upon cytoplasmic genes and reinforce the validity of the species concept when based on variation in the nuclear genome. Perhaps more interestingly, if species introgression of the mitochondrial genome is found to be common our estimates of species divergence times could be conservative and interpretation would be largely population-specific.

VI. CONCLUSIONS

Present appreciation of the mechanism of mtDNA evolution is based largely upon data obtained from comparatively few groups of organisms. In animals mtDNA evolution is characterized by features such as a very high rate of base pair substitution with a preponderance of transitional over transversional changes. Evolution through changes of genome length are mainly confined to small oligonucleotide variations and these are most frequent in the D-loop and other noncoding regions. Modeling and theoretical expectations of variation in the mitochondrial as compared to the nuclear genome has been attempted. While the high rate of evolution in animal mtDNA is in part related to a high mutation rate, it is apparent that the dynamics of mtDNA transmission in germinal cells and molecular mechanisms of replication and repair need to be more fully investigated. More generally, evolutionary mechanism in mtDNA appears to vary considerably on a phylogenetic scale. The relative constancy of genome size seen in mammalian evolution is much less constrained in higher plant and fungal species. Indeed, large length mutations, the presence of introns in gene coding sequences, and recombination make our conceptions of mtDNA evolution in some fungi different to those gained from the mammalian mitochondrial genome. Additionally, nuclear variation in *Aspergillus* species promises to be greater than estimates from mtDNA restriction site variation, indicating a substantial conservation and lower rate of evolution in the mitochondrial genome compared to the nuclear genome of these species.

Models of mitochondrial DNA variation offer a comparative treatment of evolution with that of the nuclear genome. The models account for variation attributable to reproductive behavior, the relative contribution of maternal and paternal mitochondrial genomes, and differential female and male migration rates between subpopulations. At present our understanding of mtDNA microevolution at a within-species level relies heavily on studies of populations of higher animals. Similar studies of microevolution in other taxa, for example, fungi, higher plants, algae, and protozoa, with their widely differing modes of reproduction may further our understanding of the comparative evolution of nuclear and mitochondrial genomes. Lastly, the investigation of hybrid zones between related species is worthy of study. Mitochondrial gene flow and nuclear gene isolation between species, while emphasizing the genetical control of reproductive isolation in nuclear genes, may present problems for mtDNA-based phylogenies. Evolutionary relationships based upon the nuclear genome may vary in time and order of branching with those based upon the mitochondrial genome. Hence, our present understanding of evolutionary mechanisms in the mitochondrial genome show them to be phylogenetically very diverse. Some aspects of mtDNA evolution may be accountable by molecular mechanism and functional constraint alone. While the neutral gene theory provides a basis for modeling mtDNA evolution it is hoped that further research will offer an experimental approach for the testing of this evolutionary hypothesis.

REFERENCES

1. **Ephrussi, B., Hottinguer, H., and Chimenes, A. M.,** Action de l'acriflavine sur les levures. I. La mutation "petite" colonie, *Ann. Inst. Pasteur (Paris)*, 76, 351, 1949.
2. **Ephrussi, B.,** *Nucleo-Cytoplasmic Relations in Micro-Organisms,* Clarendon Press, Oxford, 1953.
3. **Nass, M. M. K. and Nass, S.,** Intra-mitochondrial fibres with DNA characteristics. I. Fixation and electron staining reactions, *J. Cell Biol.,* 19, 593, 1963.
4. **Nass, M. M. K. and Nass, S.,** Intra-mitochondrial fibres with DNA characteristics. II. Enzymatic and other hydrolytic treatments, *J. Cell Biol.,* 19, 613, 1963.
5. **Wallace, D. C.,** Structure and evolution of organelle genomes, *Microbiol. Rev.,* 46, 208, 1982.
6. **Clark-Walker, G. D. and Sriprakash, K. S.,** Size diversity and sequence rearrangements in mitochondrial DNAs from yeasts, in *Mitochondrial Genes,* Slonimski, P., Borst, P., and Attardi, G., Eds., Cold Spring Harbor Laboratory, Cold Spring Harbor, N.Y., 1982, 349.
7. **McArthur, C. R. and Clark-Walker, G. D.,** Mitochondrial size diversity in the Dekkera/Brettanomyces yeasts, *Curr. Genet.,* 7, 29, 1983.
8. **Bendich, A. J.,** Plant mitochondrial DNA: the last frontier, in *Mitochondrial Genes,* Slonimski, P., Borst, P., and Attardi, G., Eds., Cold Spring Harbor Laboratory, Cold Spring Harbor, N.Y., 1982, 477.
9. **Leaver, C. J. and Gray, M. W.,** Mitochondrial genome organisation and expression in higher plants, *Ann. Rev. Plant Physiol.,* 33, 373, 1982.
10. **Goldbach, R. W., Arnberg, A. C., van Bruggen, E. F. J., Defize, J., and Borst, P.,** The structure of *Tetrahymena pyriformis* mitochondrial DNA. I. Strain differences and occurrence of inverted repetitions, *Biochim. Biophys. Acta,* 477, 37, 1977.
11. **Goldbach, R. W., Bollen-deBoer, J. E., van Bruggen, E. F. J., and Borst, P.,** Replication of the linear mitochondrial DNA of *Tetrahymena pyriformis, Biochim. Biophys. Acta,* 562, 400, 1979.
12. **Goddard, J. M. and Cummings, D. J.,** Structure and replication of mitochondrial DNA from *Paramecium aurelia, J. Mol. Biol.,* 97, 593, 1975.
13. **Goddard, J. M. and Cummings, D. J.,** Mitochondrial DNA replication in *Paramecium aurelia.* Cross linking of the initiation end, *J. Mol. Biol.,* 109, 327, 1977.

14. **Weslowski, M. and Fukuhara, H.,** Linear mitochondrial deoxyribonucleic acid from the yeast *Hansenula mrakii, Mol. Cell Biol.,* 1, 387, 1981.

15. **Grant, D. and Chiang, K. S.,** Physical mapping and characterization of *Chlamydomonas* mitochondrial DNA molecules: their unique ends, sequence homogeneity, and conservation, *Plasmid,* 4, 82, 1980.

16. **Quetier, F. and Vedel, F.,** Heterogeneous population of mitochondrial DNA molecules in higher plants, *Nature (London),* 268, 365, 1977.

17. **Levings, C. S. and Pring, D. R.,** The mitochondrial genome of higher plants, *Stadler Symp.,* 10, 77, 1978.

18. **Synenki, R. M., Levings, C. S., and Shah, D. M.,** Physiochemical characterisation of mitochondrial DNA from soybean, *Plant Physiol.,* 61, 460, 1978.

19. **Levings, C. S., Shah, D. M., Hu, W. W. L., Pring, D. R., and Timothy, D. H.,** Molecular heterogeneity among mitochondrial DNAs from different maize cytoplasms, in *Extrachromosomal DNA. ICN-UCLA Symposium on Molecular and Cellular Biology,* Cummings, D. J., Fox, C. F., Borst, P., Dawid, I. G., and Weissman, S. M., Eds., Academic Press, New York, 1979, 63.

20. **Sparks, R. B. and Dale, R. M. K.,** Characterization of ³H-labelled supercoiled mitochondrial DNA from tobacco suspension culture cells, *Molec. Gen. Genet.,* 180, 351, 1980.

21. **Spruill, W. M., Levings, C. S., and Sederoff, R. R.,** Recombinant DNA analysis indicates that the multiple chromosomes of maize mitochondria contain different sequences, *Dev. Genet.,* 1, 363, 1980.

22. **Dale, R. M. K.,** Sequence homology among different size classes of plant mtDNAs, *Proc. Natl. Acad. Sci. U.S.A.,* 78, 4454, 1981.

23. **Powling, A.,** Species of small DNA molecules found in mitochondria from sugar-beet with normal and male sterile cytoplasms, *Mol. Gen. Genet.,* 183, 82, 1981.

24. **Brennicke, A. and Blanz, P.,** Circular mitochondrial DNA species from *Oenothera* with unique sequences, *Mol. Gen. Genet.,* 187, 461, 1982.

25. **Dale, R. M. K.,** Structure of plant mitochondrial DNAs, in *Mitochondrial Genes,* Slonimski, P., Borst, P., and Attardi, G., Eds., Cold Spring Harbor Laboratory, Cold Spring Harbor, N.Y., 1982, 471.

26. **Palmer, J. D. and Shields, C. R.,** Tripartite structure of *Brassica campestris* mitochondrial genome, *Nature (London),* 307, 437, 1984.

27. **Pring, D. R., Levings, C. S., Hu, W. W. L., and Timothy, D. H.,** Unique DNA associated with mitochondria in the "S" type cytoplasm of male-sterile maize, *Proc. Natl. Acad. Sci. U.S.A.,* 74, 2904, 1977.

28. **Pring, D. R. and Levings, C. S.,** Heterogeneity of maize cytoplasmic genomes among male-sterile cytoplasms, *Genetics,* 89, 121, 1978.

29. **Lonsdale, D. M., Fauron, C. M.-R., Hodge, T. P., Pring, D. R., and Stern, D. B.,** Structural alterations in the mitochondrial genome of maize associated with cytoplasmic male sterility, in *Genetic Rearrangement. The 5th John Innes Symposium,* Chater, K. F., Cullis, C. A., Hopwood, D. A., Johnston, A. W. B., and Woolhouse, H. W., Eds., Croom Helm, London, 1983, 183.

30. **Levings, C. S., Kim, B. D., Pring, D. L., Conde, M. F., Mans, R. J., Laughnan, J. R., and Gabay-Laughnan, S. J.,** Cytoplasmic reversion of cms-S in maize: association with a transpositional event, *Science,* 209, 1021, 1980.

31. **Levings, C. S., Sederoff, R. R., Hu, W. W. L., and Timothy, D. H.,** Relationships among plasmid-like DNAs of the maize mitochondria, in *Structure and Function of Plant Genomes,* Ciferri, O. and Dure, L., Eds., Plenum Press, New York, 1982, 363.

32. **Thompson, R. D., Kemble, R. J., and Flavell, R. B.,** Variations in mitochondrial DNA organisation between normal and male-sterile cytoplasms of maize, *Nucleic Acids Res.,* 8, 1999, 1980.

33. **Lonsdale, D. M., Thompson, R. D., and Hodge, T. P.,** The integrated forms of the S-1 and S-2 DNA elements of maize male sterile mitochondrial DNA are flanked by a large repeated sequence, *Nucleic Acids Res.,* 9, 3657, 1981.

34. **Bernardi, G., Baldacci, G., Bernardi, G., Faugeron-Fonty, G., Gaillard, C., Goursot, R., Huyard, A., Mangin, M., Morotta, R., and de Zamaroczy, M.,** The petite mutation: excision sequences, replication origins and suppressivity, in *The Organization and Expression of the Mitochondrial Genome,* Kroon, A. M. and Saccone, C., Eds., Elsevier, Amsterdam, 1980, 21.

35. **Cummings, D. J., Belcour, L., and Grandchamp, C.,** Etude au microscope électronique du DNA mitochondrial de *Podospora anserina* et présence d'une série multimérique de molécules circulaires de DNA dans des cultures senescentes, *C. R. Acad. Sci. Ser. D.,* 287, 157, 1978.

36. **Stahl, U., Kück, U., Tudznyski, P., and Esser, K.,** Characterization and cloning of plasmid-like DNA of the ascomycete *Podospora anserina, Mol. Gen. Genet.,* 178, 639, 1978.

37. **Jamet-Vierney, C., Begel, O., and Belcour, L.,** Senescence in *Podospora anserina:* amplification of a mitochondrial DNA sequence, *Cell,* 21, 189, 1980.

38. **Belcour, L., Begel, O., Mossé, M. O., and Vierney, C.,** Mitochondrial DNA amplification in senescent cultures of *Podospora anserina:* variability between the retained amplified sequences, *Curr. Genet.,* 3, 13, 1981.

39. **Kück, U., Stahl, U., and Esser, K.,** Plasmid-like DNA is part of mitochondrial DNA in *Podospora anserina, Curr. Genet.,* 3, 151, 1981.

40. **Wright, R. M., Horrum, M. A., and Cummings, D. J.,** Are mitochondrial structural genes selectively amplified during senescence in *Podospora anserina? Cell,* 29, 505, 1982.

41. **Lazarus, C. M., Earl, A. J., Turner, G., and Küntzel, H.,** Amplification of a mitochondrial DNA sequence in the cytoplasmically inherited "ragged" mutant of *Aspergillus amstelodami, Eur. J. Biochem.,* 106, 633, 1980.

42. **Lazarus, C. M. and Küntzel, H.,** Amplification of a common mitochondrial DNA sequence in three new "ragged" mutants, in *The Organization and Expression of the Mitochondrial Genome,* Kroon, A. M. and Saccone, C., Eds., Elsevier, Amsterdam, 1980, 87.

43. **Lazarus, C. M. and Küntzel, H.,** Anatomy of amplified mitochondrial DNA in "ragged" mutants of *Aspergillus amstelodami:* excision points within protein genes and a common 215 bp segment containing a possible origin of replication, *Curr. Genet.,* 4, 99, 1981.

44. **Küntzel, H., Köchel, H. G., Lazarus, C. M., and Lünsdorf, H.,** Mitochondrial genes in *Aspergillus,* in *Mitochondrial Genes,* Slonimski, P., Borst, P., and Attardi, G., Eds., Cold Spring Harbor Laboratory, Cold Spring Harbor, N.Y., 1982, 391.

45. **Bertrand, H., Collins, R. A., Stohl, L. L., Goewert, R. R., and Lambowitz, A. M.,** Deletion mutants of *Neurospora crassa* mitochondrial DNA and their relationship to the "stop-start" growth phenotype, *Proc. Natl. Acad. Sci. U.S.A.,* 77, 6032, 1980.

46. **de Vries, H., de Jonge, J. C., van't Sant, P., Agsteribbe, E., and Arnberg, A.,** A "stopper" mutant of *Neurospora crassa* containing two populations of aberrant mitochondrial DNA, *Curr. Genet.,* 3, 205, 1981.

47. **Collins, R. A., Stohl, L. L., Cole, M. D., and Lambowitz, A. M.,** Characterization of a novel plasmid DNA found in mitochondria of *N. crassa, Cell,* 24, 443, 1981.

48. **Stohl, L. L., Collins, R. A., Cole, M. D., and Lambowitz, A. M.,** Characterization of the two new plasmid DNAs found in mitochondria of wild-type *Neurospora intermedia* strains, *Nucleic Acids Res.,* 10, 1439, 1982.

49. **Stern, D. B. and Lonsdale, D. M.,** Mitochondrial and chloroplast genomes of maize have a 12 kb DNA sequence in common, *Nature (London),* 299, 698, 1982.

50. **Lonsdale, D. M., Hodge, T. P., Howe, C. J., and Stern, D. B.,** Maize mitochondrial DNA contains a sequence homologous to the ribulose-1, 5-bisphosphate carboxylase large subunit gene of chloroplast DNA, *Cell,* 34, 1007, 1983.

51. **van den Boogaart, P., Samallo, J., and Agsteribbe, E.,** Similar genes for a mitochondrial ATPase subunit in the nuclear and mitochondrial genomes of *Neurospora crassa, Nature (London),* 298, 187, 1982.

52. **Brown, T. A., Ray, J. A., Waring, R. B., Scazzocchio, C., and Davies, R. W.,** A mitochondrial reading frame which may code for a second form of ATPase subunit 9 in *Aspergillus nidulans, Curr. Genet.,* 8, 489, 1984.

53. **Farrelly, F. and Butow, R. A.,** Rearranged mitochondrial genes in the yeast nuclear genome, *Nature (London),* 301, 296, 1983.

54. **Gellissen, G., Bradfield, J. Y., White, B. N., and Wyatt, G. R.,** Mitochondrial DNA sequences in the nuclear genome of a locust, *Nature (London),* 301, 631, 1983.

55. **Jacobs, H. T., Posakony, J. W., Roberts, J. W., Xin, J., Britten, R. J., and Davidson, E. H.,** Mitochondrial DNA sequences in the nuclear genome of *Strongylocentrotus purpuratus, J. Mol. Biol.,* 165, 609, 1983.

56. **Lewin, R.,** Promiscuous DNA leaps all barriers, *Science,* 219, 478, 1983.

57. **Butow, R. A., Farrelly, F., Zassenhaus, H. P., Hudspeth, M. E. S., Grossman, L. I., and Perlman, P. S.,** var 1 determinant region of yeast mitochondrial DNA, in *Mitochondrial Genes,* Slonimski, P., Borst, P., and Attardi, G., Eds., Cold Spring Harbor Laboratory, Cold Spring Harbor, N.Y., 1982, 241.

58. **Macino, G. and Tzagoloff, A.,** Assembly of the mitochondrial membrane system: the DNA sequence of a mitochondrial ATPase gene in *Saccharomyces cerevisiae, J. Biol. Chem.,* 254, 4617, 1979.

59. **Hensgens, L. A. M., Grivell, L. A., Borst, P., and Bos, J. L.,** Nucleotide sequence of the mitochondrial structural gene for subunit 9 of yeast ATPase complex, *Proc. Natl. Acad. Sci. U.S.A.,* 76, 1663, 1979.

60. **Hack, E. and Leaver, C. J.,** Synthesis of a dicyclohexylcarbodiimide-binding proteolipid by cucumber (*Cucumis sativus* L.) mitochondria, *Curr. Genet.,* 8, 537, 1984.

61. **Gray, M. W. and Spencer, D. F.,** Is wheat mitochondrial 5S ribosomal RNA prokaryotic in nature? *Nucleic Acids Res.,* 9, 3523, 1981.

62. **Chao, S., Sederoff, R., and Levings, C. S.,** Partial sequence of the 5S-18S gene region of the maize mitochondrial genome, *Plant Physiol.,* 71, 190, 1983.

63. **Hack, E. and Leaver, C. J.,** The α-subunit of the maize F_1-ATPase is synthesized in the mitochondrion, *EMBO J.,* 2, 1783, 1983.

64. **Boutry, M., Briquet, M., and Goffeau, A.,** The α-subunit of a plant mitochondrial F_1-ATPase is translated in the mitochondrion, *J. Biol. Chem.,* 258, 8524, 1983.

65. **Leaver, C. J. and Gray, M. W.,** Mitochondrial genome organization and expression in higher plants, *Ann. Rev. Plant Physiol.,* 33, 373, 1982.

66. **Dawson, A. J., Jones, V. P., and Leaver, C. J.,** The apocytochrome *b* gene in maize mitochondria does not contain introns and is preceded by a potential ribosome binding site, *EMBO J.,* 3, 2107, 1984.

67. **Anderson, S., Bankier, A. T., Barrell, B. G., de Bruijn, M. H. L., Coulson, A. R., Drouin, J., Eperon, I. C., Nierlich, D. P., Roe, B. A., Sanger, F., Schreier, P. H., Smith, A. J. H., Staden, R., and Young, I. G.,** Sequence and organization of the human mitochondrial genome, *Nature (London),* 290, 457, 1981.

68. **Anderson, S., de Bruijn, M. H. L., Coulson, A. R., Eperon, I. C., Sanger, F., and Young, I. G.,** Complete sequence of bovine mitochondrial DNA: conserved features of the mammalian mitochondrial genome, *J. Mol. Biol.,* 156, 683, 1982.

69. **Anderson, S., Bankier, A. T., Barrell, B. G., de Bruijn, M. H. L., Coulson, A. R., Drouin, J., Eperon, I. C., Nierlich, D. P., Roe, B. A., Sanger, F., Sehreier, P. H., Smith, A. J. H., Staden, R., and Young, I. G.,** Comparison of the human and bovine mitochondrial genomes, in *Mitochondrial Genes,* Slonimski, P., Borst, P., and Attardi, G., Eds., Cold Spring Harbor Laboratory, Cold Spring Harbor, N.Y., 1982, 5.

70. **Bibb, M. J., van Etten, R. A., Wright, C. T., Walberg, M. W., and Clayton, D. A.,** Sequence and gene organization of mouse mitochondrial DNA, *Cell,* 26, 167, 1981.

71. **Fauron, C. M. R. and Wolstenholme, D. R.,** Structural heterogeneity of mitochondrial DNA molecules within the genus *Drosophila, Proc. Natl. Acad. Sci. U.S.A.,* 73, 3623, 1976.

72. **Wolf, K., Lang, B., del Giudice, L., Anziano, P. Q., and Perlman, P. S.,** *Schizosaccharomyces pombe:* a short review of a short mitochondrial genome, in *Mitochondrial Genes,* Slonimski, P., Borst, P., and Attardi, G., Eds., Cold Spring Harbor Laboratory, Cold Spring Harbor, N.Y., 1982, 355.

73. **Borst, P. and Grivell, L. A.,** The mitochondrial genome of yeast, *Cell,* 15, 705, 1978.

74. **Clark-Walker, G. D., McArthur, C. R., and Sriprakash, K. S.,** Partial duplication of the large ribosomal RNA sequence in an inverted repeat in circular mitochondrial DNA from *Kloeckera africana, J. Mol. Biol.,* 147, 399, 1981.

75. **Hudspeth, M. E. S., Shumard, D. S., Bradford, C. J. R., and Grossman, L. I.,** Organisation of *Achlya* mtDNA: a population with two orientations and a large inverted repeat containing the rRNA genes, *Proc. Natl. Acad. Sci. U.S.A.,* 80, 142, 1983.

76. **Scazzocchio, C., Brown, T. A., Waring, R. B., Ray, J. A., and Davis, R. W.,** Organization of the *Aspergillus nidulans* mitochondrial genome, in *Mitochondria 1983,* Schweyen, R. J., Wolf, K., and Kaudewitz, F., Eds., de Gruyter, Berlin, 1983, 303.

77. **Macino, G.,** Mapping of mitochondrial structural genes in *Neurospora crassa, J. Biol. Chem.,* 255, 10563, 1980.

78. **Kück, U. and Esser, K.,** Genetic map of mitochondrial DNA in *Podospora anserina, Curr. Genet.,* 5, 143, 1982.

79. **Heckman, J. E., Sarnoff, J., Alzner-de Weerd, B., Yin, S., and RajBhandary, U. L.,** Novel features in the genetic code and codon reading patterns in *Neurospora crassa* mitochondria based on sequences of six mitochondrial tRNAs, *Proc. Natl. Acad. Sci. U.S.A.,* 77, 3159, 1980.

80. **Barrell, B. G., Anderson, S., Bankier, A. T., deBruijn, M. H. L., Chen, E., Coulson, A. R., Drouin, J., Eperon, I. C., Nierlich, D. P., Roe, B. A., Sanger, F., Schreier, P. H., Smith, A. J. H., Staden, R., and Young, I. G.,** Different pattern of codon recognition by mammalian mitochondrial tRNAs, *Proc. Natl. Acad. Sci. U.S.A.,* 77, 3164, 1980.

81. **Bonitz, S. G., Berlani, R., Coruzzi, G., Li, M., Macino, G., Nobrega, F. G., Nobrega, M. P., Thalenfeld, B. E., and Tzagoloff, A.,** Codon recognition rules in yeast mitochondria, *Proc. Natl. Acad. Sci. U.S.A.,* 77, 3167, 1980.

82. **Roe, B. A., Wong, J. F. H., Chen, E. Y., Armstrong, P. W., Stankiewicz, A., Ma, D. P., and McDonough, J.,** A modified nucleotide 3′ to the anticodon may modulate their codon response, in *Mitochondrial Genes,* Slonimski, P., Borst, P., and Attardi, G., Eds., Cold Spring Harbor Laboratory, Cold Spring Harbor, N.Y., 1982, 45.

83. **Yin, S., Heckman, J., and RajBhandary, U. L.,** Highly conserved GC-rich palindromic DNA sequences flank tRNA genes in *Neurospora crassa* mitochondria, *Cell,* 26, 325, 1981.

84. **Yin, S., Burke, J., Chang, D. D., Browning, K. S., Heckman, J. E., Alzner-de Weerd, B., Potter, M. J., and RajBhandary, U. L.,** *Neurospora crassa* mitochondrial tRNAs and rRNAs: structure, gene organization and DNA sequences, in *Mitochondrial Genes,* Slonimski, P., Borst, P., and Attardi, G., Eds., Cold Spring Harbor Laboratory, Cold Spring Harbor, N.Y., 1982, 361.

85. **Crick, F. H. C.,** Codon-anticodon pairing: the wobble hypothesis, *J. Mol. Biol.,* 19, 548, 1966.

86. **Fox, T. D. and Leaver, C. J.,** The *Zea mays* mitochondrial gene coding cytochrome oxidase subunit II has an intervening sequence and does not contain TGA codons, *Cell,* 26, 315, 1981.

87. **Browning, K. S. and RajBhandary, U. L.,** Cytochrome oxidase subunit III in *Neurospora crassa* mitochondria, *J. Biol. Chem.,* 257, 5253, 1982.

88. **Jukes, T. H.,** Amino acid codes in mitochondria as possible clues to primitive codes, *J. Mol. Evol.*, 18, 15, 1981.
89. **Gray, M. W.,** Mitochondrial genome diversity and the evolution of mitochondrial DNA, *Can. J. Biochem.*, 60, 157, 1982.
90. **Sederoff, R. R.,** Structural variation in mitochondrial DNA, *Adv. Genet.*, 22, 1, 1984.
91. **Gray, M. W. and Doolittle, W. F.,** Has the endosymbiont hypothesis been proven? *Microbiol. Rev.*, 46, 1, 1982.
92. **Gray, M. W.,** The bacterial ancestry of plastids and mitochondria, *Bioscience*, 33, 693, 1983.
93. **Shah, D. M. and Langley, C. H.,** Inter- and intra-specific variation in restriction maps of *Drosophila* mitochondrial DNAs, *Nature (London)*, 281, 696, 1979.
94. **Fauron, C. M. R. and Wolstenholme, D. R.,** Extensive diversity among *Drosophila* species with respect to nucleotide sequences within the adenine + thymine-rich region of mitochondrial DNA molecules, *Nucleic Acids Res.*, 8, 2439, 1980.
95. **Fauron, C. M. R. and Wolstenholme, D. R.,** Intraspecific diversity of nucleotide sequences within the adenine + thymine-rich region of mitochondrial DNA molecules of *Drosophila mauritiana*, *Drosophila melanogaster* and *Drosophila simulans*, *Nucleic Acids Res.*, 8, 5391, 1980.
96. **Reilly, J. G. and Thomas, C. A.,** Length polymorphisms, restriction site variation, and maternal inheritance of mitochondrial DNA of *Drosophila melanogaster*, *Plasmid*, 3, 109, 1980.
97. **Goddard, J. M., Fauron, C. M. R., and Wolstenholme, D. R.,** Nucleotide sequences within the A + T-rich region of the large-rRNA gene of mitochondrial DNA molecules of *Drosophila yakuba*, in *Mitochondrial Genes*, Slonimski, P., Borst, P., and Attardi, G., Eds., Cold Spring Harbor Laboratory, Cold Spring Harbor, N.Y., 1982, 99.
98. **Gillum, A. M. and Clayton, D. A.,** Displacement-loop replication initiation sequence in animal mitochondrial DNA exists as a family of discrete lengths, *Proc. Natl. Acad. Sci. U.S.A.*, 75, 677, 1978.
99. **Walberg, M. W. and Clayton, D. A.,** Sequence and properties of the human KB cell and mouse L cell D-loop regions of mitochondrial DNA, *Nucleic Acids Res.*, 9, 5411, 1981.
100. **Sekiya, T., Kobayashi, M., Seki, T., and Koike, K.,** Nucleotide sequence of a cloned fragment of rat mitochondrial DNA containing the replication origin, *Gene*, 11, 53, 1980.
101. **Ferris, S. D., Wilson, A. C., and Brown, W. M.,** Evolutionary tree for apes and humans based on cleavage maps of mitochondrial DNA, *Proc. Natl. Acad. Sci. U.S.A.*, 78, 2432, 1981.
102. **Upholt, W. B. and Dawid, I. G.,** Mapping of mitochondrial DNA of individual sheep and goats: rapid evolution in the D-loop region, *Cell*, 11, 571, 1977.
103. **Goldbach, R. W., Bollen-de Boer, J. E., van Bruggen, E. F. J., and Borst, P.,** Conservation of the sequence and position of the ribosomal RNA genes in *Tetrahymena pyriformis* mitochondrial DNA, *Biochim. Biophys. Acta*, 521, 187, 1978.
104. **Hauswirth, W. W., van de Walle, M. J., Laipis, P. J., and Olivo, P. D.,** Heterogeneous mitochondrial DNA D-loop sequences in bovine tissue, *Cell*, 37, 1001, 1984.
105. **Pring, D. R., Conde, M. F., and Gengenbach, B. G.,** Cytoplasmic genome variability in tissue culture derived plants, *Environ. Exp. Bot.*, 21, 369, 1981.
106. **McNay, J. W., Chourey, P. S., and Pring, D. R.,** Molecular analysis of genomic stability of mitochondrial DNA in tissue cultured cells of maize, *Theor. Appl. Genet.*, 67, 433, 1984.
107. **Nagy, F., Török, I., and Maliga, P.,** Extensive rearrangements in the mitochondrial DNA in somatic hybrids of *Nicotiana tabacum* and *Nicotiana knightiana*, *Mol. Gen. Genet.*, 183, 437, 1981.
108. **Ward, B. L., Anderson, R. S., and Bendich, A. J.,** The size of the mitochondrial genome is large and variable in a family of plants (Curcurbitaceae), *Cell*, 25, 793, 1981.
109. **Timothy, D. H., Levings, C. S., Pring, D. R., Conde, M. F., and Kermicle, J. L.,** Organelle DNA variation and systematic relationships in the genus *Zea*: teosinte, *Proc. Natl. Acad. Sci. U.S.A.*, 76, 4220, 1979.
110. **Weissinger, A. K., Timothy, D. H., Levings, C. S., and Goodman, M. M.,** Patterns of mitochondrial DNA variation in indigenous maize races of Latin America, *Genetics*, 104, 365, 1983.
111. **Sederoff, R. R., Levings, C. S., Timothy, D. H., and Hu, W. W. L.,** Evolution of DNA sequence organization in mitochondrial genomes of *Zea*, *Proc. Natl. Acad. Sci. U.S.A.*, 78, 5953, 1981.
112. **Lazowska, J., Jacq, C., and Slonimski, P. P.,** Sequence of introns and flanking exons in wild-type and *box-3* mutants of cytochrome *b* reveals an interlaced splicing protein coded by an intron, *Cell*, 22, 338, 1980.
113. **Jacq, C., Pajot, P., Lazowska, J., Dujardin, G., Claisse, M., Groudinsky, O., de la Salle, H., Grandchamp, C., Labouesse, M., Gargouri, A., Guiard, B., Spyridakis, A., Dreyfus, M., and Slonimski, P.,** Role of introns in the yeast cytochrome-*b* gene: cis- and trans-acting signals, intron manipulation, expression and intergenic communications, in *Mitochondrial Genes*, Slominski, P., Borst, P., and Attardi, G., Eds., Cold Spring Harbor Laboratory, Cold Spring Harbor, N.Y., 1982, 155.

114. **Waring, R. B., Davies, R. W., Lee, S., Grisi, E., McPhail Berks, M., and Scazzocchio, C.,** The mosaic organization of the apocytochrome *b* gene of *Aspergillus nidulans* revealed by DNA sequencing, *Cell*, 27, 4, 1981.

115. **Waring, R. B., Davies, R. W., Scazzocchio, C., and Brown, T. A.,** Internal structure of a mitochondrial intron of *Aspergillus nidulans, Proc. Natl. Acad. Sci. U.S.A.,* 79, 6332, 1982.

116. **Collins, R. A. and Lambowitz, A. M.,** Structural variations and optional introns in the mitochondrial DNAs of *Neurospora* strains isolated from nature, *Plasmid,* 9, 53, 1983.

117. **Davies, R. W., Waring, R. B., Ray, J. A., Brown, T. A., and Scazzocchio, C.,** Making ends meet: a model for RNA splicing in fungal mitochondria, *Nature (London),* 300, 719, 1982.

118. **Waring, R. B. and Davies, R. W.,** Assessment of a model for intron RNA secondary structure relevant to RNA self-splicing — a review, *Gene,* 28, 277, 1984.

119. **Lazowska, J., Jacq, C., and Slonimski, P. P.,** Splice points of the third intron in the yeast mitochondrial cytochrome *b* gene, *Cell,* 27, 12, 1981.

120. **Waring, R. B., Brown, T. A., Ray, J. A., Scazzocchio, C., and Davies, R. W.,** Three variant introns of the same general class in the mitochondrial gene for cytochrome oxidase subunit 1 in *Aspergillus nidulans, EMBO J.,* 3, 2121, 1984.

121. **Lang, B. F.,** The mitochondrial genome of the fission yeast *Schizosaccharomyces pombe:* highly homologous introns are inserted at the same position of the otherwise less conserved *coxl* genes in *Schizosaccharomyces pombe* and *Aspergillus nidulans, EMBO J.,* 3, 2136, 1984.

122. **Nobrega, F. G. and Tzagoloff, A.,** Assembly of the mitochondrial membrane system. DNA sequence and organization of the cytochrome *b* gene in *Saccharomyces cerevisiae* D 273-10B, *J. Biol. Chem.,* 255, 9828, 1980.

123. **Grivell, L. A., Arnberg, A. C., Hensgens, L. A. M., Roosendaal, E., van Ommen, G. J. B., and van Bruggen, E. F. J.,** Split genes on yeast mitochondrial DNA: organization and expression, in *The Organization and Expression of the Mitochondrial Genome,* Kroon, A. M. and Saccone, C., Eds., Elsevier, Amsterdam, 1980, 37.

124. **van Ommen, G. J. B., de Boer, P. H., Groot, G. S. P., de Haan, M., Roosendaal, E., Grivell, L. A., Haid, A., and Schweyen, R. J.,** Mutations affecting RNA splicing and the interaction of gene expression of the yeast mitochondrial loci *cob* and *oxi-3, Cell,* 20, 173, 1980.

125. **Bonitz, S. G., Coruzzi, G., Thalenfeld, B. E., and Tzagoloff, A.,** Assembly of the mitochondrial membrane system: structure and nucleotide sequence of the gene coding for subunit 1 of yeast cytochrome oxidase, *J. Biol. Chem.,* 255, 11927, 1980.

126. **Grivell, L. A., Hensgens, L. A. M. Osinga, K. A., Tabak, H. F., Boer, P. H., Crusius, J. B. A., van der Laan, J. C., de Haan, M., van der Horst, G., Evers, R. F., and Arnberg, A. C.,** RNA processing in yeast mitochondria, in *Mitochondrial Genes,* Slonimski, P., Borst, P., and Attardi, G., Ed., Cold Spring Harbor Laboratory, Cold Spring Harbor, N.Y., 1982, 225.

127. **Croft, J. H.,** Unpublished data, 1984.

128. **Croft, J. H. and Jinks, J. L.,** Aspects of the population genetics of *Aspergillus nidulans,* in *Genetics and Physiology of Aspergillus,* Smith, J. E. and Pateman, J. A., Eds., Academic Press, New York, 1977, 339.

129. **Heyting, C. and Menke, H. H.,** Fine structure of the 21 S ribosomal RNA region on yeast mitochondrial DNA. III. Physical location of mitochondrial genetic markers with molecular nature of ω, *Mol. Gen. Genet.,* 168, 279, 1979.

130. **Faye, G., Dennebouy, N., Kujawa, C., and Jacq, C.,** Inserted sequence in the mitochondrial 23S ribosomal RNA gene of the yeast *Saccaromyces cerevisiae, Mol. Gen. Genet.,* 168, 101, 1979.

131. **Labouesse, M. and Slonimski, P. P.,** Construction of novel cytochrome *b* genes in yeast mitochondria by subtraction or addition of introns, *EMBO J.,* 2, 269, 1983.

132. **Rowlands, R. T. and Turner, G.,** Maternal inheritance of mitochondrial markers in *Aspergillus nidulans, Genet. Res.,* 28, 281, 1976.

133. **Jadayel, D. M. and Croft, J. H.,** unpublished results, 1984.

134. **Earl, A. J., Turner, G., Croft, J. H., Dales, R. B. G., Lazarus, C. M., Lünsdorf, H., and Küntzel, H.,** High frequency transfer of species specific mitochondrial DNA sequences between members of the Aspergillaceae, *Curr. Genet.,* 3, 221, 1981.

135. **Belliard, G., Vedel, F., and Pelletier, G.,** Mitochondrial recombination in cytoplasmic hybrids of *Nicotiana tabacum* by protoplast fusion, *Nature (London),* 281, 401, 1979.

136. **Galun, E., Arzee-Gonen, P., Fluhr, R., Edelman, M., and Aviv, D.,** Cytoplasmic hybridization in *Nicotiana* mitochondrial DNA analysis in progenies resulting from fusion between protoplasts having different organelle constitutions, *Mol. Gen. Genet.,* 186, 50, 1982.

137. **Croft, J. H. and Dales, R. B. G.,** Mycelial interactions and mitochondrial inheritance in *Aspergillus,* in *The Ecology and Physiology of the Fungal Mycelium,* Jennings, D. H. and Rayner, A. D. M., Eds., Cambridge University Press, London, 1984, 433.

138. **Brown, T. A., Davies, R. W., Ray, J. A., Waring, R. B., and Scazzocchio, C.,** The mitochondrial genome of *Aspergillus nidulans,* contains reading frames homologous to the human URFs 1 and 4, *EMBO J.,* 2, 427, 1983.

139. **Spooner, R. A. and Turner, G.,** unpublished data, 1984.

140. **Osiewacz, H. D. and Esser, K.,** The mitochondrial plasmid of *Podospora anserina:* a mobile intron of a mitochondrial gene, *Curr. Genet.,* 8, 299, 1984.

141. **Aoki, K., Yoshio, T., and Takahata, N.,** Estimating evolutionary distance from restriction maps of mitochondrial DNA with arbitary G + C content, *J. Mol. Evol.,* 18, 1, 1981.

142. **Lansman, R. A., Shade, R. O., Shapira, J. F., and Avise, J. C.,** The use of restriction endonucleases to measure mitochondrial DNA sequence relatedness in natural populations, *J. Mol. Evol.,* 17, 214, 1981.

143. **Cann, R. L. and Wilson, A. C.,** Length mutations in human mitochondrial DNA, *Genetics,* 104, 699, 1983.

144. **Brown, W. M. and Vinograd, J.,** Restriction endonuclease cleavage maps of animal mitochondrial DNAs, *Proc. Natl. Acad. Sci. U.S.A.,* 11, 4617, 1974.

145. **Nathans, D. and Smith, H. O.,** Restriction endonucleases in the analysis and restructuring of DNA molecules, *Ann. Rev. Biochem.,* 44, 273, 1975.

146. **Aquadro, C. F. and Greenberg, B. D.,** Human mitochondrial DNA variation and evolution: analysis of nucleotide sequences from seven individuals, *Genetics,* 103, 287, 1983.

147. **Kimura, M.,** Evolutionary rate at the molecular level, *Nature (London),* 217, 624, 1968.

148. **Birky, C. W., Jr., Maruyama, T., and Fuerst, P.,** An approach to population and evolutionary genetic theory for genes in mitochondrial and chloroplasts, and some results, *Genetics,* 103, 513, 1983.

149. **Chapman, R. W., Clairborne-Stephens, J., Lansman, R. A., and Avise, J. C.,** Models of mitochondrial DNA transmission genetics and evolution in higher eukaryotes, *Genet. Res.,* 40, 41, 1982.

150. **Gresson, R. A. R.,** Presence of the sperm middle-piece in the fertilised egg of the mouse *Mus musculus,* Nature (London), 145, 425, 1940.

151. **Friedlander, M.,** Monospermic fertilization in *Chrysopa carnia* (Neuroptera; Chrysopidae): behaviour of the fertilizing spermatozoa prior to syngamy, *Int. J. Insect Morphol. Embryol.,* 9, 53, 1980.

152. **Dawid, I. B. and Blackler, A. W.,** Maternal and cytoplasmic inheritance of mitochondrial DNA in *Xenopus, Dev. Biol.,* 29, 152, 1972.

153. **Piko, L. and Matsumato, L.,** Number of mitochondria and some properties of mitochondrial DNA in the mouse egg, *Dev. Biol.,* 49, 1, 1976.

154. **Michaelis, G. S., Hauswirth, W. W., and Laipis, P. J.,** Mitochondrial DNA copy number in bovine oocytes and somatic cells, *Dev. Biol.,* 99, 246, 1982.

155. **Mitchell, M. B. and Mitchell, H. K.,** A case of "maternal" inheritance in *Neurospora crassa, Proc. Natl. Acad. Sci. U.S.A.,* 38, 442, 1952.

156. **Levings, C. S. and Pring, D. R.,** Restriction endonuclease analysis of mitochondrial DNA from normal and Texas cytoplasmic male sterile maize, *Science,* 193, 158, 1976.

157. **Hutchinson, C. A., Newbold, J. E., Potter, S. S., and Edgell, M. M.,** Maternal inheritance of mammalian mitochondrial DNA, *Nature (London),* 251, 536, 1974.

158. **Avise, J. C., Lansman, R. A., and Shade, R. O.,** The use of restriction endonucleases to measure mitochondrial DNA sequence relatedness in natural populations. I. Population structure and evolution in the genus *Peromyscus, Genetics,* 92, 279, 1979.

159. **Lansman, R. A., Avise, J. C., and Huettel, M. D.,** Critical experimental test of the possibility of "paternal leakage" of mitochondrial DNA, *Proc. Natl. Acad. Sci. U.S.A.,* 80, 1969, 1983.

160. **Borkhardt, B. and Olsen, L. W.,** Paternal inheritance of the mitochondrial DNA in interspecific crosses of the aquatic fungus *Allomyces, Curr. Genet.,* 7, 403, 1983.

161. **Hauswirth, W. W. and Laipis, P. J.,** Mitochondrial DNA polymorphism in a maternal lineage of Holstein cows, *Proc. Natl. Acad. Sci. U.S.A.,* 79, 4686, 1982.

162. **Birky, C. W.,** Transmission genetics of mitochondria and chloroplasts, *Ann. Rev. Genet.,* 12, 471, 1978.

163. **Engels, W. R.,** Estimating genetic divergence and genetic variability with restriction endonucleases, *Proc. Natl. Acad. Sci. U.S.A.,* 78, 6329, 1981.

164. **Ewens, W. J., Spielman, R. S., and Harris, H.,** Estimation of genetic variation at the DNA level from restriction endonuclease data, *Proc. Natl. Acad. Sci. U.S.A.,* 78, 3748, 1981.

165. **Hudson, R. R.,** Estimating genetic variability with restriction endonucleases, *Genetics,* 100, 711, 1982.

166. **Ewens, W. R.,** The role of models in the analysis of molecular genetic data, with particular reference to restriction fragment data, in *Statistical Analysis of DNA Sequence Data,* Weir, B., Ed., Marcel Dekker, New York, 1983, 45.

167. **Kaplan, N.,** Statistical analysis of restriction map data and nucleotide sequence data, in *Statistical Analysis of DNA Sequence Data,* Weir, B., Ed., Marcel Dekker, New York, 1983, 75.

168. **Nei, M. and Li, W.-H.,** Mathematical model for studying genetic variation in terms of restriction endonucleases, *Proc. Natl. Acad. Sci. U.S.A.,* 76, 5269, 1979.

169. **Nei, M. and Tojima, F.,** DNA polymorphism detectable by restriction endonucleases, *Genetics,* 97, 145, 1981.

170. **Cummings, D. J.,** Evolutionary divergence of mitochondrial DNA from *Paramecium aurelia, Mol. Gen. Genet.,* 180, 77, 1980.

171. **Bogler, S. A., Zarley, C. D., Burkianek, L. L., Fuerst, P. A., and Byers, T. J.,** Interstrain mitochondrial DNA polymorphism detected in *Acanthamoeba* by restriction endonuclease analysis, *Mol. Biochem. Parasitol.,* 8, 145, 1983.

172. **Avise, J. C. and Lansman, R. A.,** Polymorphism of mitochondrial DNA in populations of higher animals, in *Evolution of Genes and Proteins,* Nei, M. and Koehn, R. M., Eds., Sinauer Associates, Sunderland, Mass., 1983, 147.

173. **Ferris, S. D., Brown, W. M., Davidson, W. S., and Wilson, A. C.,** Extensive polymorphism in the mitochondrial DNA of apes, *Proc. Natl. Acad. Sci. U.S.A.,* 78, 6319, 1981.

174. **Brown, W. M.,** Polymorphism in mitochondrial DNA of humans as revealed by restriction endonuclease analysis, *Proc. Natl. Acad. Sci. U.S.A.,* 77, 3605, 1980.

175. **Langley, C. H., Montgomery, E. A., and Quattlebaum, W. F.,** Restriction map variation in the *Adh* region of *Drosophila, Proc. Natl. Acad. Sci. U.S.A.,* 79, 5631, 1982.

176. **Leigh-Brown, A. J. and Ish-Horowitz, D.,** Evolution of the 87A and 87C heat-shock loci in *Drosophila, Nature (London),* 290, 677, 1981.

177. **Birley, A. J.,** Restriction endonuclease map variation and gene activity in the *Adh* region in a population of *Drosophila melanogaster, Heredity,* 52, 103, 1984.

178. **Avise, J. C., Giblin-Davidson, C., Laerm, J., Patton, J. C., and Lansman, R. A.,** Mitochondrial DNA clones and matriarchal phylogeny within and among geographic populations of the pocket gopher, *Geomys pinetum, Proc. Natl. Acad. Sci. U.S.A.,* 76, 6694, 1979.

179. **Selander, R. K., Smith, M. M., Yang, S. Y., Johnson, W. E., and Gentry, J. B.,** Biochemical polymorphism and systematics in the genus *Peromyscus.* I. Variation in the old-field mouse (*Peromyscus polionotus*), in *Studies in Genetics. VI,* Univ. Texas Publ. No. 7103, Austin, Tex., 1971, 49.

180. **Ferris, S. D., Sage, R. D., Prager, E. M., Ritte, U., and Wilson, A. C.,** Mitochondrial DNA evolution in mice, *Genetics,* 105, 681, 1983.

181. **Selander, R. K.,** Genic variation in natural populations, in *Molecular Evolution,* Ayala, F. J., Ed., Sinauer Associates, Sunderland, Mass., 1976, 21.

182. **Jeffreys, A.,** DNA sequence variants in the ^GY, ^GY, $\delta-$ and β-globin genes of man, *Cell,* 18, 1, 1979.

183. **Brown, W. M., George, M., Jr., and Wilson, A. C.,** Rapid evolution of animal mitochondrial DNA, *Proc. Natl. Acad. Sci. U.S.A.,* 76, 1967, 1979.

184. **Brown, G. G. and Simpson, M. V.,** Novel features of animal mtDNA evolution as shown by sequences of two rat cytochrome oxidase subunit II genes, *Proc. Natl. Acad. Sci. U.S.A.,* 79, 3246, 1982.

185. **Lansman, R. A., Avise, J. C., Aquadro, C. F., Shapira, J. F., and Daniel, S. W.,** Extensive genetic variation in mitochondrial DNA's among geographic populations of the deer-mouse, *Peromyscus maniculatus, Evolution,* 37, 1, 1983.

186. **Brown, W. M., Prager, E. M., Wang, A., and Wilson, A. C.,** Mitochondrial DNA sequences of primates: tempi and mode of evolution, *J. Mol. Evol.,* 18, 225, 1982.

187. **Miyata, T., Hayashida, H., Kikuno, R., Hasegaura, M., Kobayachi, M., and Koike, K.,** Molecular clock of silent substitution: at least six-fold preponderance of silent changes in mitochondrial genes over those in nuclear genes, *J. Mol. Evol.,* 19, 28, 1982.

188. **Miyata, T. and Yasunaga, T.,** Rapidly evolving mouse α-globin-related pseudogene and its evolutionary history, *Proc. Natl. Acad. Sci. U.S.A.,* 78, 450, 1981.

189. **Kunkel, T. A. and Loeb, L. A.,** Fidelity of mammalian DNA polymerases, *Science,* 213, 765, 1981.

190. **Seitz-Mayr, G. and Wolf, K.,** Extrachromosomal mutator inducing point mutations and deletions in mitochondrial genome of fission yeast, *Proc. Natl. Acad. Sci. U.S.A.,* 79, 2618, 1982.

191. **Kimura, M.,** Estimation of evolutionary distances between homologous nucleotide sequences, *Proc. Natl. Acad. Sci. U.S.A.,* 78, 454, 1981.

192. **Kimura, M.,** A simple method for estimating evolutionary rates of base substitutions through comparative studies of nucleotide sequences, *J. Mol. Evol.,* 16, 111, 1980.

193. **Kimura, M.,** *The Neutral Theory of Molecular Evolution,* Cambridge University Press, London, 1983.

194. **Prager, E. M. and Wilson, A. C.,** Construction of phylogenetic trees for proteins and nucleic acids: empirical evaluation of alternative matrix methods, *J. Mol. Evol.,* 11, 129, 1978.

195. **Fitch, M.,** On the problem of discovering the most parsimonious tree, *Am. Nat.,* 111, 223, 1977.

196. **Fitch, W. M.,** Estimating the total number of nucleotide substitutions since the common ancestor of a pair of homologous genes: comparison of several methods and three beta haemoglobin messenger RNA's, *J. Mol. Evol.,* 16, 153, 1980.

197. **Felsenstein, J.,** Evolutionary trees from DNA sequences: a maximum likelihood approach, *J. Mol. Evol.,* 17, 368, 1981.

198. **Felsenstein, J.,** Inferring evolutionary trees from DNA sequences, in *Statistical Analysis of DNA Sequence Data,* Weir, B. S., Ed., Marcel Dekker, New York, 1983, 133.
199. **Kaplan, N. and Risko, K.,** A method for estimating rates of nucleotide substitution using DNA sequence data, *Theor. Pop. Biol.,* 21, 318, 1982.
200. **Cann, R. L., Brown, W. M., and Wilson, A. C.,** Polymorphic sites and the mechanism of evolution in human mitochondrial DNA, *Genetics,* 106, 479, 1984.
201. **Hayashi, J. I., Yonekana, H., Gotoh, O., Tagashira, Y., Moriwaki, K., and Yosida, T. H.,** Evolutionary aspects of variant types of rat mitochondrial DNA's *Biochim. Biophys. Acta,* 564, 202, 1979.
202. **Osheroff, N., Speck, S. H., Margoliash, E. C. I., Veerman, J., Vilmo, B. W., König, B. W., and Muyses, A. O.,** The reaction of primate cytochrome c with cytochrome c oxidase, *J. Biol. Chem.,* 258, 5731, 1983.
203. **Brown, W. M. and Wright, J. W.,** Mitochondrial DNA analyses and the origin and relative age of parthenogenetic lizards (genus *Cnemidophorus*), *Science,* 203, 1247, 1979.
204. **Powell, J. R.,** Interspecific cytoplasmic gene flow in the absence of nuclear gene flow: evidence from *Drosophila, Proc. Natl. Acad. Sci. U.S.A.,* 80, 492, 1983.
205. **Ferris, S. D., Sage, R. D., Huang, C.-M., Nielson, J. T., Ritte, U., and Wilson, A. C.,** Flow of mitochondria DNA across a species boundary, *Proc. Natl. Acad. Sci. U.S.A.,* 80, 2290, 1983.
206. **Dobzhansky, Th. and Epling, C.,** Contributions to the genetics, taxonomy and ecology of *Drosophila pseudoobscura* and its relatives, *Carnegie Inst. Wash. Publ.,* 554, 1, 1944.
207. **Ayala, F. J. and Powell, J. R.,** Allozymes as diagnostic characters of sibling species of *Drosophila, Proc. Natl. Acad. Sci. U.S.A.,* 69, 1094, 1972.
208. **Coyne, J.,** Lack of genetic similarity between two sibling species of *Drosophila* as revealed by varied techniques, *Genetics,* 84, 593, 1976.
209. **Anderson, W. W., Ayala, F. J., and Michod, R. E.,** Chromosomal and allozymic diagnosis of three species of *Drosophila, J. Hered.,* 68, 71, 1977.
210. **Kimura, M. and Maruyama, T.,** Pattern of neutral polymorphism in a geographically structured population, *Genet. Res.,* 18, 125, 1971.

Chapter 6

EVOLUTION OF HISTONE GENES

William F. Marzluff

TABLE OF CONTENTS

I. INTRODUCTION

Histones are the basic proteins which package the DNA in eukaryotic chromosomes. The pioneering fractionation work of Johns and co-workers[1] simplified the histone complement to five classes of molecules. A different fractionation scheme of Luck and co-workers[2] led to similar results: five classes of histone proteins. Four of these, now called H2a, H2b, H3, and H4, appeared to be homogeneous in any one species, while the fifth H1 was obviously heterogeneous. The sequencing work of DeLange and Smith and co-workers[3-6] demonstrated that histones H3 and H4 were homogeneous proteins and highly conserved during evolution, with only 2 to 4% divergence in amino acid sequence between mammals and plants. These studies led to the idea that there were a limited number of different histone molecules in any given organism. At about the same time the first reports of histone microheterogeneity at the level of amino acid sequence appeared, indicating the possibility of several genes necessary to code for each of the different classes of histone proteins.[6-8] A much larger number of variants of histone H1 were found than of the other histones, a total of 5 different H1 proteins present in mammalian somatic cells.[9] More recently, the studies of Zweidler and co-workers[10,11] have demonstrated that both mammals and birds contain a limited number of nonallelic histone variants. In addition, Cohen and Zweidler and co-workers[12,13] described a number of different histone protein variants which are differentially expressed during early sea urchin development. More than one type of histone protein for a particular histone has been reported in the primitive eukaryotes, yeast,[14,15] and *Tetrahymena*[16] as well. This microheterogeneity at the protein level is a common feature of histone genes.

While the studies on histone protein chemistry were in progress, the role of histones in forming the structure of the chromosomes was elucidated. The basic unit of all chromosomes is the nucleosome, composed of 146 bp of DNA coiled around the histone octamer, 2 molecules each of histone H2a, H2b, H3, and H4. This basic structural unit is found in all eukaryotes from yeast and *Tetrahymena* to man and higher plants. Linking two nucleosomes is a small stretch of DNA to which the fifth histone, histone H1, is bound. The nucleosome core structure has been highly conserved in eukaryotes.[17]

The sequences of the histone proteins show only tentative homologies which may point to a single precursor. However, all the core histones show the same general theme of a basic amino terminal domain and a hydrophobic central region, followed by a short basic C-terminal domain.[18] Thus, it is possible to imagine a primordial histone with this general structure that could selfassociate. There are histone-like DNA binding proteins present in *Escherichia coli*[19] and an archaebacter, *Thermoplasma acidophilum*.[20] This gene could then have duplicated to ultimately give rise to the four core histone genes. Histone H1 with its different structural properties may well have originated independently. Once the four individual core histone proteins were present, the rate at which they have accumulated amino acid substitution has been very slow. H2a and H2b have changed more rapidly than H3 and H4. Presumably, gene duplication led occasionally to new histone genes which were fixed in a species giving rise to protein microheterogeneity. In addition to the conservation of structure at the protein levels many of the histone genes may share common regulatory sequences, since they are coordinately expressed. These sequences include sequences in the mRNA itself involved in regulation of mRNA halflife and in the regions flanking the genes presumably involved in regulation of transcription. Thus, the histones form a unique multigene set, comprised of genes coding for several types of proteins, with microheterogeneity within each type. The genes share common regulatory properties and, as an additional complexity in many organisms, there are tissue or stage-specific gene sets. Because of the extremely high conservation of the protein sequences, presumably due to selective pressure on the precise multicomponent nucleosome core structure, it is relatively easy to compare DNA sequences across broad phylogenetic ranges. The common regulatory property of specific alteration of mRNA halflife also places additional selective pressures on the DNA structure coding for the mRNA, including possibly the portion of the DNA coding for the histone protein sequence.

In this chapter, I review what is known about histone gene organization, mRNA structure, protein and DNA sequences, and regulation of histone mRNA levels. Then I speculate as to how this might fit into an evolutionary scheme for histone genes with emphasis on the data still needed to fill in the model. My own work is concerned solely with mouse histone genes and mRNAs, so this chapter borrows heavily on the work of others, particularly Drs. Lynna Hereford, Mitchell Smith, and Michael Gruenstein on yeast histone genes; Drs. Larry Kedes, Max Birnstiel, Eric Weinberg, and Geoff Childs on sea urchin histone genes; Drs. Oliver Destree and Hugh Woodland on *Xenopus* histone genes; Drs. Doug Engel, Jerry Dodgson, and Julian Wells on chicken histone genes; and Drs. Nat Heintz, Bob Roeder, and Gary Stein on human histone genes. In addition, there are several excellent recent reviews by Hentschel and Birnstiel,[21] Kedes,[22] Isenberg,[18] and Von Holt et al.[23] to which the reader is referred for additional information.

II. ORGANIZATION OF HISTONE GENES

A clue to the pattern of evolution of the histone genes may lie in the organization of the genes. In all species studied thus far at least some, if not all, of the histone genes are tightly clustered with genes for other histones. The different types of repeat units reported thus far are shown in Figure 1. These fall into two main classes, a repeated organization often but not necessarily tandemly repeated, and a jumbled cluster without repeating units.

In the primitive eukaryote, yeast, the genes are clustered in closely linked pairs H2a, H2b and H3, H4. It is not known if yeast contains an H1 histone. Yeast has the lowest copy number of any known organism, two copies of each gene.[24,25] There are two clusters of each pair, located on at least three different chromosomes. The pairing of H2a with H2b and H3 with H4 is consistent with the protein pairs found in nucleosomes. The pairs are in an inverted orientation with the 5′ ends juxtaposed and about 600 to 800 bp apart. The different gene pairs have obviously diverged from each other long ago (see below). The other primitive eukaryotes in which some histone genes have been studied are *Tetrahymena*[26] and *Neurospora*.[27] While there is not yet complete data, there are obviously independent copies of genes for H3 and H4 histones in *Tetrahymena*.[26] In contrast, *Neurospora* may have only a single copy each of H3 and H4 genes which are linked in a similar fashion to yeast histone genes.[27] The *Neurospora* genes differ from the other histone genes in that they have intervening sequences. The general theme of the organization of histone genes in yeast, a dispersed group of genes which has maintained some clustering of genes, is typical of histone genes in many organisms.

The other typical pattern found in several species is that the genes for all five histones are closely linked and tandemly repeated (Figure 1). The first genes coding for protein isolated from a eukaryote were the genes coding for the histones synthesized in the cleavage stage of sea urchin embryos.[28] There are several hundred gene copies for each histone in the sea urchin which code for the early histone genes. All the genes in the unit are transcribed from the same DNA strand. The same gene order has been maintained over the 180 million years of evolution of the sea urchin species.[22] These copies are tandemly repeated and are highly conserved within a species, suggesting that they are constantly undergoing correction by gene conversion. These genes are expressed in early development when large amounts of histone mRNA are required. In addition, there are a number of genes coding for the histones expressed later in development.[22] These are organized in a manner reminiscent of the yeast and mammalian histones with the genes dispersed but still linked.[29,30] There are a limited number (10 to 20) of genes in this class coding for each histone. The H2a and H2b

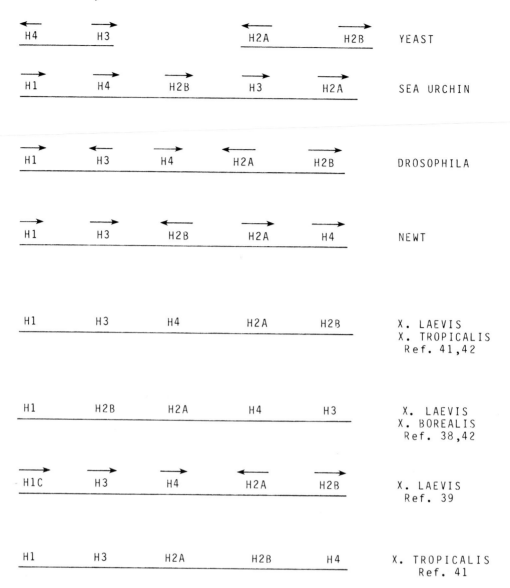

FIGURE 1.   Examples of repeated histone gene clusters. The repeated histone gene clusters from different species are shown. The direction of transcription is shown where it is known. The H1 gene is placed arbitrarily at the far left of each repeat. The distances between genes are not drawn to scale. Where there are identifiable gene pairs like the H2a-H2b or H3-H4 pairs in yeast and *Drosophila*, they are indicated by brackets.

proteins differ in several amino acids among the early and late proteins but the H3 and H4 proteins have the same amino acid sequence.[29,31]

In *Drosophila* the genes are also present in tandemly repeated units.[32] These repeats differ from the sea urchin in that both strands contain coding regions, as if the two different pairs, H2a-H2b, H3-H4, present in yeast had been linked to each other and then tandemly repeated. Unlike the sea urchin these genes seem to code for all the histone mRNAs in all *Drosophila* cells. There have been no other histone genes isolated and these genes are expressed both in early *Drosophila* embryos[33] and in *Drosophila* tissue culture cells.[34] Cytogenetic studies localize all the *Drosophila* histone genes to a single locus,[35] suggesting all the repeats are present in one large tandem structure.

Histone genes have been studied in two amphibian species, newts and frogs. The newt,

Notophthalmus viridescens, is another organism with several hundred homogeneous repeated gene copies for histones,[36] although here the repeats are interspersed with satellite DNA with the result that some repeats are very far (>20 kb) from others.[37]

The organization in the frog *Xenopus,* another primitive vertebrate which has been extensively studied, is more complex. The repeat structure in *Xenopus* is a relatively rapid evolving structure. In *Xenopus laevis,* there are at least three types of repeat, with different gene orders.[38-42] There are substantial differences in repeat length and restriction maps between individuals.[39,40] In contrast, in a closely related species, *Xenopus borealis,* there is a single major repeat which is found in all individuals.[39] The repeat unit is large (>16 kb) and contains 1 copy each of the 5 histone genes. Interestingly, the gene order is not the same in the newt[36] and the frog (Figure 1). In addition to the major repeat there are other genes which are separate from the repeat, probably in dispersed clusters similar to that found in the late sea urchin histone gene and mammalian histone genes.[39] Because of the size of the repeat it is not yet known whether these are tandemly repeated units.

Woodland and co-workers[39] have suggested that the major repeats in *Xenopus* are evolving rapidly, presumably by amplification of a cluster containing five histone genes which has diverged in the spacer region to give altered restriction enzyme maps in some clusters. Different clusters (with different gene order) may have been amplified and selected for since they each contain one copy of each histone gene. In some species (e.g., *Xenopus laevis*) more than one cluster has been amplified, while in others (e.g., *Xenopus borealis*) only one cluster was amplified. Since these two species diverged recently (8 to 10 million years ago), the amplification and subsequent divergence must have occurred recently. This is in direct contrast to the situation in sea urchins where the same gene order and cluster size has been maintained over a long period of time and the repeated cluster itself is probably evolving by mutation coupled with gene conversion. There is one sea urchin species, *Lytechinus pictus,* which has three types of repeat unit;[43] however, the gene order is the same in all repeats, although the spacer regions have diverged considerably and the repeats probably represent a splitting of the original repeat in *L. pictus* rather than independent duplication of another cluster of histone genes.

The repeated genes in *Xenopus* are probably expressed during oogenesis when there is an accumulation of large amounts of histone mRNA and hence serve the same purpose as the repeated genes in sea urchins: to provide histones for the rapidly cleaving early embryo. The low copy number genes are probably expressed at other developmental stages, although the time of expression for any of the genes has not been determined.

The dispersed cluster type of organization is typical of that found in many higher eukaryotes. The histone gene organization in mouse,[44,45] humans,[46,47] and chickens[48,49] is similar; there are clusters of genes probably containing all five histone genes. In humans[50,51] and chickens[52] H1 genes in the histone clusters have been isolated, but they have not yet been conclusively identified in mouse histone gene clusters. These clusters are characterized by a random gene arrangement. There is no pattern of neighboring histone genes and the genes may be coded by either strand of the DNA. For example, in the mouse, there is an adjacent H2a and H3 transcribed from the same strand close to an H3 and H2b, which are oriented divergently. In some human clusters there are H3-H4 pairs and in the chicken H2a-H2b pairs similar to those found in yeast. However, genes for these same proteins are also found associated with genes for other histone proteins (e.g., H2b-H3 pairs or H2b-H4 pairs). A similar disjointed gene organization is also found in other eukaryotes, including the late histone genes in sea urchins[29] and the histone genes in a higher plant, wheat.[53]

In addition to the jumbled orientation, at least in the mouse, there is more than one cluster which contains closely related histone genes. One is located on chromosome 3 and the other on chromosome 13.[54] Both clusters presumably contain genes for all the histones. All the histone genes in these clusters have not yet been isolated from any species so the overall organization is not yet known and it is possible a pattern will emerge once the overall organization is known.

A common feature of all these species is that the multiple genes for the same histone code for identical protein sequences but nonidentical mRNAs. Thus, the untranslated regions both 5' and 3', vary among the mRNAs for the same protein. In addition, there is microheterogeneity in some of the protein sequences due to single amino acid changes which have not been detected by protein sequencing, presumably because any one gene only codes for a small minority of the histone protein.[44,55]

This difference in mRNA sequences among the genes distinguishes the repeated gene sets from the jumbled clusters. The early sea urchin genes are found in about 7 kb tandem repeats with all 5 genes in the same orientation and gives rise to 5 mRNAs, 1 for each histone. Not only the mRNA sequences but the spacer sequences are highly conserved within the repeat among individuals in any sea urchin species.[56] It should be noted that the jumbled organization of the late genes is present in the same organism. The repeated structure implies that gene conversion involving the entire repeat unit must occur at a relatively high frequency to maintain homogeneity within a species. The jumbled clusters obviously do not undergo the same gene conversion events and have evolved separately from each other.

The gene organization described here covers only some of the histones in each species. The other genes have not been isolated. Thus, there are tissue-specific histone sets (e.g., those active in spermatogenesis) in sea urchins, mammals, and presumably other organisms which have not yet been identified. The histone genes isolated from mice and humans code for the replication set of histone variants, while those from chickens presumably do. There are also examples of genes which have been isolated for replacement or tissue-specific variants. The genes may not be linked tightly with other histone genes. Histone H5 (an H1-like histone) found in nucleated erythrocytes is not closely linked with other histone genes,[57] while the H1 genes are present in the jumbled clusters and tandem repeats. A "replacement" variant H3.3 gene in the chicken is also not closely linked with other histone genes.[58]

Obviously, selective pressure has maintained close linkage of histone genes in all species studied thus far. In higher eukaryotes and in the late sea urchin genes the linkage of the different clusters isolated thus far has not been conclusively demonstrated but seems likely. This is in contrast to the situation in other coordinately expressed genes (e.g., α- and β-globin and egg white proteins) where the genes have not maintained close linkage. While the tandemly repeated structure is used in some species it seems to be a special case which evolved to allow very rapid synthesis of large amounts of mRNAs necessary in early development of some species. The maintenance of close linkage between genes for different histones probably reflects the coordinate control of these genes, possibly by a single enhancer element in the cluster.

Characteristic of histone genes is the lack of intervening sequences. There are two examples of histone genes which contain intervening sequences, the H3 and H4 histone genes of *Neurospora*,[27] which may code for all the *Neurospora* H3 and H4 protein, and the H3.3 replacement variant gene from chickens.[58] The intervening sequences in the *Neurospora* and chicken H3 genes are located at different sites of the protein.

Histones are obviously very ancient genes and since it is thought that during evolution genes may lose intervening sequences it is possible that the primordial histone genes did contain intervening sequences. These have since been lost. Since histone genes have to be expressed in high levels in all cells it is possible that the genes may be expressed more efficiently and rapidly if they have lost the intervening sequences (and the polyadenylation sites). The finding of intervening sequences in an H3.3 gene, which does not have to be expressed in large amounts, is consistent with this possibility. It will be of great interest to examine the gene structure of replacement variant and minor histone variant genes in other species to see if they have also retained intervening sequences.

III. EXPRESSION OF HISTONE GENES

The histone genes are transcribed by RNA polymerase II.[59,60] None of the genes described thus far, except the histone genes of *Neurospora*,[27] and the gene for the chicken H3.3,[58] a replacement variant histone, contain intervening sequences. The histone mRNAs of yeast[61] and *Tetrahymena*[26] are polyadenylated, while those of other species, with the possible exception of the histone mRNAs found in *Xenopus* eggs which contain short A stretches,[62] are not polyadenylated. The histone mRNAs which are not polyadenylated all have a characteristic stem-loop structure at the 3′ end of the mRNA.[21] The genes share the 5′ promoter elements common to other genes transcribed by RNA polymerase II, the TATA, cap and "CAAT" boxes.[21] Generally, the mRNAs contain short untranslated regions less than 50 nucleotides at the 5′ end and less than 80 at the 3′ end, which, coupled with the lack of poly A, makes both the histone mRNAs and their genes among the smallest in the cell.

The principal advantage of having the genes organized in a repeat structure is that the gene number is kept the same for each of the histone genes and the sequence can be maintained by gene conversion. In addition, the genes may be more readily coordinately controlled. Repeat structures are common for many genes which are present in multiple copies in the genome. In yeast, with only two copies of each histone gene, selection can maintain the protein sequences, but even here the protein sequences of the H2b and H2a genes differ slightly and either one of the H2b genes or H2a genes (and hence proteins) is dispensable.[63,64]

In the higher eukaryotes, with jumbled clusters, the genes are also coordinately expressed. It seems likely that if gene conversion operates on these genes it occurs much less frequently and only involves the small coding region targets. As a result the proteins are probably not as homogeneous as once thought. As more genes are sequenced, amino acid substitutions are found which presumably are present only in a small fraction of that histone protein. These alterations probably lead to functional proteins and some of them must be selected for, since the same replacement protein sequence variants are found in many different species, including mammals and birds. It seems likely that the histone coding regions are still under extreme selection even in the case of genes which are apparently expressed at very low levels.

IV. PROBLEMS IN THE STUDY OF HISTONE GENE EVOLUTION

The time course of divergence of different genes can be estimated using an evolutionary clock.[65] At the DNA level a gene will typically show two classes of mutations: (1) those that may not be under selection (3rd base changes and changes in intervening sequences or flanking regions) and (2) those that lead to amino acid sequence changes and hence are under strong selective pressure in the case of histones. The rate of each of these changes treated separately may be related to the separation in evolutionary time of the sequences. Nonselective (or neutral) mutations in 3rd bases or in noncoding regions occur at the same rate per nucleotide as changes in the amino acid sequence, but may not be selected against and thus accumulate faster.[66,67] The noncoding sequence changes which are truly neutral should be useful for getting distance on a relatively short evolutionary time scale and may be similar for all genes. (Of course, some of the noncoding sequences may have a function subject to selection and hence vary at a different rate from the "neutral" changes and changes in the coding regions.) The noncoding regions are also subject to other changes such as deletions and additions which may lead to large divergence in a rather short time. Mutations in the coding region which do not alter the amino acid sequence may be useful for following the evolution of genes over a long time frame, making interpretation more reliable since the basic structure of the gene imposed by the triplet code allows one to identify all the changes. However, there is much more scatter in the accumulated data for the silent changes, than in that for replacement changes.[66,67] The degree to which these may be sensitive to selection is not known. Changes in amino acid sequence occur at a different rate for each protein based on the structural requirements of the protein.[65] This rate is essentially 0 for histones (<1 amino acid substitution per 100 million years[70]), so replacement changes cannot be used as a clock. There is as much variation within a species as between species.

Evolutionarily related genes can be divided into two categories: orthologous genes and paralogous genes.[65] Genes that are orthologous have a direct evolutionary relationship, while those that are paralogous are related but are on two different branches of the gene family. Depending on the context of the comparison however, two genes may either be orthologous or paralogous to each other. Thus, when comparing the hemoglobin genes among species, the α-like genes form one orthologous gene set while the β-like genes from another orthologous gene set. Any β gene is paralogous to any α gene. However, if one were comparing the hemoglobin genes from a number of species with the myoglobin genes, then the α genes and βgenes would be orthologous in this context and any myoglobin gene would be paralogous to an α or β hemoglobin gene.

A major problem with evolution of a multigene family is the ability to identify which genes are orthologous and which are paralogous between species. Thus, within sea urchin species the early genes are orthologous to each other and an evolutionary relationship can be defined by comparing these sequences.[71] The late genes are also orthologous to each other and a similar relationship can be defined by comparing the coding regions of these genes.[72] However, individual late genes may not necessarily be compared between species (except in the coding regions) until the orthologous pairs can be identified. Thus, evolution of a number of divergent gene copies could proceed by each copy evolving independently. In this case two closely related species might each have 20 copies of a gene, each copy having its counterpart in the other species which would be more similar than other genes in the same species. Alternatively, the 20 genes in each species could undergo conversion followed by divergence, in which case all the genes in one species would be more similar to each other than to the genes in the other species. Therefore, defining the pattern of evolution of the late sea urchin histone genes may only be feasible once all the late genes are isolated and one can get an idea of whether the late genes spread separately within a species or each has its orthologue in other sea urchin species. Similar problems exist interpreting the evolution of histone genes in all higher eukaryotes. Among rodents, S1 nuclease mapping of histone genes using mouse probes reveals that there are similar genes in the hamster and rat which share similarity to a particular mouse gene in the untranslated regions, suggesting that each mouse histone gene may have its individual counterpart in the other species.[73]

The problem is more acute comparing sequences across different phyla. With which of the sea urchin genes, if any, are the genes in the frog or chicken orthologous? Did two or more independent gene sets evolve before sea urchin and are their descendents present in all higher animals? Were these two sets extremely ancient (e.g., are they descended from the genes of the most primitive eukaryotes as reflected in yeast) and are their descendents in all eukaryotes? Or did the splitting into two or more gene sets occur independently over and over again in evolution? Since we only have a small portion of the total histone gene complement of all organisms except yeast, it is impossible to definitively answer these questions. I will try to analyze the available evidence in terms of the questions I have asked above.

V. EVOLUTION OF DNA SEQUENCES FOR THE CODING REGION

The histone genes provide an excellent opportunity to study the changes in nucleotide sequences that can occur without altering the protein sequence. Since H3 and H4 proteins vary by no more than 4% in amino acid sequences among different species,[5,6] the comparison of nucleic acid coding sequences may show the range of sequence changes at the nucleotide level allowed in evolution. It may be possible to identify sequences necessary for mRNA function. In addition, since there are multiple nonallelic genes in many species, the changes that may occur within a single group of coordinately expressed genes may be defined. Of particular interest is the probability of their being a large number of paralogous histone genes which have been evolving independently for a long time within a single species. One should also be able to trace an orthologous set of histone genes as it has evolved in much the same way that the globin gene structure has been analyzed.[67]

There is concern that the great stability in histone protein sequences might be due to special mechanisms which restrict the mutation rate in these genes. The studies of the sea urchin early histone genes by Birnstiel and Weinberg and their co-workers[56,71] has definitely ruled out this possibility. They have shown that even in a tandemly repeated gene cluster (subject to gene conversion to maintain stability within a species) there is the same rate of nucleotide change ($0.5\%/10^6$ years) as is found for globin genes in mammals or total unique sea urchin DNA.[71] Yager et al.[56] have shown the expected amount of polymorphism in nucleotide sequence in both spacer and coding regions among individuals within a species in the *S. purpuratus* early histone genes. Thus, the histone genes do not exhibit stability due to a low mutation rate, but must maintain the stability due to selection.

Since deducing the evolutionary relationships of the histone genes will rely primarily on silent nucleotide changes, the information available for the sea urchin genes also provides critically important assurance that this method will be applicable. Birnstiel and associates have collected the data to show that this is true for the sea urchin early genes over a period of 180 million years (Figure 2).[71] In addition, the late histone genes, which are present in each urchin species, vary from the early genes by much more within a species than the early genes do between species[72] (Tables 1 and 2).

The divergence in silent substitutions is not as consistent as the divergence in coding sequences. In the globin sequences, for example, the silent changes between human and rabbit was much lower than any other mammalian pair, although all the replacement substitutions gave similar results.[66] This variability makes conclusions drawn by comparing any two sequences suspect. Unfortunately in some cases this is all the data available; however, in some cases there are multiple sequences for the same gene within a species or sequences of clearly linked genes (e.g., late sea urchin H3 and H4) between species. Averaging across this group of data will give much more reliable results. Variability (scatter in the data) could be due simply to statistical variation or to other factors operating convergently in two different species, (e.g., similar codon preferences) as seems to be the case in the chicken and mouse. Any conclusions drawn will be based on a network of consistent data as often as possible. In no case is the data as yet complete enough to rule out alternative explanations.

A number of sequences of histone genes from a diverse set of species have been determined and in several cases independent genes from the same species have been sequenced. The lack of extensive data from more closely related species and the existence of multiple, independent (paralogous) gene sets prevents a detailed map of histone gene evolution. Nonetheless, the data available confirm the existence of independently evolving gene sets in a single species. The constancy of the amino acid sequences also allows one to identify nucleotide sequences which have been conserved, implying selection operating on the mRNA structure. The conclusions are limited by the lack of knowledge of the expression of many of these genes, whether they are expressed only in particular tissues or particular develop-

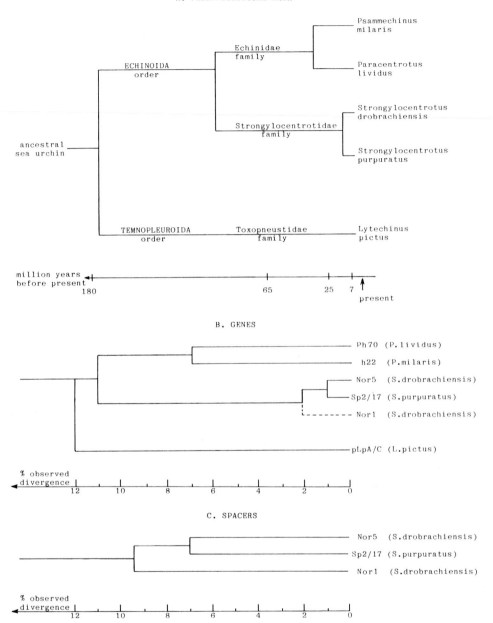

FIGURE 2.   Evolution of early sea urchin histone genes. The divergence of the early histone genes of sea urchin is plotted for five sea urchin species. The fossil record is shown in (A). The percent nucleotide change for coding and spacer DNA is shown in B and C. The data is taken from Busslinger et al.,[71] except for the data on *L. pictus* which is taken from Roberts et al.[72a]

mental stages. The sole exception is the sea urchin where the paralogous early and late histone genes have been well-documented at the protein, RNA, and DNA levels.

Many single-base mutations can occur which do not alter the amino acid sequence of the protein. There are four classes of these.[66] The most common are 3rd base changes in codon sets in which the amino acid is determined by the first two nucleotides and the third nucleotide can be any of the four. A related situation is for isoleucine where there are three codons,

Table 1
SILENT SITE DIVERGENCE IN DIFFERENT GENES IN
SAME SPECIES

			Divergence (%)	A.A. changes
Mouse H3	MH3-1	MH3-2	13	1
	MH3-1	MH3-3	19	1
	MH3-3	MH3-2	15	0
Chicken H2b				
CKH2B-1	CKH2B-2		9	1
Chicken H3				
CKH3-1	CKH3-2		116	4
Xenopus H4				
XEN H4-1	XENH4-2		49	0
Human H4				
HU4-1	HU4-2		98	0
Human H2B				
HuH2B-1	HuH2B-2		53	3
S. purpuratus H3 and H4				
H3 early	H3 late		>100	0
H4 early	H4 late		>100	0
Yeast (all 4)				
YH2A-1	YH2A-2		39	2
YH2B-1	YH2B-2		59	4
YH3-1	YH3-2		34	0
YH4-1	YH2		7	0
Xenopus H1			2	11

Note: The sequences of the coding regions of histone genes isolated from the same species are compared to get the percent divergence in the replacement (silent) sites as described by Perler et al.[66] For consistency the genes are denoted by an abbreviation for the species, H1, etc., for the type of histone and number 1, 2, or 3 to denote independently isolated genes. The MH3-1 and MH3-2 genes are the mouse H3.1 and H3.2 genes described by Sittman et al.[44] The chicken H2B-1 gene is from Grandy et al.[76] and the CKH2B-2 gene was isolated by Harvey et al.[55] The CKH3-1 gene codes for the replication variant gene and the CKH3-2 gene for the replacement (somatic) gene isolated by Engel et al.[58] The *Xenopus* XenH4-1 gene was isolated by Moorman et al.[81] and the XenH4-2 gene was isolated by Turner et al.[82] The human HuH4-1 gene was isolated by Sierra et al.[78] and the HuH4-2 gene was isolated by Zhong et al.(79). The human HuH2B-1 gene was isolated by Zhong et al.[79] and the HuH2B-2 gene was isolated by Marashi et al.[80] The sequence of the early H3 *S. purpuratus* genes was from Sures et al.[87] and the early H4 gene from Grunstein and Grunstein.[88] The sequences of the late H3 and H4 genes are from Kaumeyer and Weinberg.[72] The yeast sequences are from Wallis et al.[14] (H2A) and Choe et al.[15] (H2B) and Smith and Andresson[74] (H3 and H4). The *Xenopus* H1 genes were isolated by Turner et al.[39] Most of these genes code for nonallelic protein variants and the number of amino acid changes within a species is also shown.

AUA, AUU, AUC, with the variation again in the third base. The other two classes are situations in which only two nucleotides can be found in a particular position and still specify the same amino acid. This occurs in a number of codons where the third base can be either purine (AAA, AAG-lysine) or either pyrimidine (AAU, AAC-asparagine). The first base of some codons can be changed without changing the amino acid in two cases (UUG, UUA, and CUX for leucine and AGA, AGG, and CGX for arginine). Finally, there is one case in which one cannot reach the alternate codons for the same amino acid by a series of single nucleotide changes without altering the amino acid in the process. This is in serine codons (UCX, AGC, and AGU). Therefore, if the protein sequence is totally conserved, a change from a UC to an AG codon should not occur. This conclusion must be tempered by the observation that a possible pathway for this change would involve an intermediate ACX codon and hence a serine → threonine substitution which could be conservative. With this

Table 2
SILENT SITE
DIVERGENCE IN H3
GENES

Mouse — CKH3-1	64
Mouse — HU	90
Mouse — Xen	57
Mouse — *L. pictus* (late)	67
Xenopus-CKH3-1	106
HU — CKH3-1	116
L. pictus-CKH3-1	136
HU — Xen	118
L. pictus — HU	124
L. pictus — Xen	57
S. purpuratus (early) — Mouse	135
S. purpuratus (early) — CKH3-1	150
S. purpuratus — Ps. mil. (early)	63
S. purpuratus — L. pictus (late)	30

Note: The replacement site divergence for the H3 genes from different species was calculated as described by Perler et al.[66] For the mouse sequences, the divergence is the average divergence seen for the three different mouse H3 genes compared with the indicated gene. The human gene was from Zhong et al.,[79] the *Xenopus* gene from Moorman et al.,[81] the *L. pictus* gene from Childs et al.,[29] and the *Psammechinus milaris* (Ps. mil.) early H3 gene from Schaffner et al.[89] The other genes are the same as in Table 1.

background in mind I will review the current published sequences and compare them with regard to evolutionary divergence using the method of Perler et al.[66] The assumptions behind the calculations include the postulate that there are random mutations and that mutations that do not change the amino acid sequence are neutral. Thus, other factors, including mRNA structure and codon preference will alter the results, particularly in the class I (two nucleotide possibilities) codons. The percent divergence (assuming a Poisson distribution of random mutations) after correcting for multiple events has been calculated by the method of Perler et al.[66] Note that this method often yields values greater than 100%, and for completely random variation the divergence would be infinite. Comparison of sequences among the vertebrates should give some indications of possible orthologous genes from different species. The studies of Efstradiatis and co-workers[67] on the globin and insulin genes[66] have shown that the silent site replacements can serve as measures of distance in evolution with a large scatter in the data. Silent sites diverge in an apparent biphasic way over time. First, there is a rapid divergence probably due to variation of many sites (no selection), followed by a slower divergence due to selective pressure to maintain the mRNA sequence. It is not possible to know *a priori* the rate of the replacement site or silent site divergence. However, the silent site divergence for globin genes extends to about 100% at 500 million years. There is reason to expect that the possible relationships between chicken and vertebrate histone genes and possibly as far back as *Xenopus* could be detected by these comparisons. The significance of these possible relationships would be strengthened if linked genes in both species show similar divergence. However, since the protein sequence is conserved identification of orthologous genes is difficult.

VI. COMPARISON OF CODING REGION SEQUENCES WITHIN A SPECIES

Table 1 compares the changes in silent nucleotides in the cases where more than one gene for the same histone protein have been isolated in a single species. The amount of divergence varies dramatically from a few percent for chicken H2b and mouse H3 genes to extensive variation between the two human H4 genes isolated and the early vs. late sea urchin genes.

In one species, yeast, all of the histone genes have been sequenced[14,15] and there are two copies of each gene. The histone genes of this simple eukaryote, with a total of eight histone genes, exhibit many of the properties of histone genes in all organisms. There are two nonallelic copies of each of the four histones. The gene sequences are sufficiently different in nucleotide sequence to conclude that these genes have been separated in evolution for a long time. The H2b genes differ in over 10% of the nucleotides, including 6 replacement changes leading to 4 amino acid changes and an AG → UC change in a serine codon. The H2a genes differ in a smaller number of nucleotides, about 5% of the total, including two replacement changes. The H2b genes are more divergent than the H2a genes, although both sets of genes have diverged substantially (Table 1).

The yeast H3 and H4 genes present a different picture (Table 1). The amino acid sequences predicted from the DNA sequence are identical. The H3 genes show substantial silent nucleotide changes as did the H2a and H2b genes. The H4 genes vary dramatically from the other genes. There are only 4 nucleotide differences among the 2 H4 genes, which are tightly linked to the H3 genes). Thus, the H4 genes either have a stronger selective pressure on the nucleotide sequence or have recently been homogenized by gene conversion. Gene conversion seems the most likely explanation although the gene conversion "unit" must be small since the H3 gene, which is within 800 bases, was not affected.[74]

These nonallelic variant H2a and H2b genes apparently do not have separate functions since yeast grow normally with only one of the two gene copies.[63,64] This situation is similar to that emerging for higher eukaryotes. The presence of nonallelic genes (some with amino acid variants) and conserved coding regions, suggesting possible gene conversion, and different degrees of variation among copies of genes coding for the same protein, are all common features of the histone genes of higher eukaryotes (see below).

In the vertebrate histone genes, sequences are available for at least one of each core histone gene from *Xenopus*, chicken, human, and mouse; for some histones from several species more than one sequence is known. To give an idea of the scarcity of data, there are probably at least 20 to 50 copies of each gene in each species, and with the possible exception of *Xenopus*, these genes are not in a cluster obviously subject to gene conversion. Hence each gene may well have a different sequence. I will first review the multiple sequences for a gene within a species and then compare the core histone sequences across species. Three examples of obviously closely related genes within a species are known.

Three H3 genes from the mouse are compared in Table 1. Two of these genes are closely linked (5 kb apart) on chromosome 13 and the third (H3[3,75]) is present on chromosome 3.[54] The sequences vary 13 to 19% in the coding region (replacement sites). The mouse H3 genes differ in one amino acid, the reported H3.1 and H3.2 protein variants.[7,8] A similar degree of divergence is seen among two chicken H2b genes. The chicken H2b genes also differ in a single amino acid.[55,76] This amount of divergence in the mouse H3 and chicken H2b genes is similar to that seen in other genes which have arisen by recent duplication within a species (the human β- and δ globin genes,[67] the mouse β-major and β-minor,[67] and the rat preproinsulin genes[66].) The percent change in these genes suggests that these genes have diverged (within the species) within the last 40 million years. This interpretation leads to several paradoxes. First, the mouse genes are in different clusters on different chromosomes. This would determine the time when the chromosomal rearrangement occurred, making the shift of some of the mouse histone genes to another chromosome a recent event. The genes on chromosome 13 are not significantly more divergent from the genes on chromosome 3 than they are from each other. The three genes must have started diverging from each other at about the same time. Secondly, these genes code for different proteins, H3.1 and H3.2, each of which is found in all mammals. The rodents have more H3.2 than H3.1 protein while other mammals have more H3.1 than H3.2. It is hard to imagine a mechanism whereby H3.1 genes continue to evolve from H3.2 genes (or vice-versa) on an evolutionarily rapid scale. Finally, while the coding regions of these genes are very similar, the untranslated regions and flanking regions are totally dissimilar, except for the 3' end of the mRNAs which are similar among all mouse histone mRNAs. This is not the case in the rat preproinsulin gene[66] or in the β-major, β-minor, or β- and σ globin genes,[67] where the flanking and intervening sequences show homology consistent with the inferred divergence time of the coding regions. Either the mouse genes are subject to gene conversion limited exactly to the coding region (a model hard to reconcile with the different protein sequence) or the selective pressures on the nucleotides in the coding region are extreme. If this is the case and these genes have been evolving separately since the mammalian radiation, then very similar genes should exist in other mammalian species. As yet no such H3 gene has been isolated from humans. However, S1 nuclease analysis of human mRNA using mouse H3 probes detects some human mRNAs with very similar sequences in the coding region to the mouse genes, so some genes of this type may exist in human DNA.[77] S1 nuclease mapping of hamster and rat mRNAs using mouse histone probes definitely detects genes orthologous to one of the mouse histone H3 and H2a genes even in noncoding regions. Among these rodents at least individual genes are evolving independently.[73]

Less definitive conclusions can be made with regard to the chicken H2b genes; however, they show the similar characteristic to mouse H3 genes of very little flanking region homology.[55] There is not enough sequence information on chicken proteins to know the significance of the single amino acid change predicted by these sequences. It is also not clear if both these genes are expressed.

A situation exactly analogous to the mouse H3 and chicken H2b genes exists in the late histone H3 and H4 genes of the sea urchin *Lytechinus pictus*.[72a] There are multiple copies of the genes. The coding regions of two of these genes are very similar to each other (<5% change). The flanking and untranslated regions of the gene are extremely divergent. Thus, the pattern of a set of genes with highly conserved coding regions and very divergent flanking and untranslated regions is present in three widely varied species (mouse, chicken, sea urchin) and will probably be characteristic of at least one set of histone genes in all higher eukaryotes.

There are other examples of more than one gene being isolated in a single species, but in these cases there are not large homologies between the genes. In sea urchins the genes expressed in early and late stages have been isolated. These H3 and H4 proteins are identical in sequence, although the mRNAs are very different.[29] These genes provide an excellent

Table 3
SILENT SITE
DIVERGENCE IN H4 GENES

Mouse — HU4-1	47
Mouse — HU4-2	86
Mouse — Xen 4-1	62
Mouse — Xen 4-2	90
CK — Mouse	66
CK — Xen 4-1	62
CK — Xen 4-2	102
CK — HU4-1	87
CK — HU4-2	55
Mouse — Wheat	73
Mouse — *L. pictus* (late)	90
Mouse — *S. purpuratus* (early)	152
L. pictus — *S. purpuratus* (late)	40
L. pictus — *S. purpuratus* (early)	40

Note: The silent site divergence for the H4 genes from different species was calculated as described for Perler et al.[66] The genes used are those in Table 1. In addition, the mouse gene sequenced by Seiler-Tuyns and Birnstiel,[84] the chicken gene isolated by Sugarman et al.,[52] the wheat gene sequenced by Tabata et al.,[53] and the sea urchin genes from Grunstein and Grunstein[89] (early genes), Childs et al.[29] (late *L. pictus*), and Kaumeyer and Weinberg[72] (late *S. purpuratus*) were analyzed. The divergence of the late sea urchin gene has been estimated from the nucleotide differences.

example of genes which have evolved separately for a very long time. The late H3 and H4 genes from *Strongylocentrotus purpuratus* differ from the early genes by 17.3 (H3) and 18.2% (H4) of total nucleotides, which yield a corrected percent divergence of silent sites of about 100%, as large as is observed in any pairwise comparison of histone H4 gene sequences (e.g., sea urchin—plant, human—frog) (see Table 3). Comparing early genes of one sea urchin species with early genes of another sea urchin species show divergence of about 8 to 10% of total nucleotides and similar numbers are found (6.1% -H3, 8.1% -H4) for the differences among late genes of *Strongylocentrotus purpuratus* and *Lytechinus pictus*.[72] In addition to the nucleotide changes the sea urchin early and late H3 and H4 genes show 3 AG → TC codon changes in serine codons, further suggesting the different ancient lineage of these gene sets.

The expression of the genes isolated in other species has not been as well documented so that different functions cannot be ascribed to these genes. Two H4 genes have been isolated from human DNA[78,79] and these have diverged from each other in 98% of the silent sites, as much as the early and late H4 genes of sea urchins. Two human H2b genes[79,80] have also been sequenced and these also differ substantially (55%) from each other in the replacement sites as well as having several amino acid changes. Two complete H4 genes, XenH4-1 and XenH4-2, have been isolated from *Xenopus*.[81-83] These differ from each other in 48% of the replacement changes (silent substitutions). A third H4 gene (XenH4-3) from *Xenopus laevis* has also been partially sequenced[38] and varies from the other two in 15 (XenH4-1) and 30% (XenH4-2) of silent sites. Two of these genes, XenH4-1 and XenH4-3, are the most similar with differences like that of the mouse H3 and chicken H2b genes. XenH4-1 and XenH4-2 are somewhat similar and XenH4-2 and XenH4-3 are the most distant. The function of these genes is not known, although *Xenopus* presumably has different histone genes expressed in eggs and adult tissues, in addition to other possibly developmentally regulated histone genes. These three H4 genes are much more similar to each other than are the early and late sea urchin histone genes.

Two chicken H3 genes which are different in structure and function have been sequenced.[58] One has two intervening sequences and codes for a protein with similar amino acid sequence to mammalian H3.3, the replacement variant histone. The other lacks intervening sequences and codes for the replication variant histone. As expected these coding regions have diverged as much, if not more, than the early and late sea urchin histone genes. However, the position of the serine codons has been maintained. Since the metabolism of the replacement mRNAs is regulated differently from the replication variant histone mRNAs, there may be different selective pressures on the "silent" nucleotides and hence more changes have accumulated. It is more likely, however, that these genes are very ancient. This is indicated particularly by the different gene structure. In this respect it will be of great interest to compare sequences of the H3.3 genes from various species.

A final pair of genes for the same histone in the same species are the two H1 genes from *Xenopus laevis*.[39] These genes differ the least among sequenced histone genes (2% in the coding region). In addition, the flanking regions have been highly conserved (4% different in 400 nucleotides), directly flanking the genes. These differences are not much greater than those observed between individuals in the early histone genes of *Strongylocentrotus purpuratus*.[56] Despite this high degree of nucleic acid sequence homology, the proteins differ in 11 amino acids (5 single base substitutions, and 2 deletions or additions of 1 and 5 amino acids). There are only four silent substitutions. Variability in amino acid sequence is expected for the H1 protein, but this high degree of amino acid changes between very similar genes is again paradoxical. These H1 genes are of the same H1 subtype as assayed by gel electrophoresis. If this is a general finding then the H1 protein subtypes themselves may be quite heterogenous in protein sequence.

VII. COMPARISON OF HISTONE GENE SEQUENCES AMONG SPECIES

There are a number of different histone gene sequences known from different species. The comparison of these may help to identify genes which are related in different species. Since most of the mouse genes (all except H4) and most of the *Xenopus* genes (all except H4-2 and H4-3) sequenced are linked to each other or clearly from closely related clusters, it is possible to compare several genes from a cluster in each species. I argue that these two clusters may be orthologous. Comparing the H4 genes, most of the H4 genes are widely divergent (80 to 100% total change) when different species are compared (Table 2). This includes the expected widely divergent sea urchin and vertebrate genes and sea urchin and plant H4 genes.

Examination of silent changes among the H4 genes demonstrates the problem of interpreting the divergence due to silent site mutations. While most of the genes show very large divergence, there are several examples where the divergence is much less. For example, the mouse[84] and human gene H4-1 show only 47% divergence of silent sites, while the mouse and human H4-2 have diverged 98%. The chicken and mouse gene show only 66%

divergence; however, the human H4-1 and chicken gene show 87% divergence, while the chicken and human H4-2 gene show 55% divergence. These results suggest an evolutionary tree where the chicken and human H4-2 are orthologous and the mouse and human H4-1 are orthologous. Comparing the chicken, human, and mouse genes with the *Xenopus* genes shows that the one *Xenopus* gene, XenH4-2, has diverged over 100% from all the other H4 genes, while another, XenH4-1, is only 62% different from mouse and chicken and 85% different from the two human genes. These data imply that the mouse and chicken H4 genes are orthologous to the *Xenopus* XenH4-1 gene and hence to each other, in direct contradiction to the conclusion reached by comparing chicken, mouse, and human genes with each other. A possible answer is that none of these genes can be judged orthologous simply based on these sequence comparisons. These genes represent too small a sample and there is not enough data from intermediate species to draw firm conclusions. Nevertheless, the silent changes among the mouse, chicken, and human H4-2 are of the same magnitude as those observed for the orthologous globin genes, so some of these genes may be orthologous. In particular, since the linked *Xenopus* H2b genes are orthologous with the mouse genes (see below), the mouse and *Xenopus* XenH4-1 gene may be orthologous. This conclusion depends on the mouse H4 gene being closely related to the other mouse histone genes, an excellent possibility since the codon usage is so similar.[44] The one definitive conclusion that can be drawn is the two human H4 genes are extremely divergent from each other (much more than pairwise comparisons of either human gene with mouse or chicken), proving that they have been evolving independently for a long time and must come from two paralogous gene sets.

The H3 genes present a similar picture (Table 3). Comparing all the H3 genes among species the three mouse genes are similarly distant from the H3 genes from each of the other species, as expected for the recently diverged mouse histone genes. They differ 64 ± 6% from the chicken gene, 90 ± 10% from the human gene, and 57 ± 3% from the *Xenopus* gene. The other pairwise comparisons show that the chicken and human genes are both very distant from the *Xenopus* genes and from each other, as distant as the early and late sea urchin genes or the replication and replacement chicken genes. These data suggest that the mouse and *Xenopus* H3 genes are orthologous and that the rate of substitution of silent changes may be relatively low for these genes because of common selective pressures, possibly due to common regulatory properties. This conclusion must be considered tentative, but is consistent with the relationships between the other mouse and *Xenopus* genes. In contrast, the other H3 genes are not orthologous with the mouse or with each other.

There is less data available for the H2a and H2b genes, but the data show a similar pattern (Table 4). Here one can clearly detect an orthologous set of genes, the sequenced H2b genes from all the vertebrates. The identification of these as an orthologous set of genes is strengthened by the fact that linked H2a-H2b genes have been sequenced in two species and show similar divergence and that the linked *Xenopus* genes and the mouse genes all show similar divergence. The mouse H2a gene[85] is very similar to the chicken H2a gene, a maximum of 39% variation in silent sites. These genes are more similar than the orthologous early sea urchin genes from two different species which vary by 55% and about as similar as the late histone genes from *Lytechinus pictus* and *Strongylocentrotus purpuratus*.[72] Because of the extreme similarity in codon usage in the chicken and mouse genes (see below), this is certainly a minimum divergence; however, the H2a and H2b genes are much more similar between chicken and mouse than the H3 and H4 genes which are not orthologous, but show the same codon bias. In contrast the other pairs of H2a genes (mouse-human,[79] mouse-*Xenopus*,[86] chicken-human,[55] chicken-*Xenopus*, or human-*Xenopus*) all show 80 to 100% divergence. In each case the divergence of the chicken and mouse from human or *Xenopus* is the same as expected for othologous genes. In contrast the human-*Xenopus* comparison shows more divergence from the mouse or chicken-*Xenopus* comparison, consistent with the human H2a gene not being orthologous with the mouse or chicken H2a gene.

Table 4
SILENT SITE DIVERGENCE IN H2a AND H2b
GENES

	H2b	Genes	H2a	Genes
CK — mouse	42	(25)	39	(23)
Mouse — HU-1	41		81	
Mouse — HU-2	44			
CK — HU-1	54		88	
CK — HU-2	58			
MU — XEN	75		97	
XEN — HU-1	78		99	
XEN — HU-2	118		153	
CK-XEN	80		99	
S. purpuratus — Ps. mil.			55	
(early)				

Note: The silent site divergence for the H2a and H2b genes from different species was calculated as described by Perler et al.[66] For the chicken genes the value given is the average of the divergence from the H2b-1 and H2b-2 genes. The actual divergence for the mouse and chicken genes is given in parenthesis and the divergence just considering the changes in the class of codons of the type GCX is given. Because of the extreme codon preference in these species there is only a single change in the AAPu or AAPy type codons. The mouse H2b gene was isolated by Sittman et al.,[44] the mouse H2a gene was sequenced by Specher and Marzluff,[85] the human H2a gene was sequenced by Zhong et al.,[79] and the *Xenopus* H2a and H2b genes were sequenced by Moorman et al.[86] The sequence of the early sea urchin genes was taken from Sures et al.[87] *S. purpuratus* and Schaffner et al.[89] *(Ps. mil.)*.

The H2b genes present a similar situation except all the sequenced H2b genes may be from a single orthologous set. Again the mouse[44] and chicken[55,76] genes are highly homologous with only about 40% silent substitutions. The mouse H2b,[44] human H2b-1,[79] and human H2b[80] genes are also more homologous than any other mouse-human gene pair, (41%) and the chicken H2b-human H2b genes show similar divergence (54%). The chicken genes may not have diverged from the mouse genes as much due to the selective pressure of codon usage. In addition, comparing four genes (two chicken, 1 human H2b-1, and 1 mouse) with the *Xenopus* H2b gene[86] shows that they are all $78 \pm 3\%$ divergent from the *Xenopus* gene. The human H2b-2 gene is more divergent from the *Xenopus* gene, but this may be due to the variation found in the "clock" defined by the silent site mutations. It may also be a member of the same orthologous set, since it is clearly as closely related to the mouse and chicken H2b gene as to the human H2b-1 gene.

Therefore, these five genes may share a "recent" (in vertebrates) common ancestral gene. The human H2b-2 gene has diverged equally from the mouse and chicken genes as the human H2b-1 gene. The H2b-2 is also as divergent from the human H2b-1 gene as it is from the mouse gene. It is possible that these two human genes duplicated around the time of the mammalian radiation and have evolved separately since. Thus, both these genes may be members of the same orthologous set of H2b genes viewing vertebrates as a whole, although they have diverged within humans and are paralogous within that species. Alternatively, there may not be individual human histone genes which are very similar to each other as has been found in chicken and mouse. These data strongly suggest that one will be able to detect orthologous gene sets in vertebrates and they certainly indicate that the lack of homology among the other genes is due to their arising from paralogous genes which had diverged from each other prior to the amphibians.

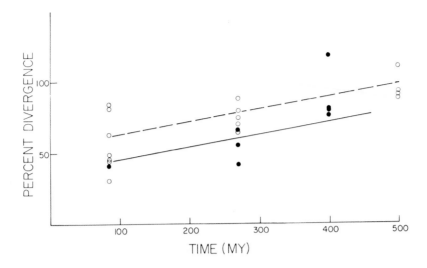

FIGURE 3.   Silent site divergence of vertebrate H2b genes. The divergence of the vertebrate H2b (Table 4) gene is plotted vs. the time of divergence. The percent divergence of the globin gene is plotted for comparison purposes. The line for the histone H2b genes was deliberately drawn parallel to the globin genes. Data points for histone genes (●) are taken from Table 4 and for globin genes (○) from Perler et al.[66]

The silent site divergence for the H2b genes is plotted in Figure 3. For comparison purposes the silent site divergence for globin genes is also plotted. While a number of lines could be drawn due to the scatter in the data, I have chosen to draw parallel lines. The globin gene divergence extrapolates back to 60% divergence, while the H2b genes extrapolate back to 35% divergence. One interpretation of this is that there are a large number (60%) of "neutral" silent sites in globin genes and a much smaller number of "neutral" silent sites in the histone genes. The other silent sites are under selective pressure.

The vertebrate histone genes sequenced thus far could have arisen from a single set of ancient histone genes, which have given rise to a number of gene sets which we now view as paralogous. The only data to examine in this regard are the serine codons. There are 24 serine codons in the 4 core histone genes; 19 are TCX and 5 are AGPy. In all the vertebrate genes sequenced, including the chicken replacement variant gene, the same type of codon is present for a particular serine; there are no TC → Ag transitions. In contrast, in the sea urchin there are 3AG → TC changes between the early and late H3 and H4 genes and in the yeast H2b genes there is an AG → TC change. To alter these codons requires two independent changes. There is an example of a serine → threonine change (ACX) in the H2a gene of many species[10,89] H2a.1 vs. H2a.2 protein variants, but with no example of the second substitution having occurred. It seems very unlikely that this pattern of serine codons would have arisen randomly many times independently, and thus there may have been a single primordial gene set for these histone genes. Regardless, some of the genes in a particular species have been separated from each other for a very long time, as long or longer than the time since the divergence of frogs from birds and mammals.

Despite the fact that one can identify orthologous genes, there are still paradoxes in some of the sequence data. Genes coding for nonallelic variants H3.1 and H3.2[44] and H2a.1 and H2a.2 have been isolated from the mouse.[85] These are very similar in their coding region but contain amino acid substitutions. The three changes which have been sequenced in the two sets of genes are an arg → lys in H2a.1 and H2a.2, a pro → ala in H2a.1 and H2a.2, and a cys → ser in H3.1 and H3.2 All could involve a single nucleotide change. Yet in all three cases there are multiple nucleotide changes, and in one case all three bases are changed.[44,85]

Table 5
YEAST HISTONE CODON USAGE[84]

TTT	6	TCT	50	TAT	12	TGT	0
TTC	14 Phe	TCC	29 Ser	TAC	16 Tyr	TGC	0 Cys
TTA	20 Leu	TCA	5	TAA	6 Term	TGA	1 Term
TTG	56	TCG	0	TAG	1	TGG	0 Trp
CTT	0	CCT	6	CAT	5	CGT	0
CTC	0 Leu	CCC	0 Pro	CAC	13 His	CGC	0 Arg
CTA	9	CCA	26	CAA	38 Gln	CGA	0
CTG	5	CCG	0	CAG	2	CGG	0
ATT	26	ACT	36	AAT	6	AGT	0
ATC	32	ACC	19 Thr	AAC	20 Asn	AGC	1 Ser
ATA	0 Ile	ACA	8	AAA	42 Lys	AGA	0 Arg
ATG	8 Met	ACG	0	AAG	73	AGG	0
GTT	27	GCT	72	GAT	18	GGT	82
GTC	23 Val	GCC	41 Ala	GAC	8 Asp	GGC	0 Gly
GTA	0	GCA	3	GAA	47 Glu	GGA	0
GTG	2	GCG	1	GAG	1	GGG	0

Note: The codon usage in the yeast histone genes is presented.

Around the change in H3.1 and H3.2 three of four nucleotides are changed (one in an adjacent codon).[44] Despite this divergence the genes are extremely homologous elsewhere and other amino acid changes are not found. Within a species (only one — the mouse has been studied enough thus far) the replication-type nonallelic variants show as great nucleotide homology within a variant (H3.2 vs. H3.2) as between variants (H3.1 vs H3.2). Presumably the same homology will be found within another species (e.g., human H3.1 vs. human H3.2) and these genes will be equally divergent from the mouse genes. How this situation has been maintained across a large number of species remains a mystery.

VIII. CODON USAGE IN HISTONE GENES

The analysis of nucleotide divergence presented above assumes that mutations are random. With respect to silent mutations this implies that there is no selection in favor or against certain codons. This is clearly not the case for any of the species, although different species show different stringencies in their codon preferences. In yeast there is a fairly strict set of codon preferences[74] which differ from those of the vertebrates, making the yeast genes AT rich while the vertebrate and sea urchin genes are all GC rich (Table 5). The yeast histone genes show a similar codon selection to other abundant yeast mRNAs.[74] This codon usage is very different from that seen in the *Neurospora* histone genes.[27] The codon usage in different sea urchin species is similar for the early and late histone genes in different species (Table 6). Thus, the extensive divergence between these two gene sets in the silent changes is not due to selective pressure of differential codon usage.

The mouse and chicken histone genes show a particularly nonrandom pattern of codon usage (Table 7). There is almost exclusive use of codons ending in G or C. As a result the H2a and H2b genes differ in only 2 nucleotides in the 100 codons of class I (e.g., lysine AAG or AAA). There is more extensive variation in codons where all four bases can be found in the 3rd base. The codon preference is not absolute, although codons ending in G or C are highly preferred and codons ending in A are very rare. The mouse and chicken histone genes show more homology to each other than expected by comparing these two genes with other histone genes. For example, the group of H2b genes discussed above may

Table 6
SEA URCHIN HISTONE CODON USAGE

	E	L		E	L		E	L		E	L
TTT	6	0	TCT	15	4	TAT	5	1	TGT	2	1
TTC	20	12 Phe	TCC	13	4 Ser	TAC	28	13 Tyr	TGC	3	3 Cys
TTA	0	0 Leu	TCA	7	0	TAA	3	4 Term	TGA	1	0 Term
TTG	9	4	TCG	2	0	TAG	4	0	TGG	0	0 Trp
CTT	15	10	CCT	15	1	CAT	15	4	CGT	27	40
CTC	33	21 Leu	CCC	14	6 Pro	CAC	9	4 His	CGC	44	18 Arg
CTA	14	0	CCA	16	7	CAA	15	3 Gln	CGA	19	1
CTG	30	5	CCG	3	0	CAG	39	17	CGG	7	0
ATT	12	1	ACT	11	2	AAT	3	2	AGT	14	2
ATC	57	25	ACC	44	44 Thr	AAC	23	7 Asn	AGC	19	6 Ser
ATA	1	0 Ile	ACA	16	0	AAA	45	8 Lys	AGA	14	1 Arg
ATG	22	10 Met	ACG	12	0	AAG	103	40	AGG	24	5
GTT	16	2	GCT	37	15	GAT	12	4	GGT	25	14
GTC	31	28 Val	GCC	72	34 Ala	GAC	15	8 Asp	GGC	29	8 Gly
GTA	6	0	GCA	29	2	GAA	23	5 Glu	GGA	37	26
GTG	21	2	GCG	9	1	GAG	40	19	GGG	16	0

Note: The codon usage in the sea urchin early (E) and late (L) histone genes is presented. The data for the early genes is for *S. purpuratus* and *Ps. milaris* (taken from Kedes[22] and *L. pictus* (taken from Roberts et al.[72] The data for the late genes are for *L. pictus* and are taken from Roberts et al.[72a]

all be orthologous. Yet the chicken and mouse genes are more similar than the mouse and human even though they should be farther apart in evolution. This is probably a manifestation of the similar codon usage in these two genes. The codon usage alone does not make all chicken genes similar to mouse histone genes. The H3 and H4 genes are much more divergent between these two species, although they show the same codon preference. Similar effects are seen when comparing mouse H4 and wheat H4 genes. These are more similar than most other pairs of H4 genes, although they must be separated by over 1 billion years of evolution. Again the codon usage in these two genes is very similar, although there is not enough data to establish whether the same codon usage extends across many wheat histone genes. The higher degree of similarity in these genes may be due to selection to maintain the codon usage pattern. Thus, many of the random mutations (e.g., to give A as the 3rd base of a codon) would be eliminated by selection at the codon usage level even though the protein sequence would not be affected.

The *Xenopus* and human histone genes sequenced thus far do not show the extreme codon usage pattern of the mouse and chicken genes. They do show the same general trend of preferred codons (Table 8). There is very little use of codons ending in A as in the mouse and chicken genes and a preference for CTX codons for leucine. There is more extensive use of AGPu codons for arginine and less use of codons which end in CG, consistent with the low content of CG doublets in eukaryotic DNA. However, in none of the vertebrate histone genes is there a significantly lower frequency of CG doublets than would be expected from the base composition. There is more variability between the orthologous mouse and human genes than between the mouse and chicken genes due to the effect of codon usage.

The codon usage preference must be maintained by selection throughout evolution. The result of this is that the chicken and mouse histone genes and mRNAs are very GC-rich. It is conceivable that the selection acts to maintain the high percentage of GC rather than to maintain a particular codon usage.

Table 7
HISTONE CODON USAGE

Mouse[a]

TTT	2	TCT	2	TAT	0	TGT	1
TTC	10 Phe	TCC	9 Ser	TAC	15 Tyr	TGC	2 Cys
TTA	0 Leu	TCA	0	TAA	3 Term	TGA	1 Term
TTG	4	TCG	8	TAG	1	TGG	0 Trp
CTT	1	CCT	4	CAT	4	CGT	10
CTC	4 Leu	CCC	11 Pro	CAC	8 His	CGC	42 Arg
CTA	2	CCA	0	CAA	1 Gln	CGA	0
CTG	34	CCG	7	CAG	22	CGG	5
ATT	0	ACT	4	AAT	0	AGT	0
ATC	28	ACC	31 Thr	AAC	9 Asn	AGC	9 Ser
ATA	0 Ile	ACA	0	AAA	2 Lys	AGA	1 Arg
ATG	11 Met	ACG	2	AAG	61	AGG	1
GTT	0	GCT	13	GAT	1	GGT	8
GTC	11 Val	GCC	35 Ala	GAC	13 Asp	GGC	26 Gly
GTA	0	GCA	2	GAA	0 Glu	GGA	6
GTG	21	GCG	9	GAG	25	GGG	3

Chicken[b]

TTT	0	TCT	2	TAT	1	TGT	1
TTC	11 Phe	TCC	9 Ser	TAC	19 Tyr	TGC	0 Cys
TTA	0 Leu	TCA	0	TAA	2 Term	TGA	2 Term
TTG	0	TCG	20	TAG	1	TGG	0 Trp
CTT	2	CCT	1	CAT	0	CGT	6
CTC	11 Leu	CCC	16 Pro	CAC	13 His	CGC	39 Arg
CTA	2	CCA	1	CAA	1 Gln	CGA	0
CTG	35	CCG	6	CAG	19	CGG	9
ATT	2	ACT	2	AAT	2	AGT	0
ATC	32	ACC	20 Thr	AAC	13 Asn	AGC	10 Ser
ATA	0 Ile	ACA	2	AAA	2 Lys	AGA	1 Arg
ATG	12 Met	ACG	14	AAG	78	AGG	1
GTT	0	GCT	2	GAT	2	GGT	4
GTC	16 Val	GCC	37 Ala	GAC	14 Asp	GGC	35 Gly
GTA	0	GCA	1	GAA	1 Glu	GGA	3
GTG	22	GCG	28	GAG	29	GGG	9

[a]   The codon frequencies for the mouse histone genes sequenced by Sittman et al.[44] and Seiler-Tuyns and Birnstiel[84] are presented.

[b]   The codon usage in the chiken histone genes is presented. The genes analyzed are all the genes described in Tables 1 to 4 except the replacement variant H3 gene,[52] which is not included.

IX. OTHER SELECTIVE PRESSURES ON HISTONE mRNA STRUCTURE

Since there is clearly selection operating to maintain a particular codon usage, it is of interest to ask whether there are other constraints on mRNA structure which are also maintained by selection. Histone mRNAs are a good candidate for this type of selection since the mRNAs participate directly in one type of control, the control of mRNA degradation. When DNA synthesis is inhibited, the halflife of histone mRNA changes dramatically to

Table 8
FURTHER HISTONE CODON USAGE

Human[a]

TTT	3	TCT	7	TAT	4	TGT	1
TTC	8 Phe	TCC	10 Ser	TAC	15 Tyr	TGC	1 Cys
TTA	2 Leu	TCA	0	TAA	3 Term	TGA	1 Term
TTG	10	TCG	4	TAG	1	TGG	0 Trp
CTT	8	CCT	7	CAT	5	CGT	10
CTC	10 Leu	CCC	2 Pro	CAC	10 His	CGC	38 Arg
CTA	1	CCA	4	CAA	4 Gln	CGA	4
CTG	19	CCG	6	CAG	15	CGG	8
ATT	11	ACT	8	AAT	7	AGT	0
ATC	21	ACC	23 Thr	AAC	6 Asn	AGC	6 Ser
ATA	0 Ile	ACA	4	AAA	17 Lys	AGA	2 Arg
ATG	11 Met	ACG	2	AAG	49	AGG	1
GTT	3	GCT	18	GAT	3	GGT	14
GTC	9 Val	GCC	23 Ala	GAC	12 Asp	GGC	35 Gly
GTA	5	GCA	5	GAA	2 Glu	GGA	5
GTG	23	GCG	19	GAG	26	GGG	7

Xenopus[b]

TTT	3	TCT	6	TAT	8	TGT	0
TTC	11 Phe	TCC	12 Ser	TAC	15 Tyr	TGC	2 Cys
TTA	0 Leu	TCA	1	TAA	4 Term	TGA	1 Term
TTG	4	TCG	3	TAG	1	TGG	0 Trp
CTT	1	CCT	6	CAT	1	CGT	7
CTC	16 Leu	CCC	10 Pro	CAC	12 His	CGC	36 Arg
CTA	3	CCA	3	CAA	3 Gln	CGA	2
CTG	34	CCG	1	CAG	19	CGG	13
ATT	3	ACT	11	AAT	3	AGT	1
ATC	33	ACC	34 Thr	AAC	14 Asn	AGC	5 Ser
ATA	0 Ile	ACA	2	AAA	21 Lys	AGA	6 Arg
ATG	12 Met	ACG	0	AAG	56	AGG	15
GTT	9	GCT	28	GAT	9	GGT	5
GTC	21 Val	GCC	33 Ala	GAC	9 Asp	GGC	30 Gly
GTA	0	GCA	8	GAA	4 Glu	GGA	26
GTG	2	GCG	3	GAG	29	GGG	13

[a] The codon usage for the human histone genes is presented. The genes used are the ones sequenced by Zhong et al.[79] and Sierra et al.[78]

[b] The codon usage in the *Xenopus* histone genes sequenced by Moorman et al.[86] and Turner and Woodland[82] is presented.

about 10 to 15 min.[50,90,91] Part of the mechanism for this change must be due to specific recognition of the histone mRNA relative to other mRNAs by a nuclease. The histone mRNAs are distinguished by their 3' ends which are not polyadenylated but rather end in a short hairpin loop. It is possible that this loop is necessary for the degradation of the mRNA, although it is already known that this structure plays an active role in the formation of the 3' end of the mRNA and this may indeed be its main function.[92,93] The histone mRNAs have relatively short untranslated regions and it is possible that the coding regions themselves play a direct role in some aspects of histone mRNA metabolism. To test this possibility the sequences of the histone genes have been compared to see if there are any regions which were much more highly conserved than others.

The H3 genes show a highly conserved region of 63 nucleotides from codon 105 to codon 125 (Figure 4). This region is highly conserved among the three mouse genes (one change among the three genes), one change between the mouse genes and the *Xenopus* H3 gene, and only one change between the mouse gene and the late sea urchin histone genes. These are the regions of highest homology among these genes from diverse species. A second region of high homology is found from codon 73 to 90. As is true for the other histone genes, the 3' end of the mRNA shows much less variation at the nucleotide level. In contrast, the chicken and human H3 genes which are not orthologous with the mouse genes show more variation in this region, but less variation than in other regions of the protein. According to this argument the sea urchin late genes may be orthologous to or at least under the same selective pressures as the mouse gene.

A similar situation is evident in the H2b mRNA sequences. The 75 base region from codon 82 to codon 107 shows only 4 changes between mouse and either human H2b gene or mouse and chicken in this region. There are more changes among *Xenopus* genes and mouse or human genes. Overall, the region from amino acid 82 to 125 (129 nucleotides) shows only 12 changes between *Xenopus* and mouse (9%), 9 between chicken and mouse (7%), and 10 between human and mouse (8%), all less than 10% difference. In the other two thirds of the mRNA there is 16% difference between *Xenopus* and mouse, 10% between chicken and mouse, and 12% between human and mouse. Thus, the 3' end of the coding region of the H2b mRNA is much more highly conserved at the nucleotide level than the rest of the mRNA (Figure 4).

Inspection of the H4 gene sequences reveals a highly conserved sequence in the 3' end of the mRNA from amino acid 70 to codon 95 (75 nucleotides). There are only two changes in this region among the H4 gene from mouse and human (Hu4-2) and mouse and *Xenopus* (XenH4-1). This region is also relatively invariant among mouse and chicken (7 changes) and the mouse genes and the human H4B gene and the mouse and *Xenopus* H4-2 gene. Even the wheat H4 gene shows only 5 changes in this region (not shown) as does the late sea urchin histone H4 gene. Thus, this region is most likely conserved among all the H4 RNAs (Figure 4).

Similarly a more conserved region can be identified in the H2a gene from codon 60 to codon 89. There are only 2 changes between chicken and mouse and 6 between human and mouse in this region. In general, again, the 3' region of the mRNA (i.e., amino acids 60 to 119) is much more highly conserved than the 5' (amino terminal) end of the mRNA. Therefore, in all the histone genes there is a relatively conserved region near the 3' end of the coding region. The significance of this observation is unknown but it may be that the structure near the 3' end of the coding region plays an important role in histone mRNA metabolism. Assuming that mutations are random there must have been selection against changes in this region to maintain the nucleotide sequence.

X. SUMMARY AND CONCLUSIONS

The histone genes provide an excellent system to look at evolution of a gene set in which amino acid sequences are essentially invariant over long evolutionary times. The comparison of known vertebrate sequences suggest that there are a number of paralogous gene sets present in vertebrates, which could code the same histone protein variants. Whether these gene sets share common regulatory properties is not yet known. These paralogous gene sets may share a common original gene set since the serine codons have maintained their postition in all of the genes. I have tentatively identified an orthologous H2b gene set which is present

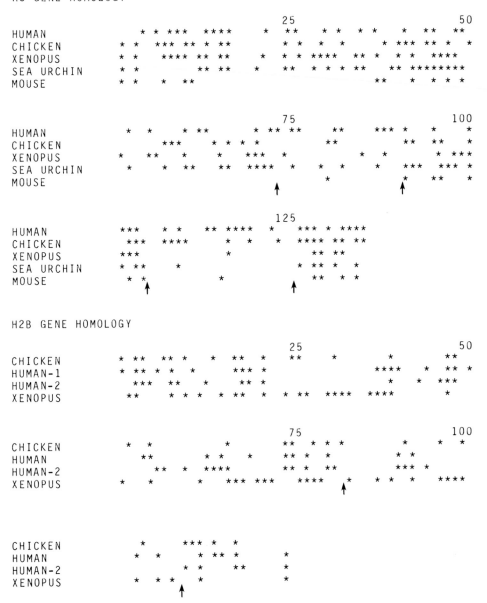

FIGURE 4.   Pattern of nucleotide changes in the coding region. The coding regions of the histone genes are compared to the mouse histone genes (an arbitrary comparison). For the H3 genes, they are compared to the H3-1 gene; the other mouse column indicates positions where there is any variation in nucleotides for any of the genes sequenced. Similarly for the H2b genes the chicken line indicates any position where the two chicken H2b genes differ from the mouse gene. The arrows mark the regions of high conservation of sequence among genes from different species. Each position indicates a codon and an asterisk indicates there is one (or more) nucleotide altered in that codon (either due to a replacement or silent substitutions).

in *Xenopus,* chicken, mouse, and human. In addition, an entire *Xenopus laevis* minor cluster may be orthologous to the mouse genes which have been isolated thus far.

In addition to selective pressures to maintain the amino acid sequence, there has also been selection to prevent extensive randomization of the nucleic acid sequence as might have been expected if silent mutations were truly neutral. This selection may be of two types:

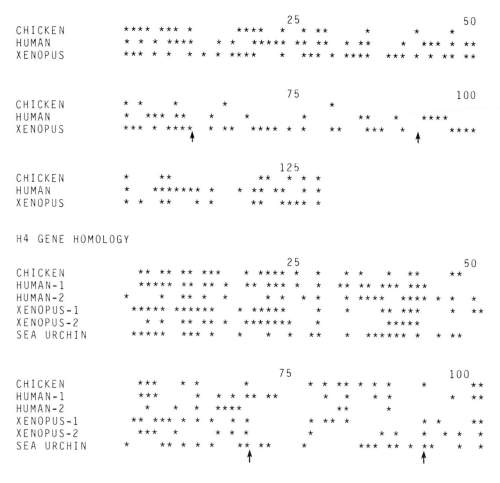

FIGURE 4. Continued.

(1) preferential codon usage particularly evident in the mouse and chicken genes and (2) conservation of nucleotide sequences in a particular region of the mRNA. These sequences may play a direct role in the mRNA metabolism and function.

Several evolutionary paradoxes have been pointed out: (1) the maintainance of coding region sequences in the presence of complete divergence of flanking regions and (2) the continued maintenance of nonallelic protein variants (e.g., H3.1 and H3.2) in various species where the nucleotide sequences within a species are very similar, while the nucleotide sequences for the same protein differ between species. This implies either a continued formation of new genes from the different protein variants or a homogenization of coding region sequences which leaves the variants intact.

The current analysis is based on very limited data. More genes must be isolated and their sequences determined to confirm what has been suggested here. In addition, it will be essential to determine when in development and in which tissues particular genes are expressed. It is apparent that the gene sets isolated from human and mouse by different investigators do not represent closely related genes. In fact these genes probably cross react very weakly with each other and will not necessarily be useful in isolating all the histone genes from a particular species. If there are orthologous genes to the mouse genes present in humans it will probably be necessary to use the mouse probes to get those human genes and vice-versa. Thus, the true gene number may be much higher than the 20 to 40 genes measured by solution hybridization (which would measure only the number in a particular family) and each gene set may have a separate function which has yet to be determined. As more data become available the evolutionary history of the histone genes should become clearer.

ACKNOWLEDGMENTS

This work was supported by NIH grant GM 29832. I am grateful to Drs. Doug Engel, Nat Heintz, Larry Yaeger, and Eric Weinberg for communicating their results prior to publication.

REFERENCES

1. **Johns, E. W., Phillips, D. M. P., Simpson, P., and Butler, J. A. V.,** Improved fractionation of arginine-rich histones from calf thymus, *Biochem. J.,* 77, 631, 1960.
2. **Luck, J. M., Rasmussen, P. S., Satake, K., and Tsvetikov, A. N.,** Further studies on the fractionation of calf thymus histone, *J. Biol. Chem.,* 233, 1407, 1958.
3. **DeLange, R. J., Fambrough, D. M., Smith, E. L., and Bonner, J.,** Calf and pea histone IV. II. The complete amino acid sequence of calf thymus histone IV: presence of E-N-acetyllysine, *J. Biol. Chem.,* 244, 319, 1969.
4. **DeLange, R. J., Fambrough, D. M., Smith, E. L., and Bonner, J.,** Calf and pea histone IV. III. Complete amino acid sequence of pea seedling histone IV: comparison with the homologous calf thymus histone, *J. Biol. Chem.,* 244, 5669, 1969.
5. **DeLange, R. J., Hooper, J. A., and Smith, E. L.,** Histone III. Sequence studies on the cyanogen bromide peptides: complete amino acid sequence of calf thymus histone III, *J. Biol. Chem.,* 248, 3261, 1973.
6. **Patthy, L., Smith, E. L., and Johnson, J.,** Histone III. V. The amino acid sequence of pea embryo histone, *J. Biol. Chem.,* 248, 6834, 1973.
7. **Marzluff, W. F., Sanders, L. A., Miller, D. M., and McCarty, K. S.,** Two chemically and metabolically distinct forms of calf thymus histone F3, *J. Biol. Chem.,* 247, 2026, 1972.
8. **Patthy, L. and Smith, E. L.,** Histone III. IV. Two forms of calf thymus histone III, *J. Biol. Chem.,* 250, 1919, 1975.
9. **Lennox, R. W. and Cohen, L. H.,** The H1 subtypes of mammals: metabolic characteristics and tissue distribution, in *Histone Genes: Structure, Organization and Regulation,* Stein, G., Stein, J., and Marzluff, W. F., Eds., Wiley & Sons, New York, 1984, 373.
10. **Franklin, S. G. and Zweidler, A.,** Non-allelic variants of histones 2a, 2b and 3 in mammals, *Nature (London),* 266, 273, 1977.
11. **Urban, M. K. and Zweidler, A.,** Changes in nucleosomal core histone variants during chicken development and maturation, *Dev. Biol.,* 95, 421, 1983.
12. **Cohen, L. M., Newrock, K. M., and Zweidler, A.,** Stage-specific switches in histone synthesis during embryogenesis of the sea urchin, *Science,* 190, 994, 1975.
13. **Newrock, K. M., Cohen, L. H., Hendricks, M. B., Donnelly, R. J., and Weinberg, E. S.,** Stage-specific mRNAs coding for subtypes of H2a and H2b histones in the sea urchin embryo, *Cell,* 14, 327, 1978.
14. **Wallis, J. H., Hereford, L. H., and Grunstein, M.,** Histone H2B genes of yeast encode two different proteins, *Cell,* 22, 799, 1980.
15. **Choe, J., Kolodrabetz, D., and Grunstein, M.,** The two yeast histone H2A genes encode similar protein subtypes, *Proc. Natl. Acad. Sci. U.S.A.,* 79, 1484, 1982.
16. **Fusauchi, Y. and Iwai, K.,** *Tetrahymena* histone H2A. Isolation and two variant sequences, *J. Biochem. (Jpn.),* 93, 1487, 1983.
17. **Kornberg, R. D.,** Structure of chromatin, *Ann. Rev. Biochem.,* 46, 931, 1978.
18. **Isenberg, I.,** Histones, *Ann. Rev. Biochem.,* 48, 159, 1979.
19. **Laine, B., Kmiecik, D., Sautiere, P., Biserte, G., and Cohn-Solul, M.,** Complete amino-acid sequences of DNA-binding proteins HU-1 and HU-2 from *Escherichia coli, Eur. J. Biochem.,* 103, 447, 1980.
20. **DeLange, R. J., Green, G. R., and Searcy, D. G.,** A histone-like protein (HTa) from *Thermoplasma acidophilum.* Purification and properties, *J. Biol. Chem.,* 256, 900, 1981.

21. **Hentschel, C. C. and Birnstiel, M. L.,** The organization and expression of histone gene families, *Cell,* 25, 301, 1980.

22. **Kedes, L. H.,** Histone messengers and histone genes, *Ann. Rev. Biochem.,* 48, 837, 1979.

23. **Von Holt, C., Strickland, W. N., Brandt, W. F., and Strickland, M. S.,** More histone sequences, *FEBS Lett.,* 100, 201, 1979.

24. **Hereford, L., Fahner, K., Woolford, J., Rosbash, M., and Kaback, D. B.,** Isolation of yeast histone genes H2a and H2b, *Cell,* 18, 1261, 1979.

25. **Smith, M. M. and Murray, K.,** Yeast H3 and H4 histone messenger RNAs are transcribed from two non-allelic gene sets, *J. Mol. Biol.,* 169, 641, 1983.

26. **Bannon, G. A., Calzone, F. J., Bower, J. K., Allis, C. D., and Gorovsky, M. A.,** Multiple independently regulated, polyadenylated messages for histone H3 and H4 in *Tetrahymena, Nucleic Acids Res.,* 11, 3903, 1983.

27. **Woudt, L. P., Patink, A., Kempers-Veenstra, A. E., Jansen, A. E. M., Mager, W. H., and Planta, R. J.,** The genes coding for histone H3 and H4 in *Neurospora crossa* are unique and contain intervening sequences, *Nucleic Acids Res.,* 11, 5347, 1983.

28. **Kedes, L. H., Chang, A. C. Y., Housman, D., and Cohen, S. N.,** Isolation of histone genes from unfractionated sea urchin DNA by subcloning in *E. coli, Nature (London),* 255, 533, 1975.

29. **Childs, G., Nocente-McGrath, C., Lieber, T., Holt, C., and Knowles, J. A.,** Sea urchin *(Lytechinus pictus)* late stage histone H3 and H4 genes: characterization and mapping of a clustered but nontandemly linked gene family, *Cell,* 31, 383, 1982.

30. **Maxson, R., Mohun, T., Gormezana, G., Childs, G., and Kedes, L. H.,** Distinct organizations and patterns of expression of early and late histone gene sets in the sea urchin, *Nature (London),* 301, 120, 1983.

31. **Von Holt, C., DeGroot, P., Schwager, S., and Brandt, W. F.,** The structure of sea urchin histones and considerations on their function, in *Histone Genes: Structure, Organization and Regulation,* Stein, G., Stein, J., and Marzluff, W. F., Eds., John Wiley & Sons, New York, 65, 1984.

32. **Lifton, R. P., Goldberg, M. L., Karp, R. W., and Hogness, D. S.,** The organization of the histone genes in *Drosophila melanogaster:* functional and evolutionary implications, *Cold Spring Harbor Symp. Quant. Biol.,* 42, 1047, 1978.

33. **Anderson, K. N. and Lengyel, J. A.,** Changing rates of histone mRNA synthesis and turnover in *Drosophila* embryos, *Cell,* 21, 717, 1980.

34. **Parker, C. A.,** personal communication.

35. **Pardue, M. L., Kedes, C. H., Weinberg, E. S., and Birnstiel, M. L.,** Localization of sequences coding for histone messenger RNA in the chromosomes of *Drosophila melanogaster, Chromosoma,* 63, 135, 1977.

36. **Stephenson, E. C., Erba, H. P., and Gall, J. G.,** Nucleic characterization of a cloned histone gene cluster of the newt *Notophthalmus viridescens, Nucleic Acids Res.,* 9, 2281, 1971.

37. **Stephenson, E. C., Erba, H. P., and Gall, J. G.,** Histone gene cluster of the newt *Notophthalmus* are separated by long tracts of satellite DNA, *Cell,* 24, 639, 1981.

38. **Zernik, M., Heintz, N., Boime, I., and Roeder, R. G.,** *Xenopus laevis* histone genes: variant H1 genes are present in different clusters, *Cell,* 22, 807, 1980.

39. **Turner, P. C., Aldridge, T. C., Woodland, H. R., and Old, R. N.,** Nucleotide sequences of H1 histone genes from *Xenopus laevis.* A recently diverged pair of H1 genes and an unusual H1 pseudogene, *Nucleic Acids Res.,* 11, 4093, 1983.

40. **Von Dongen, W. M. A. M., Moorman, A. F. M., and Destree, O. H. J.,** DNA organization and expression of *Xenopus* histone genes, in *Histone Genes: Structure, Organization and Function,* Stein, G., Stein, J., and Marzluff, W. F., Eds., John Wiley & Sons, New York, 1984, 199.

41. **Ruberti, I., Fragapane, P., Pierandrei-Amaldi, P., Beccair, E., Amaldi, F., and Bozzone, I.,** Characterization of histone genes isolated from *Xenopus laevis:* genomic libraries, *Nucleic Acids Res.,* 10, 7543, 1982.

42. **Turner, P. C. and Woodland, H. R.,** Histone gene number and organization in *Xenopus: Xenopus borealis* has a homogeneous major cluster, *Nucleic Acids Res.,* 11, 971, 1983.

43. **Cohn, R. H. and Kedes, L. H.,** Nonallelic histone gene clusters of individual sea urchins *(Lytechinus pictus):* polarity and gene organization, *Cell,* 18, 843.

44. **Sittman, D. B., Graves, R. A., and Marzluff, W. F.,** Structure of a cluster of mouse histone genes, *Nucleic Acids Res.,* 11, 6679.

45. **Sittman, D. B., Chiu, I. M., Pan, C.-J., Cohn, R. H., Kedes, L. H., and Marzluff, W. F.,** Isolation of two clusters of mouse histone genes, *Proc. Natl. Acad. Sci. U.S.A.,* 78, 8078, 1981.

46. **Heintz, N., Zernik, M., and Roeder, R. G.,** The structure of the human histone genes: clustered but not tandemly repeated, *Cell,* 24, 661, 1981.

47. **Sierra, F., Lichtler, A., Marashi, F., Rickles, R., Van Dyke, T., Clark, S., Wells, J., Stein, G., and Stein, J.,** Organization of human histone genes, *Proc. Natl. Acad. Sci. U.S.A.,* 79, 1795, 1982.

48. **Engel, J. D. and Dodgson, J. B.,** Histone genes are clustered but not tandemly repeated in the chicken genome, *Proc. Natl. Acad. Sci. U.S.A.,* 78, 2856, 1981.

49. **Bruschi, S. and Wells, J. R. E.,** Vertebrate histone gene transcription occurs from both DNA strands, *Nucleic Acids Res.,* 9, 1591, 1981.

50. **Stein, G. S., Sierra, F., Plumb, M., Marashi, R., Baumbach, L., Stein, J. L., Carozzi, N., and Prokopp, K.,** Organization and expression of human histone genes, in *Histone Genes: Structure Organization and Function,* Stein G., Stein, J., and Marzluff, W. F., Eds., John Wiley & Sons, New York, 1984, 397.

51. **Heintz, N.,** personal communication.

52. **Sugarman, B. J., Dodgson, J. D., and Engel, J. B.,** Genomic organization, DNA sequence and expression of chicken embryonic histone genes, *J. Biol. Chem.,* 258, 9005, 1983.

53. **Tabata, T., Sasaki, K., and Iwabuchi, M.,** The structural organization and DNA sequence of a wheat histone H4 gene, *Nucleic Acids Res.,* 11, 5865, 1983.

54. **Graves, R. G., Wellman, S. E., Chiu, I. M., and Marzluff, W. F.,** *J. Mol. Biol.,* 183, 179, 1985.

55. **Harvey, R. P., Robins, A. J., and Wells, J. R. E.,** Independently evolving chicken histone H2b-specific 5′ element, *Nucleic Acids Res.,* 10, 7851, 1982.

56. **Yager, L. N., Kaumeyer, J. F., and Weinberg, E. S.,** Sea urchin histone genes — nucleotide polymorphisms in the H4 gene and spacers of *Strongylocentrotus purpuratus, J. Mol. Evol.,* 20, 215, 1984.

57. **Krieg, P. A., Robins, A. J., D'Andrea, R., and Wells, J. R. E.,** The chicken H5 gene is unlinked to core and H1 histone genes, *Nucleic Acids Res.,* 11, 619, 1983.

58. **Engel, J. D., Sugarman, B. J., and Dodgson, J. B.,** A chicken histone H3 gene contains intervening sequences, *Nature (London),* 297, 434, 1982.

59. **Shutt, R. and Kedes, L. H.,** Synthesis of histone mRNA sequences in isolated nuclei of cleavage stage sea urchin embryos, *Cell,* 3, 283, 1974.

60. **Sittman, D. B., Graves, R. A., and Marzluff, W. F.,** Histone mRNA concentrations are regulated at the level of transcription and mRNA degradation, *Proc. Natl. Acad. Sci. U.S.A.,* 80, 1859, 1983.

61. **Fahrner, K., Yarger, J., and Hereford, L.,** Yeast histone mRNA is polyadenylated, *Nucleic Acids Res.,* 8, 5725, 1980.

62. **Ruderman, J. V. and Pardue, M. L.,** Cell-free translation analysis of messenger RNA in echinoderm and amphibian early development, *Dev. Biol.,* 60, 48, 1978.

63. **Rykowski, M. C., Wallis, J. M., Choe, J., and Grunstein, M.,** Histone H2B subtypes are dispensable during the yeast cell cycle, *Cell,* 25, 477, 1981.

64. **Kolodrubetz, D., Rykowski, M. C., and Grunstein, M.,** Histone H2A subtypes associate interchangeably *in vivo* with histone H2B subtypes, *Proc. Natl. Acad. Sci. U.S.A.,* 79, 7814, 1982.

65. **Wilson, A. C., Carlson, S., and White, T.,** Biochemical evolution, *Ann. Rev. Biochem.,* 46, 573, 1977.

66. **Perler, F., Efstratiadis, A., Lomedico, P., Gilbert, W., Kolodner, R., and Dodgson, J.,** The evolution of genes: the chicken preproinsulin gene, *Cell,* 20, 555, 1980.

67. **Efstratiadis, A., Posakony, J. W., Maniatis, T., Lawn, R. M., O'Connell, C., Spritz, R. A., DeRiel, J. K., Forget, B., Weissman, S. M., Sightom, J. L., Blechl, A. E., Smithies, O., Boralle, F. E., Shoulders, C. C., and Proudfoot, N. J.,** The structure and evolution of the human β-globin gene family, *Cell,* 21, 653, 1980.

68. **Martin, S. L., Zimmer, E. A., Davidson, W. S., Wilson, A. C., and Kan, Y. W.,** The untranslated regions of α-globin mRNA evolve at a functional rate in higher primates, *Cell,* 25, 737, 1981.

69. **Liebhaber, S. A. and Begley, K. A.,** Structural and evolutionary analysis of the two chimpanzee α-globin mRNAs, *Nucleic Acids Res.,* 11, 8915, 1983.

70. **Dayhoff, M. O.,** Atlas protein sequence and structure, 5 -(Suppl. 2), 1, 1976.

71. **Busslinger, M., Rusconi, S., and Birnstiel, M. L.,** An unusual evolutionary behavior of a sea urchin histone gene cluster, *EMBO J.,* 1, 27, 1982.

72. **Kaumeyer, J. F., and Weinberg, E. S.,** personal communication.

72a. **Roberts, S. B., Weisser, K. E., and Childs, G.,** Sequence comparisons of non-allelic late histone genes and their early stage counterparts, *J. Mol. Biol.,* 174, 647, 1984.

73. **Graves, R. A. and Marzluff, W. F.,** unpublished results.

74. **Smith, M. M. and Andresson, O. S.,** DNA sequences of yeast H3 and H4 histone genes from two nonallelic gene sets encode identical H3 and H4 proteins, *J. Mol. Biol.,* 169, 662, 1983.

75. **Taylor, J. D., Wellman, S., and Marzluff, W. F.,** unpublished results.

76. **Grandy, D., Engel, J. D., and Dodgson, J. B.,** Complete nucleotide sequence of a chicken H2B histone gene, *J. Biol. Chem.,* 257, 8577, 1982.

77. **Marzluff, W. F. and Graves, R. A.,** Organization and expression of mouse histone genes, in *Histone Genes: Structure, Organization and Function,* Stein, G., Stein, J., and Marzluff, W. F., Eds., John Wiley & Sons, New York, 1984, 281.

78. **Sierra, F., Stein, G., and Stein, J.,** Structure and *in vitro* transcription of a human H4 histone gene, *Nucleic Acids Res.,* 11, 7069, 1983.

79. **Zhong, R., Roeder, R. G., and Heintz, N.,** The primary structure and expression of four cloned human histone genes, *Nucleic Acids Res.*, 11, 7409, 1983.

80. **Marashi, F., Prokopp, K., Stein, J., and Stein, G.,** Evidence for a human histone gene cluster containing H2b and H2a pseudogenes, *Proc. Natl. Acad. Sci., U.S.A.*, 81, 1936, 1984.

81. **Moorman, A. F. M., DeBoer, P. A. J., deLaaf, R. T. M., VanDongen, R. N. M. A. M., and Destree, O. H. J.,** Primary structure of the histone H3 and H4 genes and their flanking sequences in a minor histone gene cluster of *Xenopus laevis, FEBS Lett.*, 136, 45, 1982.

82. **Turner, P. C. and Woodland, H. R.,** H3 and H4 histone cDNA sequences from *Xenopus:* a sequence comparison of H4 genes, *Nucleic Acids Res.*, 10, 3769, 1982.

83. **Clerc, R. G., Bucher, P., Strub, K., and Birnstiel, M. L.,** Transcription of a cloned *Xenopus laevis* H4 histone gene in the homologous frog oocyte system depends on an evolutionary conserved sequence motif in the -50 region, *Nucleic Acids. Res.*, 11, 8641, 1983.

84. **Seiler-Tuyns, A. and Birnstiel, M. L.,** Structure and expression in L-cells of a cloned H4 histone gene of the mouse, *J. Mol. Biol.*, 151, 207, 1981.

85. **Sprecher, C. and Marzluff, W. F.,** unpublished results.

86. **Moorman, A. F. M., deBoer, P. A. J., deLaaf, R. T. M., and Destree, O. H. J.,** Primary structure of the histone H2A and H2B genes and their flanking sequences in a minor histone gene cluster of *Xenopus laevis, FEBS Lett.*, 144, 235, 1982.

87. **Sures, I., Lowry, J., and Kedes, L. H.,** The DNA sequence of sea urchin *(S. purpuratus)* H2A, H2B and H3 histone coding and spacer regions, *Cell,* 15, 1033, 1978.

88. **Grunstein, M. and Grunstein, J. E.,** The histone H4 gene of Strongylocentrotus purpuratus: DNA and mRNA sequences at the 5′ end, *Cold Spring Harbor Symp. Quant. Biol.*, 42, 1083, 1977.

89. **Schaffner, W., Kunz, G., Detwyler, H., Telford, J., Smith, H. O., and Birnstiel, M. L.,** Genes and spacers of cloned sea urchin histone DNA analyzed by sequencing, *Cell,* 14, 655, 1978.

90. **Graves, R. A. and Marzluff, W. F.,** Rapid reversible changes in histone gene transcription and mRNA levels in mouse myeloma cells, *Mol. Cell. Biol.*, 4, 351, 1984.

91. **Heintz, N., Sive, H. L., and Roeder, R. G.,** Regulation of human histone gene expression: kinetics of accumulation and changes in the rate of synthesis and in the half-lives of individual histone mRNAs during the HeLa cell cycle, *Mol. Cell. Biol.*, 3, 539, 1983.

92. **Birchmeier, C., Schumperli, D., Sconzo, G., and Birnstiel, M. L.,** 3′ editing of mRNAs: sequence requirements and involvement of a 60-nucleotide RNA in maturation of histone mRNA precursors, *Proc. Natl. Acad. Sci. U.S.A.*, 81, 1057, 1984.

93. **Birchmeier, C., Folk, W., and Birnstiel, M. L.,** The terminal RNA stem-loop structure and 80 bp of spacer DNA are required for the formation of the 3′ termini of sea urchin H2A mRNA, *Cell,* 35, 433, 1983.

Chapter 7

PHYLOGENY OF NORMAL AND ABNORMAL HEMOGLOBIN GENES

William P. Winter

TABLE OF CONTENTS

I. MOLECULAR GENETICS OF HEMOGLOBIN

A. Structure of Hemoglobin

Hemoglobin, the principal respiratory protein in all vertebrates with the possible exception of a few Antarctic fish, is a tetrameric protein consisting of two pairs of polypeptide subunits or "chains". By convention, one pair is referred to as the α chains and the other as the non-α chains. In most animal species, when multiple normal hemoglobin types occur, one of the chains is common to all and that commonality is the distinguishing feature of the α chain. As will be discussed more extensively below, there is a marked degree of molecular homology between the α and non-α chains of a given species as well as interspecies homology between chains of like type (and obviously therefore, between unlike types as well). Each chain has associated with it a heme group which is a protoporphyrin IX group and which contains one iron atom. In functional hemoglobin, these iron atoms must be in the ferrous or 2^+ oxidation state. When the iron assumes the 3^+ state, the resultant hemoglobin is known as methemoglobin and is totally nonfunctional. Paradoxically then, oxygen, the principal ligand of hemoglobin, is also its worst enemy. Indeed, the internal milieu of the red cell is nearly anaerobic and one of the functions of the globin chain is the protection of the heme iron from oxidation.

Human globin chains all consist of a single polypeptide of slightly over 140 amino acids. The α chains contain 141 while the non-α chains have 146. All globin chains, including myoglobin, the oxygen-binding protein found in muscle, share a common tertiary structure and all exist under similar structural constraints. The proteins are made up primarily of α helical segments connected by short random coil segments. There is no β structure. There are eight helical regions designated A to H. Perutz[1] and others have shown that the function of hemoglobin requires a series of concerted shifts in tertiary structure which place stringent limitations on the amino acids which can occupy a given location in the molecule. This is especially true in the region of the G and H helices and at the terminal regions. Other structural restrictions include the portions of the E and F helices that form the heme binding site which must be hydrophobic and those portions of each chain which contribute to the interchain contacts that characterize the quaternary structure. A more detailed review of hemoglobin structure and function may be found in the excellent book by Dickerson and Geis.[2]

B. Normal Hemoglobin Types

In general, there are three types of hemoglobin which may be considered normal hemoglobins.[3] These are adult, fetal, and embryonic hemoglobin (Table 1). Not all types are found in all species and there may be multiple hemoglobins within a type. While a type is loosely defined by the developmental stage in which it predominates, the tissue in which it is synthesized is probably a better point of departure for a definition. Embryonic hemoglobins are synthesized by yolk sac tissue[4] and occur only transiently in development. Fetal hemoglobins are synthesized in the spleen and liver during the period when the hemopoietic cells have colonized those tissues. Adult hemoglobins are synthesized in the bone marrow spaces, although the site of synthesis alone does not determine the nature of the hemoglobin to be made. Thus, when any of certain disorders leading to the production of Hb F in adults are present, the F-cells arise from marrow cells.[5]

1. Adult Hemoglobin

In humans, there are two adult hemoglobins designated Hb A and Hb A_2. The former contains two chains designated β while the latter contains δ chains in addition to the common or α chains. The major hemoglobin type is the Hb A, comprising about 96% of the hemoglobin in a normal individual. The balance is Hb A_2. The underproduction of δ chain is due

Table 1
THE COMPOSITION OF HUMAN HEMOGLOBINS (Hb A AND A₂ ARE ADULT HEMOGLOBINS, Hb F IS FETAL, AND THE BALANCE ARE EMBRYONIC)

	β	δ	γ	ϵ
α	$\alpha_2\beta_2$	$\alpha\text{-}\delta_2$	$\alpha_2\gamma_2$	$\alpha_2\epsilon_2$
	A	A₂	F	Gower II
ζ			$\zeta_2\gamma_2$	$\zeta_2\epsilon_2$
			Portland	Gower I

to a defect in the promoter region of the δ locus.[6] Some animals, such as the cat, exhibit multiplicity of adult hemoglobins, and others, such as the sheep, have an adult hemoglobin which is not expressed except under physiological stress such as anemia.

2. Fetal Hemoglobin

The hemoglobin which supports the fetus through most of its development consists of two α chains and two γ chains. This hemoglobin has a higher oxygen affinity than adult hemoglobin, presumably enabling it to compete more effectively for oxygen than it otherwise might. Thus, this feature of fetal hemoglobin may be adaptive. There are two types of human fetal hemoglobin, one of which is characterized by the A-γ gene and the other by the G-γ gene depending on whether the chains have alanine or glycine in position 136, respectively. Since these two types do not separate electrophoretically, no distinction between them is made at the level of the tetramer.

3. Embryonic Hemoglobin

Embryonic hemoglobins[4] are synthesized in minute quantities in the first few weeks of gestation by yolk sac erythropoietic cells. For obvious reasons, it is the least accessible and the most poorly known type of hemoglobin. Three types of embryonic hemoglobin are known: Hb Gower I, Hb Gower II, and Hb Portland. These hemoglobins are combinations of the embryonic α chain known as zeta, the embryonic non-α chain known as epsilon,[7] and the γ chain. The compositions of these hemoglobins, as they are presently understood, are shown in Table 1.

C. Structural Relationships Among the Chains

As was stated above, all globin chains show a striking homology which is illustrated in Figure 1. Here, gaps have been introduced to accentuate the homology, but even without this device, the similarities are evident. The human δ chain differs from the β chain in only 10 positions out of 146 while the γ chain differs in 39 (based on G-γ). Not only is the homology evident at the level of the primary structure, but it is also reflected even more vividly in the secondary and tertiary structures. All globin chains have 7 or 8 helical regions, each of which occupies an equivalent position in the many types of globin which have been characterized.

D. Structure of the Globin Genome

The human globin genome is composed of two linkage groups or three if myoglobin is included.[8] The α linkage group is located on the short arm of chromosome 16 and includes two α genes and one ζ gene as well as two pseudogenes, a pseudo-α and a pseudo-ζ. The non-α chain genes are located on chromosome 11.[9] These include single β, δ, and ϵ genes,

```
                 1         2         3         4         5
Alpha      V-LSPADKTNVKAAWGKVGAHAGEYGAEALERMFLSFPTTKTYFPHF-DL
Zeta       S- TKTQR IIVSM A ISTQ DTI T T    L    H Q
Beta       VHLTPEEKSAVTALWGKV--NVDEVGGEALGRLLVVYPWTQRFFESFGDL
Delta          T  N                A
Gamma      G F E D ATI S         EDA   T             D   N
Epsilon      F A   A   S  S M     E A               D   N
Myoglobin  G-LSDGEWQLVLNVWGKVEADIPGHGQEVLIRLFKGHPETLEKFDKFKHL

                 6     *   7         8         9     *   0
Alpha      SH-----GSAQVKGHGKKVADALTNAVAHVDDMPNALSALSDLHAHKLRV
Zeta       HP       RELRA  S  VA VGD  KSI   IGG   K  E   YI
Beta       STPDAVMGNPKVKAHGKKVLGAFSDGLAHLDNLKGTFATLSELHCDKLHV
Delta      S                                      SQ
Gamma      SAS I              TSLG AIK    D       Q
Epsilon    S S IL             TS G AIKNM       PA   K
Myoglobin  KSEDEMKASEDLKKHGATVLTALGGILKKKGHHEAEIKPLAQSHATKHKI

                 1         2         3         4         5
Alpha      DPVNFKLLSHCLLVTLAAHLPAEFTPAVHASLDKFLASVSTVLTSKYR
Zeta                  RF SD   AEA  AW      SV   S    E
Beta       DPENFRLLGNVLVCVLAHHFGKEFTPPVQAAYQKVVAGVANALAHKYH
Delta                  RN        QM
Gamma          K       T   I     E    SW  M T   S  SSR
Epsilon        K     M II  T     E     W  L S   I
Myoglobin  PVKYLEFISECITQVLQSKHPGDFGADAQGAMNKALELFRKDMASNYKEL

                 6
Myoglobin GFQG
```

FIGURE 1.   Amino acid sequence comparison of the human globin chains.[2] Gaps, indicated by dashes, have been introduced to maximize homology. Sequence of the ζ chain is shown only where it differs from that of the α chain. The δ or γ sequences are displayed similarly with respect to the β chain. The asterisks indicate the distal and proximal histidines.

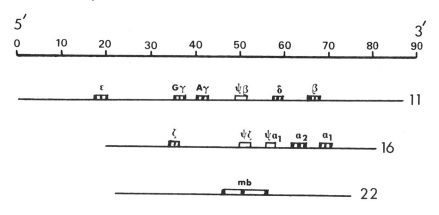

FIGURE 2.   The human globin genome. The genes and the nontranslated regions between them are shown approximately to scale. Pseudogenes are represented by empty boxes.

two closely linked γ genes, and a pseudo-β gene. The pseudogenes and other significant features of the globin genes will be discussed in more detail below. The myoglobin gene has recently been shown to be located on chromosome 22 in the region 22q11 → 22q13[10] and is thus linked neither to the α or the non-α cluster, even though it is clearly evolutionarily linked to both. The human globin genome, which is illustrated in Figure 2, appears to be generally representative of all the higher vertebrate globin genomes in the sense that the α and the non-α loci are grouped into two unlinked clusters. However, in the amphibian

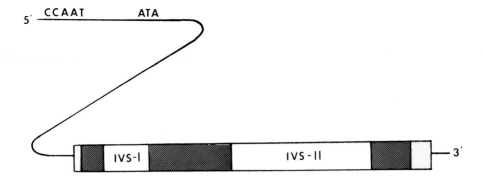

FIGURE 3.   A typical human globin chain gene. The two introns are labeled IVS-I and II. The exons are shaded darkly while the nontranslated regions 5′ and 3′ to the structural gene are shaded lightly. The control regions 5′ to the gene are shown by their respective sequences.

Xenopus tropicalis, the α and the β loci are linked.[11] Whether this is truly representative of globin phylogeny or merely a curious feature of this particular species has yet to be established. It is noteworthy that the closely related *X. laevis* has two unlinked α-β clusters as a result of tetraploidization.[11] It would be of considerable interest to characterize more amphibian and piscian globin genomes as well as those of certain invertebrates such as the midge (Insecta) or the Arc clams (Mollusca) which have hemoglobin and which, from a biological standpoint, are more primitive than reptiles, birds, and mammals.

E. Structure of the Globin Genes and Flanking Regions
One of the most remarkable and unexpected discoveries in recent years has been the finding that most if not all eukaryotic genes are made up of sections of structural information separated by nontranslated regions or intervening sequences. Moreover, regions external to the structural genes have been found to be essential to the expression of the structural genes. In this section, the principal structural features at this level of organization will be considered.

1. Structural Genes
All globin genes studied to date contain three structural segments of exons separated by two nontranslated regions, the introns. The introns are bounded by specific sequences known as splice junctions that are recognized by endonucleases which catalyze the splicing out of the intron portions of the RNA transcript to yield functional mRNA. No function is known for the nucleotide sequence of the intron, although a regulatory role has recently been proposed based on observed homology between promoters and introns.[82] The structure of the human globin genes is shown in Figure 3.

In most proteins whose genes are broken by introns, the exons code for the synthesis of a distinguishable domain of the protein.[12] The globin chain, on the other hand, shows no such domain structure. Nevertheless, it has been pointed out that the portion of the chain encoded by the second exon contains all heme-binding regions while that encoded by the third exon contains the interchain contact residues.[13-15]

2. Flanking Regions
In addition to introns and exons, the functional globin genes contain regulatory sites. All normal, functional genes appear to contain two essential regions known respectively as the ATA box and the CCAAT box.[6] The former is the RNA polymerase II attachment site and is always located 31 ± 1 bp 5′ to the cap site. The precise role of the CCAAT box, which is about 77 bp 5′ to the cap site, is not known. In the δ gene, which is translated less efficiently than the other non-α genes, the CCAAT sequence has been replaced by CCAAC,

which may in some way account for the diminished production of δ mRNA. There is also a short deletion just 5′ to the ATA box. There are other sequences in the flanking regions which are not translated but which may have some regulatory function. These include repetitive DNA sequences in the 3′ region[16] and so-called Alu family repeats in the region between the two γ genes and the δ-β genes.[17]

A 1500 bp region in the β 5′ flanking region has been sequenced and compared with similar regions flanking the δ and γ genes.[18] There was much less homology than expected, although significant homology was observed when the region was compared with mouse major (but not minor), rabbit and goat β 5′ flanking regions. Similarly, the α chain 3′ flanking in the human are unexpectedly divergent, showing less homology than expected.[19]

3. Pseudogenes

A curious feature of the globin genome is the presence of genes which bear a clear structural homology to specific functional genes but which are unable, for one reason or another, to be transcribed or translated.[20,21] Three pseudogenes are known in the human globin genome, φα, φβ, and φζ. Rabbit, mouse, and goat are also known to have pseudogenes associated with their globin genomes[20] as do the Old World monkeys.[22] Originally, a second pseudo-β gene was described but this was later shown to be an artifact of the method when an attempt was made to determine its sequence.[23] In many cases the pseudogenes have deletions which shift the reading frame so as to yield either a totally unrecognizable product or to generate terminator codons. There are also abnormalities in the 5′ flanking region.[6] One of the two loci originally identified as a second ζ gene was found by DNA sequencing to be a pseudogene by virtue of the fact that it had a terminator in the place of position 6 in the coding region.[21]

F. Developmental Aspects of Hemoglobin Genes[22]

Two developmental features of the hemoglobin genome command attention. The first is the sequential nature of the expression of the globin genes. This has been touched upon earlier in this chapter. The ζ and ε genes are active for a brief period in the first few weeks of development. They give way, by an unknown mechanism, to the synthesis of α and γ chains. The γ chains, in turn are replaced by β chains. The sequence is summarized in Figure 4. The nature of the switching mechanism is unknown, although it has been studied extensively as a potential strategy for the treatment of thalassemia and sickle cell anemia. Comparison of deletions leading to β thalassemia, δ-β thalassemia, and hereditary persistence of fetal hemoglobin (HPFH) suggests that there is a region of DNA between the γ and δ genes which is the "switch region".[24] It has been suggested that a nearby Alu region may be the switch.[25] Even if the switching can be ascribed to a specific sequence of DNA, the question remains as to how it works and how it evolved.

The second developmental feature which is striking is that the genes are arranged on the chromosome in order of expression. This is true not only for man but also for the rabbit, the mouse, various primates, sheep, and goats[4,20,22] and may therefore be a widespread and perhaps universal characteristic. The significance of this observation remains something of a mystery. One of the hoary maxims of classical descriptive biology is that "ontogeny recapitulates phylogeny" — a reference to the at least superficial resemblance between various developmental stages and the lower forms which preceded man on the evolutionary scale. Whether this observed parallel between position and development carries any as yet undeciphered message of phylogenetic importance remains to be seen.

G. Relationship of Hemoglobin to Myoglobin

Myoglobin, the oxygen-binding protein of muscle, is generally looked upon as the archtype of the globins. It appears to have diverged from the hemoglobin-related globins from between

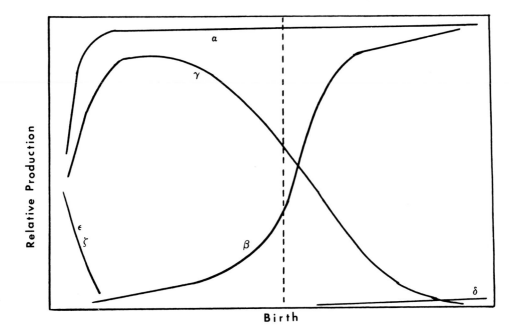

FIGURE 4.   The developmental sequence of globin chain production. The relative production of the chains is shown as a function of time from conception. The embryonic chains disappear rapidly and are replaced by α and γ chains. The latter gives way to adult hemoglobin starting in the last trimester of pregnancy.

500 to 800 million years ago.[26,27] Myoglobin is monomeric and exhibits no cooperativity in its interaction with oxygen. Its tertiary structure is virtually identical with that of the globin subunits. Few variants of human myoglobin are known,[10,28] but this is not surprising in view of the fact that no readily available tissue such as blood exists as a source of myoglobin for study. Various other animal myoglobins, which are easier to obtain, have been studied.[29] Until recently, the location of the myoglobin gene within the human genome was completely unknown as was the fine structure of the gene. Now it has been shown that like the α and β loci, the myoglobin gene contains three exons and two introns.[30,31] The gene appears to be located on chromosome 22 in the long arm between 22q11 and 22q13.

This structure of the myoglobin gene is interesting in view of some of the speculations of Gò.[13] He has postulated that the heme-binding (and therefore the oxygen-binding) segment of the molecule is the middle exon, whereas the third exon is primarily the α_1-β_1 contact region. These results suggest that even though the myoglobin gene is of much greater antiquity than the α or β globin genes, it still has a single exon for the middle portion of the molecule and not two separated by a third intron. Thus, if the Gò model is correct, the evolution of the three-exon globin gene must have predated the evolution of myoglobin. It is interesting that all the myoglobins whose sequences have been reported have the invariant or nearly invariant sequences which Gò associates with the splice junctions including the ''vestigial'' sequence in the vicinity of position 60, which he suggests is a remnant of a 4-exon primordial gene. Leghemoglobin, the oxygen-binding protein found in some root nodules, is homologous to myoglobin and as described below, has recently been found to have a four exon gene.[32]

Another interesting question raised by the analysis of Gò is the significance in myoglobin of those portions of the molecule involved in quaternary interactions. If, as proposed by Eaton[12] and by Blake[14] the α_1-β_2 and α_1-β_1 contact regions were acquired in the course of evolution by incorporating as exons, genes, or parts of genes already possessing these functions, it seems unlikely that myoglobin, a monomer, would possess homologous exons.

Thus, the presence of these exons may constitute evidence that myoglobin evolved from a primordial hemoglobin which was oligomeric and probably therefore had at least weak cooperativity.

II. PHYLOGENY OF NORMAL GLOBIN GENES

The pattern of hemoglobin quaternary structure featuring two α-like and two β-like globin chains in a tetramer with the chains encoded by unlinked, homologous genes seems to have emerged about 500 million years ago. This date roughly coincides with the date generally assigned to the emergence of the vertebrates in the Cambrian period of evolutionary history.[33] There is evidence that the primitive globin gene itself may have evolved, probably by gene duplication and divergence,[26,34,35] from some ancient heme-binding protein which also formed the evolutionary raw material for other modern heme proteins such as some of the cytochromes.[36] These and other points have been well reviewed in the vast literature on the evolution of hemoglobin[37-41] and will not be discussed further in this chapter.

The relatively recent discovery of fine-structure of the globin genes which is common to all has suggested a route by which separate genes of different functions could become integrated into one gene or how a single gene could contribute to several genes with diverse functions. The key to such a mechanism is the exon. In many recently studied gene-protein pairs, each exon corresponded to a domain of the protein with each domain connected in the protein by a short, random coil sequence. While this is not the case in the globin gene-protein pair, there are nevertheless provocative relationships evident when the globin chains are examined at the exon level. Gò[13] has applied the techniques used in domain plotting and shown that globin has three "sub-domains" or compact structures which correspond to the three exons. Of these, the third contains virtually all the α_1-β_1 contact sites, whereas the middle sub-domain contains the α_1-β_2 contact sites plus the heme binding sites. Moreover, Gò's analysis indicated that the middle exon may once have been two separate exons; i.e., the middle exon may have evolved from two exons by an unequal crossing over to eliminate the intron. Support for this prediction has come from studies of leghemoglobin, the oxygen-binding protein from the root nodules of some nitrogen-fixing plants. Soybean leghemoglobin has four exons in its gene[32,42-44] and may thus represent the ancestral four exon globin gene predicted by Gò. Consistent with Gò's prediction, the intron between the second and third exons, IVS-2, separates the distal and proximal heme-binding sites so that from a functional standpoint, the exons represent separate domains.

In addition to being homologous to myoglobin, the exons containing the heme-binding sequences show homology to some of the cytochromes,[36,45] suggesting that all heme proteins may be descended from a common ancestor. However, caution must be exercised in interpreting the homology between the plant protein and animal hemoglobins. Hyldig-Nielsen et al.[32] have calculated that if it is assumed that this homology bespeaks a common ancestor, such an ancestor would have had to have existed 1500 million years ago, which is impossible. They have therefore suggested that a virus vector may have transfected the plant genome in some more recent epoch so that its existence in the plant kingdom could be viewed as a borrowed gene rather than an evolved gene. Such an arrangement would be highly adaptive since the process of nitrogen fixation, the sole function of the root nodules, is inhibited by molecular oxygen, whereas the leghemoglobin acts as an antioxidant, rapidly and efficiently scavenging intruding oxygen molecules.

Two of the major factors which appear to have contributed to the evolution of those globin genes which we today recognize as normal are gene duplication and concerted evolution. Gene duplication was vigorously promoted 20 years ago by Ohno et al.[35] as a major mechanism of evolution. There seems to be little question that this process was indeed the mechanism by which many of the features of the globin genome arose. These features will be discussed below in the appropriate sections. The process called concerted evolution is a much newer concept.[46] It proposes that when two homologous genes lie adjacent to one another on a chromosome, they may exchange segments of DNA so that they tend to evolve as a pair rather than diverge. Hess and co-workers[83] have recently found that 5′ to 3′ homology gradients exist in the human α-chain cluster which support a polar correction process in which crossovers originate at "hotspots" and extend for random distances in the 3′ direction. Such a mechanism would clearly serve to buffer the effects of mutation since the pair could rid itself of an unfavorable mutation by replacing it with an unmutated segment from the sister gene or conversely, preserving and amplifying a desirable mutation. The situation would be somewhat analogous to the practice in computer operation whereby a backup copy of a program is made before experimenting with modification. If the modification produces the desired improvement, the backup copy can be improved by transferring the appropriate section of the program to it. If the modification proves to be undesirable, the operator can replace it from the backup copy.

There are exceptions to this mechanism. Presumably the pseudogenes are the product of duplications which have undergone mutation. It is not known how these genes escaped the protective effect of concerted evolution to diverge so extensively.

A. α-Globin Genes

1. Adult

The α-globin genes have clearly duplicated at some time in their evolutionary history. The duplicated α gene appears to be widespread in the vertebrates and is presumably of considerable antiquity. The occurrence of a duplicated α gene in the chicken would suggest that this feature must be at least 200 million years old. On the basis of mutation rate alone it would be expected that these two loci should show substantial difference, yet in man the amino acid sequences of the two loci are identical. This is most likely a consequence of concerted evolution as discussed earlier and shows the great power of this evolutionary mechanism to stabilize or temper evolutionary change. It should be noted that α chains appear to be more resistant to change than non-α chains, perhaps because they contribute to hemoglobin structure both *in utero* and at all growth stages after birth. Thus, the structure/function requirements may be more stringent for α chains than for non-α chains.

The exact extent of the α chain duplication has been determined by nucleotide sequence analysis.[46] The duplicated unit appears to extend from the poly(A) attachment site of the pseudo-α_1 gene to a point 15 bp to the 3′ side of the poly(A) addition site associated with α_1. This 50 bp segment has the sequence GCCTGTGTGTGCCTG, which can be viewed as three pentanucleotides where the first and third are identical. Proudfoot and Maniatis[47] have observed that this sequence occurs in both α genes and in the pseudo-α gene and have suggested that it may be some sort of delimiter in regulating the gene duplication process. Beyond that, unexpected divergence in the 3′ region has been observed.[18]

2. Embryonic

The ζ chain exhibits about 65% homology to the α chain from which it probably arose by gene duplication.[16] Of the 26 subunit contact sites in the α chain, 21 are conserved in the ζ and of 15 heme contact sites, 12 are conserved. It has not been possible to estimate the time of divergence of the ζ genes but it must have been quite early since the mouse, the sheep, and the rabbit all have an embryonic α gene in a homologous position on their respective α-chain chromosomes.

B. Non-α Genes

The non-α genes consist of the β, the δ, and two γ and the ε genes. Also included is the pseudo-β gene. The evolutionary relationship of these genes as deduced from structural homology is seen in Figure 5.

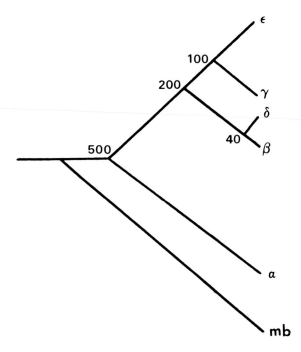

FIGURE 5. The globin family tree. The symbols γ and α represent both γ-chain genes and α-chain regnes, respectively, since it has not been possible to date their divergence. The figures represent divergence times in millions of years. Mb is myoglobin.

1. Adult

The β and δ genes appear to be the most recent divergence in the globin genome. However, the possibility that there exists in evolution an internal or selfcorrecting mechanism makes such estimations risky. There are differences between the these two genes in 10 positions which has led investigators[48,50] to calculate a time of divergence of 40 million years. This figure is in good agreement with the time as deduced from the report by Dayhoff et al.[26] that all higher primates have a δ gene since the human-higher primate divergence has been estimated to be 35 to 40 million years ago.[33]

2. Fetal

The near identity of the A-γ and G-γ chains would suggest that their gene duplication was quite recent. However, this appears not to have been the case and illustrates vividly the impact of concerted evolution. Slightom et al.[51] have compared the entire duplication, a region of about 5 kb, and have concluded that, except for a 1.5 kb region which includes the 5' two thirds of the structural gene, and which is strongly conserved, there is a 10 to 20% divergence. At 1% divergence per 10 million years[48] that corresponds to a duplication of the γ chain gene between 100 to 200 million years ago followed by a concerted mechanism to conserve the γ sequence virtually unchanged.

3. Embryonic

The embryonic non-α gene is the ε gene which, as has been pointed out above, is 5' to the γ loci. The ε chain differs in humans from the β chain at 24 positions and from the γ chain at 18 positions.[4] Estratiadis et al.[48] have placed the time of divergence of this gene from the β line at about 200 million years or about the time of divergence of birds and mammals from the reptiles.[33] This estimate is consistent with the known occurrence of an embryonic non-α gene in chickens as well as in a wide variety of mammals.

III. EVOLUTION OF ABNORMAL HEMOGLOBIN GENES

A. Types of Abnormal Globin Genes

Today over 400 abnormal or variant globin types are known.[84] The majority of the these are variants of the adult globin genes for the α and the β chains, although a growing number are variants of the γ chain; i.e., are fetal hemoglobin variants, and a few δ chain variants are known. No variants of the ε or ζ chains have been reported. Of the adult hemoglobin variants approximately one third are α chain variants while the balance are β variants. Although relatively few of these are associated with clinical manifestations, it is generally supposed that the reason for this marked imbalance is that the α chains are a component of fetal hemoglobin and must function at a time when tissue oxygenation is perhaps more critical or, more accurately, at a time when the body may be less able to adapt to compromises in the supply of oxygen. This conclusion is supported by the fact that the majority of abnormal hemoglobins which have functional abnormalities are β-chain variants.

A variety of types of structural abnormalities are known to occur in hemoglobin. These include point mutations, crossovers, deletions, insertions, and frameshifts. Double substitutions are also known. However, these almost certainly are the result of one of only two events at the level of the DNA: single base substitutions or crossovers. Similarly, all the thalassemias probably can be explained by one or the other of these two mechanisms.

1. Point Mutations

Point mutations, when they occur in one of the globin exons, lead either to amino acid substitutions (missense) or to chain terminations (nonsense). Most of the hemoglobin variants are the result of missense mutations and as a result, these are the best known consequence of point mutations and need little amplification here. Comparison of homologous sequences both between and within species suggests that this type of point mutation is a major factor in the evolution of hemoglobin genes. There is a third possibility which is often overlooked and yet, which probably also plays a significant role in providing the variability which is the raw material of evolution. That is the possibility of degenerate mutations which alter the nucleotide sequence but which in so doing generate a "synonym" condon such that there is no amino acid change. That this has happened at least once is suggested by the finding of Hb Bristol and Hb M Milwaukee I. The former is β67 val → asp whereas the latter is β67 val → glu. Thus, consideration of Hb Bristol suggests that the nucleotide sequence at position 67 of the β chain would have to be GUU/C since aspartic acid is coded for by either GAU/C, whereas Hb M Milwaukee I requires by the same argument that the sequence at position β67 be GUA/G. It is therefore evident that there is some variability within codons, even in the absence of amino acid sequence abnormalities.

The point mutations which lead to premature termination of the chain comprise the nonsense mutations. The use of the term "nonsense" to describe the codons UAG, UAA, and UGA was an unfortunate historical accident. Early students of the genetic code could assign no amino acid coding function to these codons and gave them the name "nonsense" to denote this fact since those leading to amino acid incorporation were "sense". However, the punctuation of the genetic "sentence" is anything but nonsense! Such mutations generally lead to chains shortened so drastically that they do not exhibit any detectable globin function (binding heme or forming tetramer) and so present as thalassemia.

Another manifestation of point mutations which is frequently set apart as a separate mutation type is the terminator mutation. These mutations, which are a "reverse nonsense" mutation, result from point mutations in the terminator codons which result in the loss of terminator function and the appearance of an extension variant. Since most globin genes are about the same size, it appears unlikely that either terminator or extension variants played a major role in the evolution of globin genes except possibly in very early times before the divergence of the modern chain types.

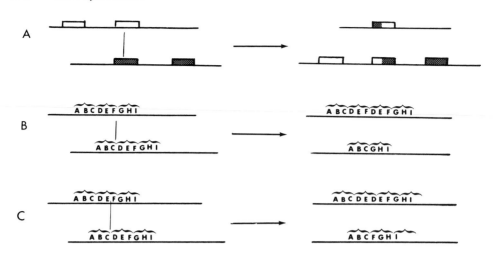

FIGURE 6.   Three crossover mechanisms. The crossovers are shown at the left and the product chromosomes to the right. The site of the cross-over is shown by the vertical line. (A) is a nonhomologous crossover resulting in fusion genes such as the Hb Lepore variants. (B) shows an in-register crossing over within a gene. This produces either an insertion with repeat (upper) or a deletion (lower). (C) is an out-of-register crossing over within a gene. The reading frame is shifted downstream from the cross-over so that no recognizable amino acid sequence remains.

When point mutations appear in introns, the results can be unpredictable. When the mutation lies at a splice junction, normal processing is blocked either wholly or partially and a thalassemia results. Point mutations within the introns are now known to activate "cryptic sites"; sequences of nucleotides which differ from normal splice junctions by only one nucleotide such that the mutation has supplied the missing nucleotide and splicing now proceeds at an inappropriate site. These mutations also tend to produce thalassemias. As in the case of terminator mutations, there is no reason to suspect that intron mutations have played a major role in the evolution of globin chains. However, as will be discussed below, thalassemias may be adaptive and some of these abnormalities may have gone to fixation in certain populations.

2. Fusion Globins, Insertions, Deletions, and Frameshifts

As was stated above, it seems likely that all mutations which are not point mutations are probably crossovers, even though the molecular manifestations may vary. There are three types of crossovers which are important in the globin genome: the classic unequal crossover involving nonhomologous pairing, in-register homologous pairing, and out-of-register homologous pairing (Figure 6). The first of these is best known and leads to the group of variants known collectively as the fusion hemoglobins. These include the Hb Lepore group which have the N-terminus of the δ chain and the C-terminus of the β chain, the anti-Lepore or Hb Miyada which begins with a β sequence and ends with a δ sequence, and Hb Kenya which is Lepore-like except that the sequence begins with a γ chain sequence. A double crossover is also known, Hb Parchman which is reminiscent of the classical geneticist's three point test cross, except on a smaller scale.[52] In all these cases there is a mispairing but the two genes are aligned such that the product gene is the same length as the "parent" genes. The second type of crossover event involves the genes aligning in such a way that the gene-to-gene pairing is correct but the codon matching is "unequal". In this case either additions or deletions will be the result but, except for the region of the abnormality, the sequences are normal. The third possibility is an out-of-register pairing involving homologous chromosomes. This will yield the frameshift mutation where all amino acid sequences on the 3' side of the mispairing will have an alternate sequence and potentially an indeterminate length because the normal terminator will be misread and passed over.

Although the evidence is not compelling, it is interesting to speculate that crossovers of the second type at least may have played a significant role in the evolution of the globin chains in their present form. Examination of the globin sequences reveals many short repeat sequences. This feature is accentuated by the fact that a disproportionate number of the deletion mutants are indeterminate. That is, one cannot say exactly which residues have been deleted because the region from which the deletion has occurred contains one or more redundant sequences. Hb Gunn Hill is one, although perhaps not the best example. There are three different pentapeptide stretches that could account for the deletion. The potential significance of this is that if there are regions that are particularly prone to crossing over, then such regions might be expected to contribute both to the evolution of the globin as well as to its spectrum of variants.

3. Nonstructural Abnormalities: The Polymorphic Frameworks

In spite of the rather large number of abnormal globins which are known, the structural portions of the globin genes are still, on the whole, highly conserved. Much of the variability which must occur in the genome at large appears to reside in the nonstructural portions — in the introns and in the flanking regions, for example. This frequently takes the form of abnormalities in restriction endonuclease sensitive sites. Such variability was first reported by Jeffreys.[53] Cloned DNA was being examined as Pst I fragments when a fragment resulting from an abnormal Pst I cleavage site was detected.[53] Kan and Dozy[54] similarly reported a Hpa I polymorphism which proved instrumental both in the prenatal detection of sickle cell anemia and in studies of the evolution of the β_S gene. Now a large number of these sites are known and as a result, it has been possible to identify groups or clusters of these restriction sites and to relate them to particular populations or ethnic groups. These are known as polymorphic frameworks and have become a sort of molecular "coat of arms" in identifying the racial and ethnographic history of a particular gene which occurs or is associated with it.

B. Mechanisms of Evolution of Abnormal Globin Genes

There are two general mechanisms by which a mutation can become established or "fixed" in a population and the hemoglobin variants appear to provide examples of both. Either the mutation has some adaptive value, i.e., it renders the carriers more fit or provides them with a selective advantage, or else it has the good fortune to find itself in an individual or population which is favored for survival for some reason entirely unrelated to the mutant gene. For a more complete discussion of the evolutionary relationship between organisms and molecules, see the article by Simpson.[55]

1. Selection and the Malaria Hypothesis

Hemoglobins S, C, and E, as well as the thalassemias, are distributed worldwide in such a way that in many areas of the world, their combined ranges virtually define the range of falciparum malaria. This observation has given rise to the "malaria hypothesis" which simply states that certain hemoglobinopathies, when present in the carrier or heterozygous condition, provide protection from falciparum malaria. In the case of sickle trait erythrocytes, the proposed mechanism for this protection is that trait cells, when occupied by the falciparum malaria merozoite, are predisposed to sickling and are thus perceived by the spleen as abnormal and destroyed. The merozoite are destroyed in the process. It has also been proposed that the sickle hemoglobin-containing cells have low intracellular potassium and provide an inhospitable environment for the parasites which have a high potassium requirement. Either way, this process appears to have provided a powerful basis for selection. Direct proof of the malaria hypothesis has been elusive at best. The coincidence of geographic distribution of the variants and the disease make a compelling albeit circumstantial case but so far may be the best evidence that malaria is the selective factor responsible for the relatively high frequency of these variants. The classical approach was to try to estimate the fitness of the various genotypes in their native setting but these attempts were less than satisfactory. Since it has become possible to culture the parasite in vitro, numerous attempts have been made to evaluate red cells of various genotypes as hosts for the organism.[56] Again, these did not provide the clear cut answer that was desired although various investigators have been able to show that the parasite grows less well in the presence of Hb S and Hb F,[57-62] suggesting that the sickling process itself is not an essential feature of selection. Durham[49] has developed a novel approach to testing the malaria hypothesis directly. He has shown that rainfall is closely linked to mosquito density which, in turn, is a good predictor of the rate of parasitemia. He has shown that there is a significant correlation between the frequency of the sickle gene and the rainfall (i.e., the malaria rate) in several populations in West Africa. There were exceptions, but these were readily explicable.

Nevertheless, it may be generally said that the strongest argument in favor of the malaria hypothesis in accounting for the geographic distribution of Hb S, Hb C, Hb E, and the thalassemias is the fact that there is no other tenable explanation for the distribution.

2. Drift and Founder Effect

The second major mechanism which can account for the distribution of abnormal hemoglobins is chance alone. This takes two closely related forms: drift in which a gene "piggybacks" on another upon which selection is acting and founder effect in which genes present by chance alone in a small population make a major contribution to the gene pool when the population expands. Both of these mechanisms are represented among the abnormal hemoglobin.

C. Evolution of Abnormal Hemoglobins

The great evolutionist George Gaylord Simpson[55] has espoused the view that molecules do not evolve, only organisms evolve. This somewhat semantic argument is challenged by many modern students of evolution who regard the stepwise response of molecules to selection pressure to be legitimate evolution.[37,39,63] It is therefore reasonable to ask the question, "have abnormal or variant hemoglobins evolved?" Depending on the definition of evolution, the answer to that question is more or less elusive. If evolution at the molecular level is thought of as a series of small discrete changes in structure arising by chance, existing in the gene pool, and becoming fixed by gradual selection then the only variant hemoglobins that can be said to have evolved are Hb S, Hb C, and Hb E. If, on the other hand, evolution is taken to mean merely the genetic history of a variant, then all abnormal hemoglobins have evolved. However, returning to the first definition, even the evolution of these three may be questioned in that the selective forces which have led to the establishment of these major polymorphisms are not selected for the function of hemoglobin as an oxygen carrier but for resistance to a disease. As hemoglobins, these appear to be no better and, in the case of Hb S and perhaps Hb C, may in fact be worse than Hb A. Falciparum malaria itself is a relative newcomer in the evolutionary picture and has established itself as a major human parasite largely as a result of increases in the population density and the relative "urbanization" of the rain forests, i.e., the practice of living in villages and of managing water in such a way as to inadvertently promote the propagation of mosquitoes. Since malaria remains a major factor in mortality and morbidity in the tropical regions of the earth, it may prevail in the time framework of evolution and the malaria-resistant hemoglobin mutants may prove to be a passing chapter in evolution rather than a milestone in the process. This is the point that Simpson[55] seeks to make when he points out that at the gene product level, evolution may appear to run backward because it is the whole organism on which selection operates.

In the following sections, the major features of the evolution of selected hemoglobin variants will be summarized. With over 400 variants known, this is by no means an exhaustive list. However, these variants have been selected as representative of the major mechanisms of evolution of abnormal hemoglobins.

1. Hb S

Hb S is the hemoglobin variant most widely studied from the evolutionary standpoint.[64] Prior to the widespread availability of gene mapping with restriction endonucleases, considerable speculation surrounded the question of how many times the gene arose in the course of human history. The presence of Hb S has been documented in India,[65] Saudi Arabia,[66] and the Mediterranean, as well as in the well-known equatorial African setting.[67] Kan and Dozy[54] conducted the first extensive study of the distribution of the β_S gene with respect to the 13/7.6 kb Hpa I marker system. They found a state of genetic disequilibrium in which the gene in West Africa and in the Mediterranean tended to be associated with the 13 kb marker, whereas Central and East African β_S as well as the Arabian and Indian genes were linked to the 7.6 kb marker. They proposed at least two mutational events, one in West Africa on a pre-existing 13 kb-carrying chromosome and the other elsewhere on a 7.6 kb-carrying chromosome with subsequent spread of the former into North Africa and the Mediterranean. The occurrence of the β_C gene on a 13 kb-bearing chromosome supports the notion that the 13 kb marker preceded the mutational events and is indigenous to West Africa. The β_C gene is assumed on good genetic grounds to have arisen independently of the β_S gene.

The dicentric origin hypothesis of Kan and Dozy was further examined by Boyer et al.,[68] who considered a second polymorphism, the Hind III site. A linkage disequilibrium similar to that for the Hpa I polymorphism was observed. They argued that while in their view, the data of Kan and Dozy were insufficient to make any judgment about the number of sites of origin of the mutation, there are theoretical reasons supported by the observed genetic disequilibrium, for believing the number of origins to be at least ten. This conclusion is consistent with recent, more extensive studies of endonuclease haplotypes in various populations.[69] Inherent in these calculations is, of course, the selection pressure of endemic falciparum malaria in those parts of the world in which the frequency of the sickle gene has reached polymorphic proportions. As discussed above, the overwhelming evidence is that malaria is the chief and probably only selective force acting to fix this otherwise deleterious gene in the population.

2. Hb C

It is still uncertain whether Hb C affords any protection from malaria and, if not, by what mechanism it became fixed in the West African population where it is found. The range of this variant is much more restricted than that for Hb S. The natural range, i.e., the range exclusive of distribution secondary to slavery, urbanization, and other nonbiological influences, is south of the Niger River in Upper Volta and the northern reaches of the Ivory Coast, Ghana, Togo, and Benin. The β_C gene is found exclusively associated with the 13-kb fragment and thus there is no evidence for its having arisen more than once.[54]

3. Hb E

Hb E is found primarily in Southeast Asia,[70] but is known to occur in India,[65] Saudi Arabia,[66] and occasionally in Europe.[71] On a worldwide basis, it is the second most common variant.[72] As is the case with Hb S, its distribution suggests that it has evolved in response to the selection pressure of malaria. The mechanism of this resistance may be the mild β-thalassemmia that accompanies this variant. This thalassemia, only recently recognized, is the result of the creation of a cryptic splice junction at position $\beta26$, the site of the substitution glu → lys. This brings about the synthesis of defective mRNA and results in the underproduction of the variant and β chains and accounts for the mild thalassemic phenotype.[73-76]

The possibility that Hb E had multiple origins was first suggested by Antonarakis et al.,[77] who found that the gene occurs in two different restriction frameworks in Southeast Asia. These same workers subsequently found that the β_E gene in Europeans occurs in yet a third framework, one that is unknown in Southeast Asia.[71] This indicates the likelihood of at least three separate mutational origins, although the authors have raised the possibility of mechanisms other than mutation to account for the origins of the gene in the respective frameworks.

4. Hb Constant Spring

Like Hb E, Hb Constant Spring is found primarily in Southeast Asia. It is an extension variant resulting from a single base substitution in the α chain terminator codon. This abnormality leads to a severe underproduction of α chains with a resultant α-thalassemia-like condition. The frequency of Hb Constant Spring has been reported to be as high as 6 to 8% in some parts of Southeast Asia, suggesting the operation of a selective force. Once again, malaria seems to be the most likely possibility because the concomitant thalassemia-like condition would seem to be capable of conferring some measure of resistance to that disease. However, there is no experimental evidence to support that conclusion.

5. Thalassemia

The term thalassemia refers collectively to abnormalities resulting in underproduction of one or more of the globin chains. The under-produced chain is named in naming the thalassemia. Thus, α-thalassemia is an underproduction of the α chain. In the early days of the study of inherited hemoglobinopathies, it was tacitly supposed that all thalassemias had the same underlying mechanism and nomenclature that described the severity of the anemia. It is now known that thalassemias are the result of an amazing array of different molecular defects, the details of which are beyond the scope of this chapter. In general, however, three types of defects can be recognized as producing the phenotype of the thalassemia: major deletions, mutations leading to deficiencies in the production of mRNA, and mutations leading to unstable or grossly abnormal globin chains.[78] All thalassemias lead to some degree of microcytic hypochromia with the exception of α-thal 2, the deletion of a single α locus where the effect is too small to be detectable. As with many of the higher frequency structural variants, the thalassemias tend to be distributed in those parts of the world where falciparum malaria is endemic. Again, the implication is that some feature of the thalassemic erythrocyte, perhaps the elevated Hb F or some membrane abnormality stemming from oxidative damage, renders it a relatively inhospitable environment to the trophozoite phase of falciparum malaria so that this widespread disorder may be the primary, if not the only, selective force active on the hemoglobin genome.

A number of elegant studies have been done showing that particular thalassemias apparently are associated with specific populations. As mentioned above, polymorphic restriction endonuclease sites can be grouped into frameworks which are in many cases characteristic of particular populations.[69] Specific defects which lead to the thalassemic phenotype have been found to be associated definite restriction frameworks.[79]

6. Hb F Sardinia

The evolution of Hb F Sardinia, γ75 ile → thr, remains one of the tantalizing mysteries of the phylogenetics of hemoglobin. This variant of the A-γ chain has a distribution that transcends both race and geography and yet there is no evidence for any functional alteration which would provide a basis for selection. However, it has been observed that although the variant is virtually absent in Ghanaians who are homozygous for sickle cell hemoglobin, it is present in significant proportion (0.150) in Ghana in persons with Sβ-thalassemia.[80] This suggests that the gene is linked to β-thalassemia and may have become fixed because of the tight linkage to the thalassemia locus which, in turn, provided for selection in the presence of malaria.

IV. EVOLUTION OF OXYGEN-BINDING PROTEINS: AN OVERVIEW

It is becoming increasingly apparent that from the standpoint of evolution, it is appropriate to view proteins in terms of a relatively small number of functional families. Thus, the oxygen-binding proteins comprise one of the major families. The origins of the heme-iron-protein complex as an oxygen-binding entity are forever lost in time, but the ubiquity of this strategy suggests that it must have been one of the earlier biochemical mechanisms to appear. Gò, Eaton, Blake, and others have examined the fine structure of modern globin genes for evidence of their evolutionary past. While these genes lack the domain structure which might have been expected, a number of revealing and exciting conclusions have been drawn by comparing the functional aspects of different regions of the molecule with the fine structure of the gene — the exons.

From this standpoint, the leghemoglobin gene may the best modern representative of the primordial globin gene. The leghemoglobin gene consists of four exons rather than three as are found in all vertebrate hemoglobins. The four exon structure was predicted by Go on the basis of his analysis of the C-C distances in three exon globins. The striking feature of this oxygen-binding protein is that each of the exons, especially exons 2, 3, and 4, appears to have a function which may have been its "reason" for being in the molecule in the first place. This modular concept of protein evolution proposes that genes evolved, in part, by incorporating intact functional units from other protein genes. Thus, the amino acid sequence of the second and third exon of leghemoglobin gene codes for amino acid sequences that bind the distal and proximal side of the heme molecule, respectively. The central exon of the vertebrate globin genes appears to have lost its intron and thus to be a fusion exon composed of primordial exons 2 and 3.

The ancestral hemoglobin is thought to have been a monomer, probably with poor cooperativity.[81] From this progenitor, through gene duplication, mutation, and crossing over, the myoglobin line appeared followed by the α/β divergence. This mechanism enabled a variety of functional properties to evolve including increased cooperativity, regulation by pH (Bohr effect), and by ambient pO_2 (organic phosphate binding).[63] Subsequent duplications have led to the modern hemoglobin genome as it occurs in the vertebrates, and occasionally in the invertebrates and plants. Viewed as a whole, the hemoglobins provide an evolutionary panorama reflecting the evolutionary biological response to the widespread need to control access to molecular oxygen. At present there appears to be no evidence to suggest that any of the 400 or so human hemoglobin variants represent further evolution of hemoglobin as a regulated oxygen carrier, but rather its adaptation to use as a "weapon" against the microscopic enemies of the red cell, sometimes to the detriment of its primary function.

REFERENCES

1. **Perutz, M. F.,** Nature of the haem-haem interaction, *Nature (London)*, 227, 495, 1972.
2. **Dickerson, R. E. and Geis, I.,** *Hemoglobin,* Benjamin Cummings, Menlo Park, Calif., 1983, chap. 3.
3. **Weatherall, D. J. and Clegg, J. B.,** Recent developments in the molecular genetics of human hemoglobin, *Cell,* 16, 467, 1979.
4. **Fantoni, A., Farace, M. G., and Gambari, R.,** Embryonic hemoglobins in man and other mammals, *Blood,* 57, 623, 1981.

5. **Dover, G. J. and Boyer, S. H.,** The cellular distribution of fetal hemoglobin: normal adults and hemoglobinopathies, *Tex. Rep. Biol. Med.,* 40, 43, 1980—81.

6. **Proudfoot, N. J., Shander, M. H. M., Manley, J. L., Gefter, M. L., and Maniatis, T.,** Structure and in vitro transcription of human globin genes, *Science,* 209, 1329, 1980.

7. **Barelle, F. E., Shoulders, C. L., and Proudfoot, N. J.,** The primary structure of the human ε-globin gene, *Cell,* 21, 621, 1980.

8. **Maniatis, T., Fritsch, E. F., Lauer, J., and Lawn, R. M.,** The molecular genetics of human hemoglobins, *Ann. Rev. Genet.,* 14, 145, 1980.

9. **Deisseroth, A., Neinhuis, A., Lawrence, J., Giles, R., Turner, P., and Ruddle, F.,** Chromosomal localization of human beta globin gene on human chromosome 11 in somatic cell hybrids, *Proc. Natl. Acad. Sci. U.S.A.,* 75, 1456, 1978.

10. **Jeffreys, A. J., Wilson, V., Blanchetot, A., Weller, P., van Kessel, A. D., Spurr, N., Solomon, E., and Goodfellow, P.,** The human myoglobin gene: a third dispersed locus in the human genome, *Nucleic Acids Res.,* 12, 3235, 1984.

11. **Jeffreys, A. J., Wilson, V., Wood, D., Simons, J. P., Kay, R. M., and Williams, J. G.,** Linkage of adult α and β globin genes in *X. laevis* and gene duplication by tetraploidization, *Cell,* 21, 555, 1980.

12. **Eaton, W. A.,** The relationship between coding sequences and function in haemoglobin, *Nature (London),* 284, 183, 1980.

13. **Gò, M.,** Correlation of DNA exonic regions with protein structural units in haemoglobin, *Nature (London),* 291, 90, 1981.

14. **Blake, C. C. F.,** Exons and the structure, function and evolution of haemoglobin, *Nature (London),* 291, 616, 1981.

15. **Blake, C. C. F.,** Exons encode protein functional units, *Nature (London),* 277, 598, 1979.

16. **Forget, B. G., Tuan, D., Biro, P. A., Jagadeeswaran, P., and Weissman, S. M.,** Structural features of the DNA flanking the human non-alpha globin genes: implications in the control of fetal hemoglobin switching, *Trans. Assoc. Am. Phys.,* 94, 204, 1981.

17. **Hess, J. F., Fox, M., Schmid, C., and Shen, C.-K., J.,** Molecular evolution of the human adult alphaglobin-like gene region: insertion and deletion of Alu family repeats and non-Alu DNA sequences, *Proc. Natl. Acad. Sci. U.S.A.,* 80, 5970, 1983.

18. **Moschonas, N., de Boer, E., and Flavell, R. A.,** The DNA sequence of the 5' flanking region of the human β-globin gene: evolutionary conservation and polymorphic differences, *Nucleic Acids Res.,* 10, 2109, 1984.

19. **Michelson, A. M. and Orkin, S. H.,** The 3' untranslated regions of the duplicated human α-globin genes are unexpectedly divergent, *Cell,* 22, 371, 1980.

20. **Little, P. F. R.,** Globin pseudogenes, *Cell,* 28, 683, 1982.

21. **Proudfoot, N. J., Gil, A., and Maniatis, T.,** The structure of the human ζ-globin and a closely linked, nearly identical pseudogene, *Cell,* 31, 553, 1982.

22. **Wood, W. G. and Weatherall, D. J.,** Developmental genetics of the human haemoglobins, *Biochem. J.,* 215, 1, 1983.

23. **Shen, S.-H. and Smithies, O.,** Human globin ψβ₂ is not a globin related sequence, *Nucleic Acids Res.,* 10, 7809, 1982.

24. **Huisman, T. H. J., Schroeder, W. A., Efremov, G. B., Duma, H., Mladenovski, B., Hyman, C. B., Rachmilewitz, E. A., Bouver, N., Miller, A., Brodie, A., Shelton, J. R., Shelton, J. B., and Apell, G.,** The present status of the heterogeneity of fetal hemoglobin in β-thalassemia: an attempt to unify some observations in thalassemia and related conditions, *Ann. N.Y. Acad. Sci.,* 232, 1107, 1974.

25. **Jagadeeswaran, P., Tuan, D., Forget, B. G., and Weissman, S. M.,** A gene duplication ending at the midpoint of a repetitive DNA sequence in one form of hereditary persistence of fetal hemoglobin, *Nature (London),* 296, 469, 1982.

26. **Dayhoff, M. O., Hunt, L. T., McLaughlin, P. J., and Jones, D. D.,** Gene duplications in evolution: the globins, in *Atlas of Protein Sequence and Structure 1972,* Dayhoff, M. O., National Biomedical Research Foundation, Washington, D.C., 1972, 17.

27. **Czelusniak, J., Goodman, M., Hewett-Emmett, D., Weiss, M. L., Venta, P. J., and Tashian, R. E.,** Phylogenetic origin and adaptive evolution of avian and mammalian hemoglobin genes, *Nature (London),* 298, 297, 1982.

28. **Dretzen, G., Bellard, M., Sassone-Corri, P., and Chambon, P.,** A reliable method for the recovery of DNA fragments from agarose and acrylamide gels, *Anal. Biochem.,* 112, 295, 1981.

29. **Hudgins, P. C., Whorton, C. M., Tomoyoshi, T., and Riopelle, A. J.,** Comparison of the molecular structure of myoglobin of fourteen primate species, *Nature (London),* 212, 693, 1966.

30. **Weller, P., Jeffreys, A. J., Wilson, V., and Blanchetot, A.,** Organization of the human myoglobin gene, *EMBO J.,* 3, 439, 1984.

31. **Blanchetot, A., Wilson, V., Wood, D., and Jeffreys, A. J.,** The seal myoglobin gene: an unusually long globin gene, *Nature (London),* 301, 732, 1983.

32. **Hyldig-Nielsen, J. J., Jensen, E. O., Paludan, K., Wilborg, O., Garrett, R., Jorgensen, P. J., and Marcker, K. A.,** The primary structure of the leghemoglobin gene from soybeans, *Nucleic Acids Res.,* 10, 689, 1982.

33. **Colbert, E. H.,** *Evolution of the Vertebrates,* John Wiley & Sons, New York, 1955, 10.

34. **Fitch, W. M.,** Evidence suggesting a partial, internal duplication in the ancestral gene for heme-containing globins, *J. Mol. Biol.,* 16, 17, 1966.

35. **Ohno, S., Wolf, U., and Atkin, N. B.,** Evolution from fish to mammals by gene duplication, *Hereditas,* 59, 169, 1968.

36. **Ozols, J. and Strittmatter, P.,** The homology between cytochrome b5, hemoglobin and myoglobin, *Proc. Natl. Acad. Sci. U.S.A.,* 58, 264, 1967.

37. **Lewin, R.,** Evolutionary history written in globin genes, *Science,* 214, 426, 1981.

38. **Coates, M. and Riggs, A.,** Perspectives in the evolution of hemoglobin, *Tex. Rep. Biol. Med.,* 40, 9, 1980—81.

39. **Hill, R. L. and Buettner-Jaunsch, J.,** Evolution of hemoglobin, *Fed. Proc. Fed. Am. Soc. Exp. Biol.,* 23, 1236, 1964.

40. **Goodman, M., Moore, G. W., and Matsuda, G.,** Darwinian evolution in the genealogy of haemoglobin, *Nature (London),* 253, 603, 1975.

41. **Pauling, L. and Zuckerkandl, E.,** Chemical paleogenetics. Molecular restoration studies of extinct forms of life, *Acta Chem. Scand.,* 17(Suppl. 1), S9, 1963.

42. **Brisson, N. and Verna, D. P. S.,** Soybean leghemoglobin gene family: normal, pseudo and truncated genes, *Proc. Natl. Acad. Sci. U.S.A.,* 79, 4055, 1982.

43. **Jensen, E. O., Paludan, K., Hyldig-Nielsen, J. J., Jorgensen, P., and Marcker, K. A.,** The structure of a chromosomal leghemoglobin gene from soybean, *Nature (London),* 291, 677, 1981.

44. **Weiberg, O., Hyldig-Nielsen, J. J., Jensen, E. O., Paludan, K., and Marcker, K. A.,** The nucleotide sequences of two leghemoglobins from soybeans, *Nucleic Acids Res.,* 10, 3487, 1982.

45. **Argos, P. and Rossman, M. G.,** Structural comparison of heme binding proteins, *Biochemistry,* 18, 4951, 1979.

46. **Michelson, A. M. and Orkin, S. H.,** Boundaries of gene conversion within the duplicated human α-globin genes. Concerted evolution by segmental recombination, *J. Biol. Chem.,* 258, 15245, 1983.

47. **Proudfoot, N. J. and Maniatis, T.,** The structure of a human α-globin pseudogene and its relationship to α-globin gene duplication, *Cell,* 21, 537, 1980.

48. **Estratiadis, A., Posakony, J. W., Maniatis, T., Lawn, R. M., O'Connell, C., Spritz, R. A., DeRiel, J. K., Forget, B., Weissman, S. M., Slightom, J. L., Blechl, A. E., Smithies, O., Barelle, F. E., Shoulders, C. C., and Proudfoot, N. J.,** The structure and evolution of the human β-globin gene family, *Cell,* 21, 653, 1980.

49. **Durham, W. H.,** Testing the malaria hypothesis in West Africa, in *Distribution and Evolution of Hemoglobin and Globin Loci,* Bowman, J., Ed., University of Chicago Sickle Cell Center Hemoglobin Symposia, Vol. 4, Elsevier, New York, 1984, 45.

50. **Spriz, R. A., DeRiel, J. K., Forget, B. G., and Weissman, S. M.,** Complete nucleotide sequence of the human δ-globin gene, *Cell,* 21, 639, 1980.

51. **Slightom, J. L., Blechl, A. E., and Smithies, O.,** Human fetal Gγ and Aγ globin genes: complete nucleotide sequences suggest that DNA can be exchanged between these duplicated genes, *Cell,* 21, 627, 1980.

52. **Adams, J. G.,** Hemoglobin Parchman: double crossover within a single human gene, *Science,* 218, 291, 1982.

53. **Jeffreys, A. J.,** DNA sequence variants in the Gγ-, Aγ-δ and β-globin genes of man, *Cell,* 18, 1, 1979.

54. **Kan, Y. W. and Dozy, A. M.,** Evolution of the hemoglobin S and C genes in world populations, *Science,* 209, 388, 1980.

55. **Simpson, G. G.,** Organisms and molecules in evolution, *Science,* 146, 1535, 1964.

56. **Livingstone, F. B.,** The malaria hypothesis, in *Distribution and Evolution of Hemoglobin and Globin Loci,* Bowman, J., Ed., University of Chicago Sickle Cell Center Hemoglobin Symposia, Vol. 4, Elsevier, New York, 1984, 15.

57. **Friedman, M. J.,** Erythrocytic mechanism of sickle cell resistance resistance to malaria, *Proc. Natl. Acad. Sci. U.S.A.,* 75, 1994, 1978.

58. **Pasvol, G., Weatherall, D. J., and Wilson, R. J. M.,** Cellular mechanism for the protective effect of haemoglobin S against *P. falciparum* malaria, *Nature (London),* 274, 701, 1978.

59. **Pasvol, G., Weatherall, D. J., and Wilson, R. J. M.,** Effects of foetal hemoglobin on susceptibility of red cells to *Plasmodium falciparum, Nature (London),* 270, 171, 1977.

60. **Roth, E. F., Friedman, M., Veda, Y., Tellez, I., Trager, W., and Nagel, R. L.,** Sickling rates human AS red cells infected in vitro with *Plasmodium falciparum* malaria, *Science,* 202, 650, 1978.

61. **Friedman, M. J., Roth, E. F., Nagel, R. L., and Trager, W.,** *Plasmodium falciparum:* physiological interactions with the human sickle cell, *Exp. Parasitol.,* 47, 73, 1979.

62. **Pasvol, G.,** The interaction between sickle haemoglobin and the malarial parasite *Plasmodium falciparum,* *Trans. R. Soc. Trop. Med. Hyg.,* 74, 701, 1980.

63. **Bunn, H. F.,** Evolution of mammalian hemoglobin function, *Blood,* 58, 189, 1981.

64. **Kan, Y. W.,** Hemoglobin abnormalities: Molecular and evolutionary studies, *Harvey Lect. Ser.,* 76, 75, 1981.

65. **Brittenham, G. M.,** The geographic and ethnographic distribution of hemoglobinopathies in India, in *Distribution and Evolution of Hemoglobin and Globin Loci,* Vol. 4, Bowman, J., Ed., University of Chicago Sickle Cell Center Hemoglobin Symposia, Elsevier, New York, 1984, 169.

66. **El Hazmi, M. A. F.,** Abnormal hemoglobins and allied disorders in the Middle East — Saudi Arabia, in *Distribution and Evolution of Hemoglobin and Globin Loci,* Vol. 4, Bowman, J., Ed., University of Chicago Sickle Cell Center Hemoglobin Symposia, Elsevier, New York, 1984, 239.

67. **Pagnier, J., Labie, D., Lachman, H. M., Dunda-Belkhodja, O., Kaptue-Noche, L., Zohoun, I., Nagel, R. L., and Mears, J. G.,** Human globin gene polymorphisms in West and Equatorial Africa, in *Distribution and Evolution of Hemoglobin and Globin Loci,* Vol. 4, Bowman, J., Ed., University of Chicago Sickle Cell Center Hemoglobin Symposia, Elsevier, New York, 1984, 145.

68. **Boyer, S. H., Panny, S. R., Smith, K. D., and Dover, G. J.,** How many ancestral mutations have led to the hemoglobin A-S polymorphism: approaches to an answer, in *Biological and Population Aspects of Human Mutation,* Hook, E. and Porter, I., Eds., Academic Press, New York, 1981, 35.

69. **Antonarakis, S. E., Boehm, C. D., Giardina, P. J. V., and Kazazian, H. H., Jr.,** Nonrandom association of polymorphic restriction sites in the β-globin gene cluster, *Proc. Natl. Acad. Sci. U.S.A.,* 79, 137, 1982.

70. **Wasi, P.,** Hemoglobinopathies in Southeast Asia, in *Distribution and Evolution of Hemoglobin and Globin Loci,* Vol. 4, Bowman, J., Ed., University of Chicago Sickle Cell Center Hemoglobin Symposia, Elsevier, New York, 1984, 179.

71. **Kazazian, H. H., Waber, P. G., Boehm, C. D., Lee, J. I., Antonarakis, S. E., and Fairbanks, V. F.,** Hemoglobin E in Europeans: further evidence for multiple origins of the βE-globin gene, *Am. J. Hum. Genet.,* 36, 212, 1984.

72. **Flatz, G.,** Hemoglobin E: distribution and population dynamics, *Humangenetik,* 3, 189, 1967.

73. **Fairbanks, V. F., Oliveros, R., Brandabur, J. H., Willis, R. R., and Fiester, R. F.,** Homozygous hemoglobin E mimics beta-thalassemia minor without anemia or hemolysis: hematologic, functional and biosynthetic studies of first North American cases, *Am. J. Hematol.,* 8, 109, 1980.

74. **Traeger, J., Wood, W. G., Clegg, J. B., and Weatherall, D. J.,** Defective synthesis of Hb E is due to reduced levels of βE mRNA, *Nature (London),* 288, 497, 1980.

75. **Benz, E. J., Jr., Berman, B. W., Tonkonow, B. L., Coupan, E., Coates, T., Boxer, L. A., Antman, A., and Adams, J. G., III,** Molecular analysis of the β-thalassemia phenotype associated with inheritance of hemoglobin E ($\alpha_2\beta_2^{26}$ glu → lys), *J. Clin. Invest.,* 68, 118, 1981.

76. **Orkin, S. H., Kazazian, H. H., Jr., Antonarakis, S. E., Ostrer, H., Goff, S. C., and Sexton, J. P.,** Abnormal RNA processing due to the exon mutation of the βE-globin gene, *Nature (London),* 300, 768, 1982.

77. **Antonarakis, S. E., Orkin, S. H., Kazazian, H. H., Jr., Goff, S. C., Boehm, C. D., Waber, P. G., Sexton, J. P., Ostrer, H., Fairbanks, V., and Chakravarti, A.,** Evidence for multiple origins of the βE-globin gene in Southeast Asia, *Proc. Natl. Acad. Sci. U.S.A.,* 79, 6608, 1982.

78. **Benz, E. J., Jr. and Forget, B. G.,** The thalassemia syndromes: models for the molecular analysis of human disease, *Ann. Rev. Med.,* 33, 363, 1982.

79. **Orkin, S. H., Kazazian, H. H., Jr., Antonarakis, S. E., Goff, S. C., Boehm, C. D., Sexton, J. P., Waber, P. G., and Giardina, P. J. V.,** Linkage of β-thalassemia mutations and β-globin gene polymorphisms in human β-globin gene cluster, *Nature (London),* 296, 627, 1982.

80. **Huisman, T. H. J.,** The occurrence of γ-chain variants and related anomalies in various populations of the world, in *Distributon and Evolution of Hemoglobin and Globin Loci,* Vol. 4, Bowman, J., Ed., University of Chicago Sickle Cell Center Hemoglobin Symposia, Elsevier, New York, 1984, 119.

81. **Furtado, M., Mathew, P. A., and Barnabas, J.,** Evolutionary analysis of functional properties in hemoglobins, *Proc. Indian Natl. Sci. Acad.,* B47, 937, 1982.

82. **Lavett, D. K.,** A model for transcriptional regulation based upon homology between introns and promoter regions and secondary structure in the promoter regions: the human globins, *J. Theoret. Biol.,* 107, 1, 1984.

83. **Hess, J. F., Schmid, C. W., and Shen, C.-K. J.,** A gradient of sequence divergence in the human adult α-globin duplication units, *Science,* 226, 67, 1984.

84. **Wrightstone, R. N.,** IHIC Hemoglobin variant list, *Hemoglobin,* 8, 243, 1984.

Chapter 8

TRANSCRIPTIONALLY REGULATORY SEQUENCES OF PHYLOGENETIC SIGNIFICANCE

P. C. Huang

TABLE OF CONTENTS

In the genetic programme, therefore, is written the results of all past reproductions, the collection of successes, since all traces of failures have disappeared. The genetic message, the programme of the present day organism, therefore, resembles a text without an author, that a proof reader has been correcting for more than two billion years continually improving, refining and completing it, gradually eliminating all imperfections.

Francois Jacob, 1973

I. INTRODUCTION

The fundamental feature of transcription, a process in which genetic information is transcribed from specific DNA sequences into discreet RNA molecules, is deceivingly simple. Biochemically, the reaction involves the initiation and elongation of a ribonucleotide chain by the formation of phosphodiester linkages using DNA as the template and the four major ribonucleoside triphosphates as the substrates. Catalyzed by DNA-dependent RNA polymerase (EC 2.7.7.6), the product is RNA. Examining the process more closely, however, one is amazed to find that transcription is a very complex and selective event, regulated precisely throughout growth, development, and differentiation, the control of which is dictated by signals encoded in DNA.

The innate differences between organisms predict a varied complexity in transcriptional regulation for prokaryotes and eukaryotes. In bacteria and bacteriophages regulation is achieved largely by interaction between proteins, such as polymerase and repressors, with specific DNA sequences, such as promoters and operators. In eukaryotic cells transcriptional control involves many different regulatory sequences such as enhancers, long terminal repeats, internal as well as distal and proximal regulators, Z-DNA, etc., in addition to the classical promoters. While the plethora of macromolecular elements participating in the recognition of and interaction with these sequences are yet to be fully identified, it is clear that as in prokaryotes, eukaryotic chromatin requires interaction among and between proteins and nucleic acids. The presence of more complex transcriptional regulation in more highly evolved systems, such as methylation of specific sequences and temporal and tissue-specific expression of different members in a gene family during development, is only beginning to be understood at the phylogenetic level.

II. DIVERGENCE OF TRANSCRIPTIONAL APPARATUS

A. RNA Polymerases Compared

Transcription of genetic information encoded in DNA into RNA involves the use of the enzyme RNA polymerase. This class of enzyme supervises the exact complementary base pairing of substrate ribonucleotides with the template DNA and catalyzes the formation of phosphodiester bonds between nucleotides during elongation of the RNA chain. To start the process of transcription precisely, the polymerase recognizes specific segments of DNA sequence upstream and/or within the gene, which are classical promoters or transcriptional regulatory elements. The termination process, in which specific sequences signal the departure of the enzyme from the template, is still largely uncharacterized.

While the DNA content of a biological system is not an index to evolutionary complexity, the simplicity of its RNA polymerase system may reflect the organization of its transcriptional units. A simple genome such as bacteriophage T7 carries a gene for its own RNA polymerase, a polypeptide of 110,000 daltons, which is transcribed by the host RNA polymerase early after infection. The transcription of other phage genes is under the control of this phage polymerase. The coding sequence for a functional form of this enzyme has recently been cloned.[1]

Escherichia coli is a natural host for bacteriophage. As has been amply documented, its RNA polymerase is larger and more complex than that of the phage, consisting of several subunits (α_2, β, β') totaling about 355,000 daltons.[2] The precise recognition of promoters by this core polymerase requires an additional polypeptide, the sigma factor. Together, these polypeptide units constitute the polymerase holoenzyme.

In *Bacillus subtilis*, as many as five holoenzyme forms of RNA polymerase have been identified; each contains a different sigma (σ) factor. These include σ^{55}, the predominant one, as well as σ^{70}, σ^{37}, σ^{32}, and σ^{28}. σ^{29} and the core enzyme constitute still another form found in the sporulating cells. These sigma factors confer promoter specificity and in so doing involve the various holoenzymes in a cascade of transcriptional control during sporulation.[3,4] Sporulation, being a form of differentiation more complex than simple growth and division, requires a more sophisticated regulatory mechanism.

Biological systems that undergo substantial differentiation apparently carry a hierarchy of transcriptional apparati, including at least three distinct forms of multimeric polymerases, each of which governs the synthesis of a different class of RNA. Thus, in all eukaryotic cells, there exist polymerases I, II, and III, responsible for the transcription of ribosomal RNA, messenger RNA, and small RNAs, respectively. The latter includes tRNA, 5S rRNA, viral RNA, human alu-family RNA, and tissue-specific identifier RNA. The subunit structure and antigenicity of polymerase II is highly conserved among many organisms ranging from the fruit fly, *Drosophila melanogaster*, to the yeast, *Saccharomyces cerevisiae*. Molecular cloning of the polymerase genes in yeast reveals that coding sequences for the largest subunit polypeptides of all three RNA polymerases are linked;[5] hence these polymerases may be evolved from each other.

Genomes of the symbiotic organelles of the eukaryotic cells also carry genes for simple RNA polymerases analogous to those of phages. Little is known about their properties.

B. Transcriptional Units

DNA evolution is often deduced from the comparison of nucleic acid or protein sequences of a given class of genes. Such comparisons are hampered by examples of appreciable sequence homology between widely divergent species and between organelles of divergent species. Wheat mitochondrial 18S genes are 80% identical to that of 16S rRNA of *E. coli* for primary sequence, 71% to *Halobacterium volcanii*, and 61% to *Xenopus* 18S rRNA.[6] Spinach, mung bean, pea, and corn mitochondrial 18S rRNA and mung bean and spinach chloroplastic 16S rRNA genes also show a high degree of homology, sharing identical 12 kb inverted repeats. In addition, their genes for the large subunits of ribulose bisphosphate carboxylase and the β subunit of chloroplast ATPase are homologous[7] as determined by both DNA hybridization and restriction enzyme mapping. These findings suggest not only a recent xenogenous origin for organelle DNA from eubacteria, but also that they may be derived through transposition from each other.

The sharing of common sequences between functionally related genes from prokaryotes and eukaryotes has also been shown. Heat shock genes produce proteins of an apparent molecular weight of 70,000 that are antigenically similar throughout eukaryotes, from man to yeast. Detailed sequence analysis of a major *Drosophila* heat shock gene, HSP 70, shows that it is 48% identical to a heat inducible gene, dnaK, from *E. coli*.[8] It is interesting to note that such a sequence is also present in *Methanosarcia barberi*, an archaebacteria. One of the *E. coli* heat shock regulatory genes, htpr, as predicted from its cloned gene sequence, actually resembles the σ factor of *E. coli* polymerase.[9]

A different way to examine phylogenetic relatedness of genes would be to compare the organization of their transcriptional units. A transcriptional unit consists of a DNA sequence from which RNA is transcribed, starting from a promoter and ending at a terminator. However, not all transcriptional units can be so clearly defined. In an increasing number of cases observed, a single nucleotide sequence may code for more than one polypeptide. This is achieved by either using different reading frames and start-terminate signals, superimposing operons of varied length, or relying on differential splicing of the primary transcripts to generate different gene products. These overlapping genes have been shown in a variety of genomes, including both DNA and RNA phages, mitochondrial DNA, insertion elements, as well as chromosomal DNA of bacteria.[10] These gene arrangements can have important regulatory implications and may serve as interesting systems to study the evolution of control signals.

Most of the simple and compact genomes use overlapping reading frames to achieve maximal information content. Bacteriophages ØX174 and G4 each encode multiple peptides in a single segment of DNA using different reading frames.[11,12] In human mitochondria the termination codon of one gene is overlapped with the initiation codon of the adjacent gene. Every sequence in its genome is encoding information.[13-15] In the Simian 40 viral genome, a single promoter directs the transcription of more than one message by a mechanism which involves differential post-transcriptional processing. The regulation of adenovirus transcription via the selective use of promoters and processing is also well established. Overlapping reading frames exist in the human hepatitis B viral genome, where both a large 832 amino acid residue surface protein and a small 226 amino acid residue putative DNA polymerase or reverse transcriptase are encoded by the same overlapping DNA sequence.[16] In these simple phage or viral genomes, many of the genes are controlled by a single or a few promoter sequences,[17] and promoters may overlap.[18]

Bacterial genomes are characteristic for the organization of genes into polycistronic operons. Functionally related genes are often linked and coordinately controlled by a single promoter-operator, the sequences of which overlap.[19] Other genes, such as the tryptophan operon, carry additional sequences which attenuate the transcriptional control, hence the name attenuator sequence.[20] The promoter region of the histidine operon has also been studied and shown to have similar control.[21,22]

Eukaryotic transcriptional units have many features which are absent in the prokaryotes. These features include:

1. Presence of tandem repeats of the same gene
2. Scattering of functionally related genes on different chromosomes
3. Inclusion of noncoding sequences within a gene (introns) or between repeating genes (spacers)
4. Multiple transcriptional regulatory elements

The divergency of transcriptional apparati depends not only upon the very nature of the polymerases but also upon the DNA sequences involved with or recognized by the regulatory elements.[23]

Eukaryotic genes with abundant transcripts usually exist in multiple copies. These include those for ribosomal RNA,[24-26] actin, tubulin, and to some extent interferon; however, not all copies of a multiple gene are functional. Analysis of the defects in these nonfunctional genes has provided additional insight into the regulatory role of DNA sequences within or flanking a gene.

The 5S rRNA genes of *Xenopus laevis* oocyte consist of multiple tandem repeats of about 700 bp; each consists of a pseudogene in addition to the gene. The sequence of these pseudogenes shows an almost perfect homology to the functional gene of 101 nucleotides; but cryptic mutations prevent their expression.[27] The mouse β-globin pseudogene lacks the two intervening sequences that interrupt all globin genes, but is otherwise homologous to the active β-globin sequence.[28] These ''processed'' sequences are not transcribed. It is postulated that these pseudogenes are evolved from sequences of DNA which were reverse transcribed from their messenger RNAs, propagated in an RNA virus, and reincorporated into the genome at novel sites.[29] Human α-globin normal and pseudogenes differ by mutation in the initiation codon and by deletion resulting in frameshift mutations which prevent its transcription into a translatable message.[30] The presence of frameshift mutations and premature termination codons of rabbit β-globin pseudogene apparently causes its transcriptional infidelity.[31] Rabbit β-2- globin, a relative of β-1, also contains a pseudo, nonfunctional gene, due to a single deletion in codon 20.[32] Not all pseudogenes are linked to their related gene sequences. While human dihydrofolate reductase gene can be assigned to chromosome 5, several of its pseudogenes generated by amplification are scattered in other chromosomes.[33]

Subsegments of certain tandemly linked repetitive genes, referred to as "orphons" have been found to be dispersed in the genome. Orphon-like elements have been found to be widespread, ranging from *Drosophila* to sea urchins and the yeast, *Saccharomyces cerevisiae*, and include both protein coding and noncoding structural gene families. In total there are more than 50 orphons in the sea urchin genome. In each of its 5 histone transcription units there are from 5 to 20 orphons. Most of the histone orphons appear to contain only one coding region. They are usually polymorphic; no two individuals have the same set of histone orphons for any coding region. The H3-coding orphon studied differs less than 2% in base sequence from the analogous region of the major histone gene clusters.[34]

Pseudogenes, processed genes, and orphons are all inactive DNA sequences which may well be the relics of evolving functional genes, resulting from nonimpeded mutations. While several mechanisms can explain their inactivity, the ones relevant to transcription may include the abolishing of signals for initiation, altered splice junctions, and improper processing of primary transcripts.

III. CIS-ACTING REGULATORY SEQUENCES

Several DNA sequences have been identified which facilitate or restrict the access of RNA polymerase to the DNA template. These sequences may form regions of special topology, allowing the polymerase to selectively open the helix at sites of transcriptional initiation.

A. Promoter Specificity

A classical promoter consists of DNA sequences which are generally located upstream from the coding sequence and are the sites of recognition and close contact by the DNA dependent RNA polymerase. In prokaryotes promoters are generally defined, by genetic evidence, as an initiation element for the expression of structural genes. RNA polymerases isolated from a wide range of bacteria such as *Mycobacterium smegmatis* and *Rhodospirillum rubrum* recognize a common promoter like that of the T7 phage. Since these bacteria are evolutionarily diversified species, certain structural elements in the protein must be so conserved that they utilize the same regulatory sequence for transcription. However, there is a high degree of heterogeneity in promoter sequences reflected by the observed variation in promoter strength. Furthermore, some promoters can be recognized only by certain polymerases. For instance, a novel form of polymerase from *Bacillus subtilis* vegetative cells recognizes only a unique promoter of T7, but not its normal promoter. Thus, polymerase and promoter interact selectively.

Prokaryotic RNA polymerase binds to DNA loosely or tightly. DNA involved in loose binding (Ka = 2×10^{11} M) remains double stranded; the halflife of dissociation of the closed E-DNA complex is about 60 min for bacteria. In the absence of σ factor, Ka is reduced to $10^7/M$ and the halflife to less than 1 sec. Sigma factor, on the other hand, promotes tight binding (Ka = $10^{14}/M$) and the halflife is increased to several hours. The binding strength varies with different promoters and is sequence-specific.[2,35] The promoters of phage T5 early genes, for instance, have been shown to be the strongest among many tested for signal strength with *E. coli* polymerase.[36]

1. Prokaryotic Promoters

Two major sequence-specific steps have been postulated for prokaryotic promoter site selection: (1) recognition — RNA polymerase holoenzyme binds to a promoter, forming a closed complex, and (2) isomerization — the closed complex is transversed to an open form.

Once an open complex is formed, RNA chain elongation is rapidly initiated in the presence of substrates. The rate of open complex formation (K_B) has been measured using an in vitro mixed transcriptional system.[37] During step 1, two classes of DNA promoter sequences are recognized, one contains the consensus hexanucleotide $T_{89}A_{89}T_{50}A_{65}A_{65}T_{100}$ and is usually situated -6 to -12 nucleotides preceding the transcription starting point.[38] The subscripts denote relative frequency of occurrence;[39] transcription start is $+1$. The second contains the consensus sextamer sequence $T_{85}T_{93}G_{81}A_{61}C_{69}A_{52}$ located about -35 nucleotides upstream from the site of transcription initiation. The distance between the Pribnow box (TTAACTA) and the sequence at -35 affects promoter strength, with optimal spacing occurring around 17 bp. High level expression in *E. coli* of genes cloned in plasmids requires proper placement of the promoter upstream. The optimal distances between Shine-Dalgarno's *AGGA* sequence in the promoter-ribosome binding site and the initiation codon *ATG* for eukaryotic and prokaryotic genes have been determined to be between 7 to 11 and some 40 bases, respectively.[40] The reason for this difference may well lie in the secondary structure of the template as well as the transcript, the former of which *E. coli* RNA polymerase must negotiate. Adequate spacing from 16 to 19 bp is apparently needed between these two promoters for efficient initiation. Neither of these two promoter sequences, however, is absolutely necessary for transcription, although very few promoters lack the -10 site. Other means most likely exist for RNA polymerase to bind and to initiate transcription. For *E. coli*, RNA polymerase also interacts with sequences upstream of specific promoters.[41] A considerable amount of information has been obtained about how chemical alteration or mutation of a single base within these prokaryotic promoters can alter their transcriptional activity as much as 100-fold.[42-44] A number of artificial promoter sequences have also been chemically synthesized and shown to be functional.

Certain prokaryotic promoters require ancillary proteins, positive regulators for the RNA polymerase to initiate transcription. A well-known example is the *lac* operon of *E. coli* which requires the putative binding of a cAMP-catabolite activator protein complex (cAMP-CAP) to an 11 bp sequence upstream (-71 to -52) of the polymerase binding site. Similarly cAMP-CAP binding between -50 and -23 and -107 to -78 are necessary for the *gal* and *arg* genes to be transcribed.[39] This catabolite activator binding protein, however, does not bind to other promoters such as T7. The operonic promoters thus represent a broader definition of transcription initiation sequences.

2. Upstream Promoters for Eukaryotic Polymerase II

The mechanistic details of eukaryotic polymerases binding to their respective promoters are not clear. Each of the multisubunit, eukaryotic RNA polymerases do recognize a set of specific promoters. Specific nucleotide sequences in eukaryotic DNA that are involved in controlling the synthesis of the primary transcript by polymerase II have been recognized indirectly, mostly through experiments in vitro. Three regions at 0, -30, and -70 upstream of the transcription initiation site may be of particular significance. They may function by controlling initiation frequency, aligning the starting point, and initiating polymerase binding, respectively.

At the point of initiation, most mRNAs seem to prefer *A* as the start, with both sides flanked by pyrimidines. There is no consensus of sequence in this region. A heptanucleotide sequence,

$$T_{82}A_{97}T_{93}A_{85}{}^{A_{63}A_{83}}_{T_{37}T_{33}}A_{50}$$

(TATA or Hogness box), similar to the prokaryotic promoter is located upstream -25 to -30 in sea urchin histone genes,[45] genes for mouse β-globin,[46,47] rat insulin I and II,[48] chicken ovalbumin and ovomucoid,[49-51] as well as mouse immunoglobulin genes.[52,53]

The transcription initiation site is often located upstream of the translational start site in eukaryotic genes. This site has a consensus sequence

$$CC_{G}^{A}CCATGG$$

as compiled from data on 211 mRNAs.[54]

The overwhelming generality of the TATAAAA sequence suggests that it might play an important role in transcription by aligning RNA polymerase. Additional support for the promoter role of this sequence is provided by comparing the transcriptional efficiencies of conalbumin and adenovirus-2 late genes which share an extensive 12 bp homology in their TATAAAA box regions,[55] and are transcribed in vitro with the same efficiency. These genes are more strongly transcribed than those of adenovirus early genes and ovalbumin, which differ in the sequence of their TATAAAA boxes.[56]

These observations suggest the existence of promoter sequences with different strengths. Wasylyk et al.[57] isolated deletion mutants of conalbumin and adenovirus late genes and used these two genes for in vitro transcription. Their studies demonstrate that a region between positions -12 and -32 upstream from the mRNA start point is essential to promote specific transcription in vitro. Not all genes are flanked by this sequence, however. The flanking sequences of the early genes of adenovirus, late transcripts of SV40,[58] and sea urchin histone H4 gene all lack this promoter sequence. Thus, the absolute requirement of this sequence is obviated.

An additional sequence further upstream, -70 to -80 to the pol II transcription initiation site, has also been observed to behave as a promoter. It has a consensus sequence, CAAT box:

$$GG_{T}^{C}CAACT$$

The precise function of this second conserved promoter needs to be more fully elucidated. Mutations and in vitro transcription studies show that this region has a strong influence on initial binding of polymerase to the template.

Although the initiation of pre mRNA transcription utilizes the TATA region around nucleotide -30 in vitro,[57,59] it requires other regions further upstream in vivo. The herpesvirus thymidine kinase gene contains several regions of DNA which are involved in transcriptional control. While sequence -16 to -32 is the proximal promoter which harbors TATA homology for initiation, sequences -47 to -61 and -80 to -105 are, respectively, the first and second distal promoters, and function in a concerted manner to control the efficiency of transcription. This has been demonstrated by altering spacing, condensing, and expanding mutations in these promoters, and assaying in oocytes.[60,61] In the chicken vitellogenin II gene there are 28 exons and 27 introns, spanning a region of 23.6 kb. Three upstream regions of interest are found. Setting the first AUG at $+14$, the region -32 to -26 contains CATAAAA; -77 to -67 contains TTGAGAATT, which is homologous to the bacterial polymerase binding site; and -101 to -90 contains a similar sequence TGTTTACATAAA.[62] The immunoglobin K gene also requires the presence of a conserved sequence element upstream to be transcribed correctly,[63] as do the heat shock protein genes in *Drosophila*.

Multiple promoters for eukaryotic polymerase II have been determined in a number of other genes. The transcription of human β interferon gene is controlled by two sequences: (1) − 77 to − 19, which includes TATAAAT and regulates constitutive and induced expression; and (2) − 210 to − 107, which suppresses the constitutive transcription level and alters kinetics of induction. The latter includes a segment which reads . . . AAAAAA AG AAAACC . . . AGTTTG TAAATC TTTTTC . . . .[64] Variant surface glycoproteins also show multiple promoters.[65]

Transcription of ε-globin gene may be initiated from at least 9 sites within 4.5 kb upstream of the canonical capping site. By primer extension and S_1 mapping the transcription boundaries have been localized at − 65 to − 250, − 900, − 1480, and − 4500.[66] Since all transcriptional products initiated at these sites are capped, alternative CAP sites upstream must also exist. In contrast, the majority of β-globin transcription initiates at only one CAP site which is − 55 to − 53 bp upstream of ATG.[67] The CAAT and TATA sequences are located at − 80 and − 30, respectively, in this gene.

Zein, a major storage protein of corn, is transcribed from two promoter regions (P_1 and P_2) into RNA of distinct sizes, ranging from 900 bp for mature mRNA to 1800 bp for the smallest precursor. These promoters, separated by about 1 kb of AT rich sequence, appear to be independently active in vivo and in vitro and may serve as double starts. When injected into *Xenopus* oocytes, however, P_2 is preferred in initiation of transcription. The sequences of these two promoters are (1) CAAT . . . TATAAT . . . P_1 . . . ATGCCTAATGG and (2) CAAAAT . . . TATATAT . . . P_2 . . . ACCTATAATATTTT.[68]

When the physical properties of several eukaryotic promoters were compared, a correlation was observed between in vivo promoter activity and denaturability, a measure of AT richness.[69] The presence of even a short stretch of AT rich region within a DNA sequence may trap an enzyme through breathing. Since some promoters are roughly 3 helix turns (27 to 34 bp) upstream the active site of the polymerase bound enzyme probably protrudes to make contact with the template downstream.

Thus, the TATA sequence which resides at a site 30 + 5 bp upstream may trap the polymerase, destabilize it, and bring the transcriptional complex into proper juxtaposition, forcing the reaction to initiate at a given site (CAP site). In the absence of the TATA sequence, such as created by mutation, a secondary CAP site + 1 to + 11 may be used for initiation. The flanking sequences may then serve as a trap. For instance, many TATA boxes are actually flanked by GC rich sequences: (1) 5′ GGGGGGTATAAAGGGGGTGG-GGGCGGG 3′ (Ad − 2 late), (2) 5′ GCCAGGGCTGCTCCTCTATAAAAGGGG 3′ (chicken conalbumin), and (3) 5′ CCGAGGTCCACTTCGCATATTAATGACGCGTGTGGCC 3′ (herpes simplex TIC). However, as shown by simultaneous alterations of the sequence at multiple sites, for example, A to C at − 28 and A to T at + 1, precise initiation of transcription requires both the TATA box and the initiation site.[70]

Thus, for eukaryotic polymerase II, at least two promoter sequences at − 10 to − 30 and − 60 to − 80, are generally involved. It is unclear whether the polymerase recognizes and binds these sites sequentially or simultaneously. It is conceivable that chromosomal proteins bring properly spaced sequences to the proximity of the enzyme in vivo. Further understanding of this coordination will become more relevant when more distant regulatory sequences are to be considered in transcriptional regulation.

While in most cases studied regulatory sequences for polymerase II lie 5′ to the initiation and capping site of transcription, recent evidence shows that sequences 3′ to the capping site may also be important. In mouse erythroleukemia cells transformed by Friend's leukemia virus complexes, cloned human α- and β-globin genes are induced differently when introduced. Analysis of the hybrid human α gene product reveals that the sequences responsible for differences in transcription of the intact α- and β-globin genes are located on the 3′ side of the mRNA capping site of the two genes. This result suggests that cis-acting regulatory sequences are also located within the structural genes.[71] A similar internal regulatory sequence is also noted in the chicken thymidine kinase gene.[72]

3. Upstream and Internal Promoters for Polymerase I

Eukaryotic polymerase I transcribes only the genes for larger ribosomal RNAs. Comparing sequences flanking these genes, no obvious promoter can be deduced;[73] however, analyses of rDNA transcripts in vitro and in vivo show that they share an initiation sequence, AGGTA, which is approximately 4.5 kb upstream from the 18S rRNA coding region of *Xenopus laevis*.[74] As deduced from 5' and 3' deletion mutant analysis, promoters for polymerase I of *Xenopus laevis* can be mapped within a 150 bp segment adjoining the initiation site. There are two major domains: (1) -7 to $+6$ (13 bp) which specifies accurate initiation in vivo, and (2) -142 to $+6$ (148 bp) which is required for maximal synthesis in vivo. Nucleotides around -75 are apparently essential for efficient initiation in *Xenopus* and in rats.[75] Additionally, several kilobases upstream a duplication of this 150 bp promoter has also been identified, which is required for in vitro transcription by polymerase I.

Sequences greater than 1150 bp upstream of the transcription initiation site for the *Xenopus laevis* ribosomal RNA gene have also been shown to affect transcription of this gene. This is demonstrated by injecting the cloned gene with its flanking spacer sequence into fertilized *Xenopus* eggs.[76,77] It is suggested that a segment of the spacer sequence may be involved in affecting the frequency with which the promoter is switched to its active conformation rather than modulating the number of polymerase loadings per gene, thus making it reminiscent of an enhancer sequence.[78]

Polymerase I promoter sequences have also been mapped by electron microscopy[79] to exist between -320 and $+115$; the presence of these sequences is critical for maximal packing density of the polymerase. Others have found the segment between -147 and -35 critical in transcriptional control. Thus, it seems that there is an hierarchy of functional domains, the function of each depending upon the transcriptional milieu.[74]

Recent study shows that in *Drosophila* one promoter for pol I is located at -43 to -27 and another lies within the first 4 nucleotides of the external, transcribed spacer.[80] The control signals for polymerase I, therefore, may well reside both upstream and within the ribosomal RNA transcription unit.

4. Internal and Downstream Promoters for Eukaryotic Polymerase III

Promoters which are a portion of a transcriptional unit have been shown for genes transcribed by eukaryotic polymerase III.[81] Internal promoters have been documented in genes for tRNA,[82-84] 5S ribosomal RNA,[85,86] as well as adenovirus VA1 gene.[87,88] For tRNA^{met} of *Xenopus laevis*, 2 sequences from $+8$ to $+30$ and from $+51$ to $+72$ are essential for transcription. For 5S genes, the promoter sequence has been deduced to lie between position $+55$ and $+80$. For the VA1 gene, two intragenic promoter boxes have been identified between $+10$ and $+69$. This "split promoter" involves the sequences . . . TCCGTGGTC . . . and . . . ACCGGGGTTCGAACCC . . . [89] At least two other downstream sequences have also been assigned to pol III promoters. These are . . . TGGCTCAGTGG . . . and . . . GGTTCGATCCC . . . [90] However, other segments of flanking sequence 5' to the tRNA genes from *Xenopus laevis* are inhibitory to the transcription of these tRNA genes in vitro.[91] A number of protein factors have been identified which allow polymerase III to recognize one or more internal sequences of certain genes, yet make contact upstream as much as 50 bp away, to initiate transcription of the gene (see later). It is interesting to note that internal control elements also exist in prokaryotic genes. The *gal* operon of *E. coli* is an example.[92] Whether control elements within a structural gene share a certain common mechanism or not is as yet unknown.

5. Possible Regulatory Role of Introns

Introns, the intervening DNA sequences between the coding regions (exons) of a gene[93] have been found in many organisms ranging from fungi to higher animals, and in viral, nuclear, as well as mitochondrial and chloroplastic DNA. Genes for 28S rRNA of *D. melanogaster*,[94] rabbit and mouse β-globin,[95,96] chicken ovalbumin,[49,50,97,98] lysozyme,[99] ovomucoid,[100] oviduct conalbumin,[55] yeast tRNA,[101,102] the late gene of adenovirus-2,[103] SV40 late gene,[104] vimentin,[105] bovine rhodopsin,[106] and metallothionein[107,108] are just a few examples. Many viral and eukaryotic genes, however, do not contain introns. Several examples are the hemagglutinin gene,[109] genes for histones,[110] gene 4 of the fowl plague virus, chlorophyll, a/b binding polypeptide, most interferons, heat shock proteins, most rRNAs and thymidine kinase of herpes simplex virus, polypeptide IX of adenovirus, hepatitis B surface antigen, and core protein.[111,112]

For two functionally related linked genes, one may have an intron but the other does not. Tobacco chloroplast genes for tRNA^{gly} (UCC) and tRNA^{arg} (UCU) show an intron of 690 nucleotides in the former and none in the latter.[113] On the other hand, isogenes that share a common promoter, such as those coding for metallothionein in mammals, are highly divergent in their intron and noncoding sequences.[107] Rat preprosomatostatin has a promoter at −31, TTTAAAA. Its primary transcript is processed stepwise as a series of polyproteins descending in size from 116 to 92 to 28 to 24 amino acids. They are coded by a gene of 1.2 kb with only one intron of 630 bp.[114]

The absence of introns from some genes, the varied number in others (up to 50), and the variation in intron size (from 10 to 1000 bp) all prompted the assumption that the intron is neither essential nor deleterious. There are at present two major trends in thought concerning evolution and functionality of intervening sequences. One invokes each exon of a gene corresponding to a unique structural domain of a protein.[115] Several exons converge to form a novel gene which codes for a multi-domain protein and which is regulated by a common transcriptional signal. The intervening sequences are thus evolutionary relics, the primary transcripts of which are recognized and processed during mRNA maturation. Strong colinear correlation[116] between protein domains and exons as in the case of immunoglobin, hemoglobin, lysozyme, and metallothionein lends credence to this thinking. The other trend invokes that intervening sequences are transposed to sites between coding sequences of a gene and that they survive evolutionary pressure by being recognized and processed during RNA splicing. Thus, introns are "selfish" disposable DNA sequences.[117] It is unclear whether intervening sequences necessarily evolve through only one of these two pathways. In fact, comparative studies show that intron sequences change more rapidly than protein coding sequences, although they are quite ancient evolutionarily, and that there are certain features in the introns suggesting that they may play a regulatory role. Introns in human globin genes, for instance, might contain regulatory elements involved in developmental switching of gene expression.[118]

a. Most Introns Share Common Splicing Signals for Post-Transcriptional Processing

It is clear from evidence obtained in studies with mRNA maturation that the intervening sequences are transcribed but later spliced out. The specific DNA sequence TACTAACA is essential for removal of the intron by splicing. By comparing the chemically synthesized decanucleotide TGTACTAACA with its variants, it was observed that alteration at positions 4 and 8 (A to C) will eliminate the splicing while at the 5th (C to T) showed no effect.[119]

b. Introns May Function as Coding Sequences

The intron of one gene may serve as the exon of another gene. Open reading frames, albeit relatively short, exist in some intron sequences. The functions for such intronic genes are as yet unclear. In yeast, additional transcripts coded for by the introns of at least one mitochondrial gene have been identified. Within the cytochrome *b* gene, a gene^s is found to code for a transacting diffusible product which can block cytochrome *b* synthesis.[119] Thus, introns may also play a role in the processing of certain yeast mitochondrial transcripts, but not all, via a Box 3 RNA maturase which is coded for by introns within the corresponding genes.

The variable splicing scheme operative during viral gene expression is one mechanism by which a genome can maximize its developmental and temporal regulation through the selective removal of introns. Since distinct exons of a gene may correspond to distinct domains of a complex protein, sequence signals in the introns between them may provide an efficient evolutionary mechanism for diversity.

c. Introns Contain Identifier Sequences for Tissue-Specific Gene Expression

The genome of each cell in an organism shares the same set of genetic information. However, only "housekeeping" genes are expressed in common. Certain genes are transcribed exclusively in specialized cells and are silent in other tissues. The elucidation of the mechanism that turns on (and off) these tissue-specific genes is a fundamental goal in the molecular study of development and differentiation.

Regulation at the gene level is evidenced by the presence of an abundant (up to 1.5×10^5 copies) class of unique sequence family in most (62%) brain-specific RNA, but very few copies (4%) in liver or kidney. These sequences are found in the introns of several cloned genes, including a few that are expressed in neuronal tissues. Termed ID (identifier) sequences, they are 82 nucleotides long and each encompasses two consensus stretches of nucleotides previously assigned to be the promoters for polymerase III. Both polymerase II and III transcripts from isolated brain nuclei show homology to ID sequences, suggesting that these are native, primary transcripts of specific cells. These data lead to a model that pol III transcription of ID sequences within the intron of brain genes activates those genes for polymerase II transcription.[120,121] The existence of various themes for post-transcriptional processing[122] suggests that there are subtle differences in the regulatory signals conveyed by the intron sequences.

6. Possible Roles of Repetitive Sequences in Regulation

One may bear in mind that evidence exists for the evolved redundancy of regulatory sequences in response to metabolic alterations. A case of note is that of the galactose operon, where the overlapping promoters TATGCTA and TATGGTT, differing in their starting point by 5 bp, are regulated by the absence or presence of CAP, respectively.

Eukaryotic genomes contain both repetitive and unique DNA sequences. While many of the unique sequences represent structural genes, some of the repetitive sequences have been postulated to play a key regulatory role in transcription.[124-126]

a. Alu Sequences

Alu sequences constitute about one third of human genomic DNA and account for one half of the highly repetitive sequences. In primates and rodents, 3 to 6% of the DNA consists of a family of 3 to 5×10^5 copies of such sequences per haploid genome. They exist as 300 nucleotide-long sequences, with an Alu 1 recognition site about 130 nucleotides from their 5' ends.[127,128] These sequences are interspersed within the genome, in satellite DNA, flanking, or within a gene. The function of Alu-family sequences is uncertain. They may be used as origins for DNA replication, in transposition, and/or in transcription.

This complex array consists of tandem and inverted repeat sequences which are interspersed among many genes.[129] A 56 kb fragment of human DNA, which carries globin genes, contains 7 copies of Alu 1 sequences. A 12 kb *onc* gene contains 2 and the 19 kb insulin gene 1 Alu 1 sequence.

Alu sequence transcripts are enriched in the nuclear heterogenous polydispersed hnRNA. The transcripts are also present in the form of cytoplasmic low molecular weight polysomal 4.5S and 7.0S RNAs; the latter is common to many eukaryotic cells and retroviruses. Cloned hamster Alu repeats were used for transcription by RNA polymerase III. The products were analyzed by two dimensional oligonucleotide analysis after T1 digestion. The results showed that they are analogous to a class of low molecular weight RNAs synthesized in growing Chinese hamster cells.[130] Several highly repeated DNA sequences were mapped near the Syrian hamster CAD gene which encodes the first three enzymes for UMP biosynthesis. This gene is amplified in a PALA resistant mutant and has been cloned. The repeated sequences show many properties of the Alu-family in humans and their transcripts in vitro by polymerase III show homology with the 7S and 4.5S cytoplasmic nonpolyadenylated RNAs.[131]

The Alu sequences in the human β-like globin gene cluster were also shown to be transcribed by RNA polymerase III in vitro.[132] Possible involvement of these pol III transcripts in pol II transcription was suggested earlier in the α-globin gene cluster. Alu family repeat sequences that are dispersed throughout the α-globin cluster are at least 20 times more efficient as templates for RNA polymerase III-dependent in vitro transcription than those in the β-globin gene cluster. Certain in vitro transcripts from the α-cluster Alu family repeats can be precipitated by Lupus antibodies which are known to interact with ribonucleoproteins (RNPs) synthesized in vivo. This suggests that the transcripts are assembled into antigenically distinguishable RNPs in vitro and exert a regulatory function in transcriptional processing events.[132a]

It is noted that some Alu sequences are located within the intron sequences of a gene, e.g., genes for mouse α-feto protein,[133] human gastrin, and gastrointestinal growth hormone.[134] Other Alu sequences, however, are located some distance from the 3' end of a gene. Human insulin gene Alu members are found 6 kb downstream from the gene.[135] The location of Alu sequences with respect to the genes they may be regulating is not obvious. Their dispersion throughout the genome and possession of direct repeats of DNA sequence suggest that Alu sequences could be important in transcription, and hence more effective on peripatetic genes.

b. LTRs

A long repetitive sequence is present in the 5' and 3' termini of retroviral RNA and its corresponding provirus. Known as long terminal repeats (LTRs), they contain the sequences necessary for transcription of the viral genome.[136-138] In addition to providing viral promoter functions, they permit activation of host genes adjacent to the viral integration site.[139]

Many retroviral LTRs end with inverted, complementary repeats of from 2 to 16 nucleotides; these repeats characteristically begin with the dinucleotide TG and end with the inverted complement CA. Also characteristic to LTRs is the presence of direct repeats of host DNA sequences at the proviral LTR/host DNA junction, presumably through the duplication of host sequences at the target site of retroviral integration. Adjacent to the 3' LTR, retroviruses contain a putative primer binding site for second strand cDNA synthesis. This site consists of a short stretch of purine nucleotides.[140] Both eukaryotic regulatory sequences for initiation, TATA at −25 and CAAT at −70, and for termination, AATAAA at the 3' end, are found in viral and proviral LTRs. There are three characteristic regions — U3, R, and U5 — within each LTR. The U3 (unique 3') region is 342 to 480 nucleotide bp in length. It contains two of the transcription signals described, the CCAAT and TATA sequences. The R region, a sequence repeated at both the 5' and 3' ends of the viral RNA, is 60 to 70 nucleotides long. It contains the polyadenylation signal AATAAA as well as a pyrimidine rich region ending with CA. The 5' region of LTRs, U5, has no demonstrated promoter sequences and varies in length from 67 to 176 nucleotides in mammalian retroviruses. The tRNAs utilized as primers in viral replication anneal to a nucleotide sequence within the viral RNA immediately adjacent to U5. This region, the primer binding site (PBS), is always complementary to the 3' terminal 16 to 19 nucleotides of a specific tRNA. The 72 bp direct repeats, containing the consensus sequence activity of the SV40 enhancers,[137] have been identified in retroviral LTRs. In some cases they have been shown to possess enhancer activity, although retroviral enhancers thus far characterized are located in the U3 region of the LTR within the long direct repeats. They are absent from the LTRs of isolates from HTLV (ATLV)[141,142] and related bovine leukemia virus. Other similar repeats of 21 bp, however, are present in the U3 region of HTLV1. They have been postulated to be involved in viral transcription in lymphoid and other target cells.[142,143]

c. IAPs

Intracisternal A particles (IAPs) are noninfectious, retrovirus-like particles localized within the endoplasmic reticulum in several mouse tumor cell lines and in normal preimplantation mouse embryos. There are about 1000 IAP genes dispersed throughout the genome, comprising over 0.2% of the total mouse DNA.[144] Restriction digests and heteroduplex mappings of cloned IAP genes reveal that DNA close to the 3' end of the IAP coding sequences is conserved, while 5' sequences diverge from one another considerably. Two major classes of these IAP genes have been studied extensively and were shown to have LTR flanking both ends characteristic of retroviral sequences.[145,146] IAP genes are transcribed in vivo and in isolated nuclei. In the latter case the transcripts constitute the majority of the RNA products. RNA polymerase II is involved in their transcription, since α-amanitin (2 $\mu g/m\ell$) inhibits their synthesis by about 90%. The site within the IAP gene for RNA initiation has been mapped within the LTR region at about 20 to 30 nucleotides 5' of a Pst I site.[147]

The origin of these LTR sequences is not known; neither is their role clear. A 1350 bp family of repetitive DNA, generated by Eco RI restriction, has been shown to be homologous to the IAP genes. This 1350 bp satellite DNA sequence may be a part of a larger family in which there are conserved Eco RI sites. Its transcriptability and LTR features suggest that its function may be to serve as transposable, movable promoters for the activation of cellular genes in the mouse.

d. Very Short Repeats

Genes, the products of which share related function, are often expressed coordinately. Coordinately inducible genes such as those coding for amino acid pathway enzymes or growth hormones, apparently share short repeating sequences unique to each gene family.[148] Consensus sequences (. . . TGACTC . . .) that have been deduced are repeatedly located 5' to the coding sequence and are distinguishable from promoters. In addition, as noted by Cheung et al.[149] the trinucleotide GTG is present frequently and symmetrically in many gene sequences of prokaryotic as well as eukaryotic DNA, ranging from the consensus *E. coli* lac repressor binding site to the transcriptionally critical sequences (upstream -87) of the human β-globin gene. This trinucleotide is also present in LTRs and in the VDJ joints of immunoglobulin genes. Potentially, such a sequence, albeit short, is important in the B to Z transition of DNA; hence it may be involved in transcriptional regulation.

B. Enhancers-Activators

Enhancers are DNA sequences which potentiate the efficient expression of certain genes by increasing the level of transcription. The prototype enhancer, the 72 bp tandem repeat of SV40 DNA, was initially identified to be situated more than 100 nucleotides upstream from the CAP site of the early genes of SV40.[150-152] Enhancers are distinct from classical promoters in that they can function upstream from, within, or downstream from an eukaryotic gene. They also function in either orientation.[153,154] More remarkably, they may act over distances as great as 10 kb, although the nature of the spacing DNA sequences play a role in their effectiveness. A basic core of 7 to 10 bp (. . . GTGGNNNG . . .) constitutes the consensus sequence of all enhancers.[137,155]

1. Viral Origin

Viral enhancers, also known as activators, have been found in many animal viruses, including simian virus, SV40, polyoma, Burkitts' lymphoma, bovine papilloma, Molony murine sarcoma, and adeno and herpes simplex viruses; they are often short tandem repeats of 50 to 100 nucleotides. These sequences have been well mapped in their respective genomes and documented for their transcriptional enhancing activity.[137,156,157] For instance, SV40 early and late promoters are located within a 342 bp sequence of noncoding region, which also contains the replication origin and signals to switch from early to late expression. It is characterized by palindromes, 1 copy of an AT rich 17 bp sequence, 3 copies of a 21 bp GC rich (GGGCGG) repeat, and 2 copies of 72 bp repeat enhancers. The precise boundaries of these control regions have been delineated.[158]

2. Cellular Origin

While cellular promoters from human, rabbit, and chicken can be enhanced with viral sequences,[154,159,160] cellular DNA with enhancer activity has been also detected recently. A 21 bp repeat of BK virus DNA was found to have a homologous counterpart in the human genome, although its enhancing activity is some 8 times less than that of the virus, as determined by a CAT assay.[161]

Sequence near the C_K gene has been shown to be essential for high expression of K chain immunoglobulin in transfected myeloma cells, although the functional similarity to viral enhancers is less clear. However, a portion of the intervening sequence between J_h (heavy chain joining region) and C_u (constant u chain region), but not between C-α and S-α- has been identified to clearly enhance transcription of the immunoglobulin heavy chain. The sequence J_h-C_u is brought into functional proximity with V_h promoters after VDJ joining in lymphocytes.[162,163]

There is recent evidence demonstrating that enhancers also exist in lower eukaryotic cells. Specifically, yeast iso-1-cytochrome *c* gene activation is shown to be enhanced by an orientation independent upstream sequence. This sequence has the characteristic enhancer domain.[164,165]

3. Activation of Oncogenes

Enhancers play a role in the expression of rearranged genes.[166] For instance, nonacute avian leukosis virus (myelocytomatosis) c-myc gene is activated by integration into a sequence close to the ALV proviral enhancers.[139,167] Indeed several oncogene activations have been associated with their translocation to other chromosomes. Murine c-myc gene is activated after recombination into the Igh (i.e., Ig heavy chain) locus in Balb/c plasmacytomas, thus creating a t(12:15) translocation.[168,169] Burkitt's lymphomas or non-Hodgkin's lymphomas undergo a t(8:14) (q24; q32) translocation, such that the c-myc gene is recombined into the Igh locus and activated.[170] C-onc genes can also be activated by translocations.[171]

4. Tissue Specificity

Activity of a specific V gene among hundreds of V genes in an individual B lymphocyte and its progeny is due to its direct rearrangement into the vicinity of an enhancer sequence located upstream of the constant (C) gene region, between the joining (J) and switch (S) segments.[172] However, the enhanced transcriptional activity is detectable only in certain tissues, suggesting that tissue-specific factors are also involved. The immunoglobulin enhancer sequence J_h - C_u, while less effective than the SV40 72 bp repeat, is not functional in CV1 cells, the nontransformed parental line of COS cells. It does, however, enhance transcription of SV40 early promoter.[173] The differential effectiveness of a given sequence of enhancer in different tissues clearly suggests that specific factors are involved in its function.[174] Steroid hormone activation is an example.[175]

A tissue-specific enhancer element has indeed been located in the major intron of a rearranged immunoglobulin heavy chain gene of mouse.[176,177] Cellular sequence homologous to the 72 bp repeat of SV40 has also been observed to enhance the efficiency of transformation in an orientation independent manner. This sequence hybridizes to many fractions of human DNA, suggesting that a family of such enhancer elements exists.

It is interesting to note that enhancer sequences in native chromatin or viral minichromosomes are hypersensitive to DNase 1 digestion. In the developmentally committed eukaryotic genes of mouse myeloma cells, enhancers upstream from the transcriptionally active immunoglobulin light chain, C_K, are DNase hypersensitive. These same genes are not hypersensitive to DNase in their nonexpressed state in the liver. Similarly, enhancer core-like sequences for the E globin gene, which is constitutively active in K562 leukemia cells, also lie within a region sensitive to DNase 1.[178] These observations suggest that the chromatin in the enhancer region changes during transcription.

The presence of an enhancer sequence, regardless of its origin in DNA, significantly increases transformation frequency. Thus, any of the 72 bp repeats of SV40 or the 73 bp repeats of Harvey Sarcoma virus, is able to enhance 10- to 100-fold the number of positive colonies when plasmids carrying a galactokinase gene are used to transform chimeric hamster cells.[179] While these results support the role of enhancer at the transcription level, they are compatible with an alternative interpretation that enhancer sequences facilitate transformation per se. In prokaryotes, transformation-specific DNA sequences are preferentially taken up by competent cells presumably through specific binding to membrane receptors.[180] While it is unlikely that a similar mechanism is operative in the eukaryotes, it is yet to be shown that the enhanced transformation is not due to an elevated level of integration.

There are other models for enhancer action.[181] Enhancers may serve as DNA attachment sites to the nuclear matrix, an environment favorable for transcription. They may serve as a ''trap'', attracting and generating more enhancer sequences for transcription.[182] They may also serve as bidirectional entry sites for RNA polymerase II, since the enhancing activities can be lowered when a functional promoter is placed between an enhancer and a transcription initiation site.[183] Enhancers may be responsible for DNA or chromatin conformational changes, as evidenced by their DNase 1 hypersensitivity in minichromosomes.[184,185]

C. Control Through DNA Topology, Promoter Conformation, and Z-DNA

While a double-stranded DNA assumes helical conformation, its topology changes upon complexing with proteins as in the case of nucleosome formation involving the histones. Certain sequences of DNA bend into a complete circle in the space of about 80 bp,[186] which is only one half of the persistent length in solution. Bends and kinks occur frequently in native trypanosome kinetoplast DNA for sequences consisting of $C(A)_{5-6}T$ repeats at 10 bp apart, in phase with the helical periodicity.[187] It is suspected that altered DNA conformations per se serve as signals for protein recognition.[188,189] Conversely, the binding of proteins, e.g., CAP, can induce DNA bending several bases toward the promoter sites located within the center of symmetry for binding. This supports the notion that DNA topology can be altered in such an effective way that the stabilization of RNA polymerase is favored and transcription promoted.[190-192] It is well recognized that initiation of transcription in vitro by both prokaryotic and eukaryotic RNA polymerases is more efficient with negatively supercoiled DNA than with free DNA; presumably less free energy is required for the initial melting of DNA in the initiation complex.[193]

A new wave of attention has been focused on sequence specific DNA conformations, especially those amenable to undergoing B to Z transitions.[194] In prokaryotes, genomic DNA is negatively supercoiled. In eukaryotes, the bulk of their DNA appears to be relaxed in vivo. Inhibition of gyrase, an enzyme which lowers the linking number of DNA, results in the depression of transcription of several operons in bacteria. Suppression of topoisomerase

I, a DNA nicking-closing enzyme, in bacteria elevates transcription. It is possible that the activity of these enzymes determines DNA helicity, the extent of which in the promoters in turn exerts the strength of polymerase binding, hence the level of transcriptional initiation. This is so because it is energetically favored.[193,195] The effects of superhelicity on transcriptional elongation and termination are yet unclear. Recent studies show that purified RNA polymerase II of yeast can selectively initiate transcription in vitro within the iso-cytochrome *c* promoter TATA box located adjacent to a sequence that can undergo B to Z transition or that can form a cruciform structure.[196] On the other hand, $Co(NH_3)_6^{2+}$ induced B to Z transitions in poly (dG-dC), poly (dG-dC), and (poly dG dm^{5}C)-(poly dG dm^{5}C)[197] result in a lowering of template activity as assayed by RNA polymerase in vitro.[198] Thus, features in the upstream sequences of DNA and the extent of supercoiling of the DNA may be important for the efficiency and specificity of transcription.

Sequences of alternating purine-pyrimidine bases such as poly (dG-dC)$_n$ are able to adopt a zigzag helical conformation, termed Z-DNA, as revealed by 0.9A resolution X-ray crystallography.[199,200] Most of the native DNA sequences exist in beta form, β-DNA, in which the guanosine (and all the other bases) is in the "anti" conformation; its C-8 is closest to the phosphodiester backbone distal to the core, while the N-6, N-1, and C-2 atoms are projected into the center of the molecule. In the Z form, however, the guanosine residues are rotated 180° about the glycosidic bond, giving the "syn" conformation. As a consequence, the base pairs in Z-DNA occupy a position at the periphery, exposing the C-5 position of cytosine and the N-7 and C-8 of guanosine to the exterior of the molecule. This results in a left-handed, zigzag sugar-phosphate backbone with only 1 groove and 12 bp per turn of the helix, in contrast with the smooth, double grooved B-DNA which has 10 bp per turn and a right-handed helix.

The presence of the Z form of DNA in nature, its unique properties, and its physical localization among DNA regulatory sequences prompted many studies aiming at the elucidation of its function.[201] There are speculations that it may be involved in transient modulations of specific proteins for regulation of transcription, replication, and recombination, as well as in transient modifications of chromatin structure. Indeed, the nucleosome is not a static structure.[202]

Water molecules in the deep groove of Z-DNA are disordered, contributing to Z-DNA instability.[203] However, Z-DNA apparently can be stabilized by at least three mechanisms: negative supercoiling, methylation and ionic conditions, and binding to proteins.

1. Negative Supercoiling

A DNA polymer is positively supercoiled when it contains more than 10 bp per right-handed helical turn and is negatively supercoiled when it contains less than 10 bp per turn. The torsional stress imparted by negative supercoiling can induce Z-DNA formation.[204-206] When a segment (call it A) of circular DNA is being transcribed or replicated it has to unwind, i.e., negatively supercoil. In unwinding, this forces a segment or segments near A to also negatively supercoil to some extent. The free energy state of negatively supercoiled DNA is higher than the normal, relaxed form of DNA, and thus it is less stable. Any process that reduces this superhelicity is therefore thermodynamically favored under these conditions. Z-DNA should behave in such a fashion, as it is thermodynamically stable at −12 bp per turn with regard to right-handed B-DNA. Thus it may provide the cushion necessary for segment A to unwind for transcription or replication while maintaining the stability of the overall structure of the chromosome. This would allow alternating purine-pyrimidine residues with the ability to flip from B to Z form to facilitate the unwinding of genes — a prelude to transcription.

2. Base Modification

Alternating purine-pyrimidine regions, such as poly (dG-dC) poly (dG-dC), undergo dramatic conformational transition at high salt or alcohol concentrations as revealed by circular dichroism.[207] Methylation or partial methylation of cytosine residues also causes the polymer to exist in Z form at di- and trivalent salt concentrations in the physiological range.[208] Since the extent of methylation correlates with transcriptional activity and since the dG m^{5}dC sequence is common in eukaryotic cells, changes between B and Z forms of DNA around a gene may serve as a switching mechanism for the turning on or off of gene activity.

It is interesting to note that alternating GC residues in supercoiled Z-DNA are not readily recognized by the Hha 1 restriction modification system in bacteria. This was shown by examining two lengths of alternating GC of 24 and 32 bp long, respectively, inserted into plasmids to generate a series of overlapping recognition sites 5'-GCGC-3'. The same sequence, however, is methylated when relaxed.[209]

3. Binding to Proteins

Z-DNA segments are also stabilized by polyamines or polyarginine. Several *Drosophila* proteins have been found to bind specifically to Z-DNA, but not to B-DNA.[204] Such proteins, ranging in size from 70 to 150 kd, have also been identified in human cancer cells, wheat germ, and *E. coli*, although in varying abundances.[210] The sequence GTG/CAC, common to DNA from a variety of sources, shows a higher imino proton exchange rate well below the thermal denaturation temperature.[211] This suggests that this type of sequence is vulnerable to binding with specific macromolecules. The regulatory function of Z-DNA may well rest on its precise interaction with these proteins.

DNA sequences enriched in $(dA-dC)_n$, $(dT-dG)_n$, and $(dG-dC)_n$ $(dC-dG)_n$, which are capable of forming Z-DNA, have been found in a variety of eukaryotic cells and animal viruses.[212-214] There are other simple purine-pyrimidine repeats such as $(GT)_n$ or $(CA)_n$ in the eukaryotic genome which show a propensity to adopt the Z-form.[215] $(CA)_n$ blocks of DNA have been recognized in specific mammalian genes such as γ-globin,[216] globin gene clusters,[217,218] within the Alu sequence,[219] the c-myc gene and in and around the actin genes,[220] and in the immunoglobulin V[210,221] and histocompatibility genes.[222] They have also been detected by hybridization with sequence probes in frogs,[223] slime mold,[220] and yeast.[213,214] Some purine-pyrimidine repeats exist with the gene. For instance, in the mouse immunoglobulin C γ H-Cγ 3 intron, there exists a sequence in which the dinucleotide CA is repeated 30 times:

$$...CC\underline{TAACTA}TT(CT)_{29}(CA)_{30}(GA)_4GC\underline{CTAACTA}...$$

Other Z-DNA sequences seem to associate with the enhancer regions, as in the case of SV40, or in long terminal repeats, as in the retroviral genome. Many cellular repeats are localized in chromosomal telomere regions. These localizations prompted further speculation that they are involved in regulation through insertion and nicking/closing of DNA helixes.[210]

The exact composition of a given Z-DNA sequence is critical to its functionality. When a specific base is replaced in the Z-DNA sequence of SV40, a significant effect on viral reproduction is observed. While a T to C change yields no ill effect, an A to T substitution results in very slow (if any) growth of the virus.

The existence of native Z-DNA has been further shown with immunochemical assays. Anti-Z-DNA antibodies were elicited by injecting rabbits with brominated Z polymer. The addition of bromine to the C-8 atom of guanosine blocks the ability of these bases to assume the "anti" conformation. Z-specific antibodies were purified by quantitative precipitation with unbrominated poly(dG-dC) poly(dG-dC), which ensured their being specific for Z-DNA.[212,224] Such anti-Z-DNA antibodies react with poly(dG-dC)$_n$ sequences inserted into pBR322 as well as with other alternating purine-pyrimidine sequences normally present in this chimeric plasmid. In a more recent study using Z-DNA left-handed helical stretches in the DNA of native rabbit β-globin cDNA sequences were shown to be induced by torsional stress.[225] This was demonstrated by inserting this cDNA into pBR322 and manipulating it with nick-closing enzymes. The Z-DNA antibody binds only to those DNA molecules that are stressed below linking number $5 = < -0.1$, but not to form I or form II (relaxed circles). The bound antibody is released when the helix is unwound. This observation is further supported by circular dichroism and electron microscopy. The bindings were shown to be specific at 3 bp stretches enriched in alternating purine-pyrimidine and GC base pairs.

At least two different idiotypes of monoclonal anti-Z-DNA antibodies have been elicited and purified:[226] one has a strong affinity for the phosphodiester backbone of Z-DNA, whereas the other has an affinity for the methylated cytosine. These findings, coupled to the finding that pBR322s Z-DNA is not all G-C, but also contains A-T, indicate that different anti-Z-DNA monoclonal antibodies might be used as probes for different sequences of Z-DNA.

Anti-Z-DNA has been used to localize Z-DNA in the interchromatin bands of *Drosophila melanogaster* polytene chromosomes by immunofluorescence techniques.[212] Since interband regions do not contain a coding sequence this observation suggests that the Z-DNA serves a stabilizing function in the transcription and replication of DNA.

Indirect immunofluorescence with the ciliate *Stylonychia mytilus* shows that only the macronucleus binds the anti-Z-DNA antibodies, whereas the micronucleus in its polytene stage does not.[227] The macronucleus is degraded upon sexual reproduction and the micronucleus replicates extensively into polyteny, and then degrades to become a new macronucleus. It is interesting to note that antibody binding to Z-DNA was decreased in the macronucleus upon DNase I treatment. The region of Z-DNA in the chromatin is therefore not extensively associated with chromosomal proteins. Because the macronucleus is the site of DNA transcription in this ciliate, a role for Z-DNA in the transcription of structural genes can then be inferred.

The studies with *Drosophilia* and *Stylonychia* point toward differences in the localization and possibly the function of Z-DNA in species of vastly differing evolutionary background. The question arises: are there similarities, e.g., evolutionarily conserved sequences of Z-DNA, in different species that are phylogenetically closely related? In a study comparing anti-Z-DNA binding between chromosomes of *Cebus albifrons* and humans, Veigas-Pequignot et al.[228] observed that in spite of the vast difference in their classical karyotype patterns, correspondingly specific regions can be detected. For instance, binding in the short arm of the 1st human chromosome and the 15th chromosome of *Cebus* is comparable. The same relationship holds for human chromosome 21 and *Cebus* chromosome 9. The significance of this study lies in the suggestion that there are conserved sequences of Z-DNA in the euchromatin of humans and *Cebus albifrons*; both are primates.

D. Terminal and Downstream Signals for Transcription

DNA sequences signaling the end of a gene function terminate transcription at the proper site of their transcriptional unit, thus allowing independent transcription of their adjacent genes. With this signal, RNA chain elongation ceases, the transcript is released, and the polymerase is dissociated from the template. At least two classes of such sequences have been identified, specific rho factor (a) dependent or (b) independent. In prokaryotes, rho factor-dependent termination occurs near the 3' end of a coding sequence which is recognized by a protein factor, rho (ρ), about 46,000 in mol wt. ρ acts through its single-stranded, RNA-dependent ATP nucleoside triphosphate phosphohydrolase activity and may assume altered conformations when bound to a nucleotide substrate or product.[229] Its functional form is probably a hexamer. The sequence of the ρ gene has recently been determined.[230]

DNA sequences at ρ-dependent termination sites, however, are heterogeneous and as more such sequences are examined, they lack a consensus feature, although between 12 to 14 residues bind to each ρ monomer.[231] It is suggested that in addition to specific binding to nascent RNA and hydrolytic activity on nucleoside triphosphates, ρ factor binding specifies sites on DNA distal to which release of RNA and termination of transcription may occur at multiple positions over a minimal of 80 to 150 bp region of the template, which is usually at the ends of operons.

Other termination signals are independent of factors such as ρ. This termination site is characterized by a GC rich region of dyad symmetry preceding the 3' terminus of the RNA, followed by an AT rich sequence interspersed with isostitches of T on the anticoding strand.[232,233] Termination efficiency, whether by pausing of the transcriptional event or cessation of elongation, at these dyad symmetry and poly pyrimidine isostitches can be correlated with the length of the stem-loop structure; a minimal length of base pairs is required. The primary event here seems to involve reformation of template configuration and displacement of the nascent RNA from it, a process applicable to that operating in the attenuator region.

As noted earlier, the attenuator region represents another regulatory sequence for the termination of transcription in bacteria which is present upstream of a structural gene. Several prokaryotic operons characteristic of autogenous regulation, such as those for histidine, phenylalanine, leucine, threonine, isoleucine, and tryptophan, each contains an overlapping set of potential stem-loop forming sequence at the end of a 100 to 200 nucleotide leader sequence. It is by mutual exclusion and competition for stem-loop formation between these potential sequences that determine whether an early termination would occur. The mechanism involving a feedback control coupling transcription and translation has been detailed for the tryp operon.[234]

For the eukaryotic polymerase III termination, such as in the 5S RNA gene, a termination sequence . . . GCCTTTTGC . . . can be recognized by purified polymerase III.[235] Transcriptional units requiring polymerase II contain intrinsic sequence signals which determine premature termination of transcription. In vivo and in isolated nuclei, transcription has been shown to pause at positions proximal to the initiation site or terminate prematurely at +175 and possibly +120. Sequence surrounding +175 in Ad2 is homologous to that of the termination site t_{R1} of λ phage.[236]

In most eukaryotic exons, a pentanucleotide AATAA is present, which is some 25 bp before the 3' end of the coding DNA region. This consensus sequence has also been suggested as a signal for termination of transcription or for polyadenylation. Immediately after the 3' end of coding sequence of several mRNAs there is often a tetranucleotide TTTT which is also a candidate signal for RNA polymerase II termination.[237]

Many mRNAs show a conserved sequence of AAUAAA in the 3' end of poly A containing mRNA. Mutation in the sequence to AAGAAA or AAUAAG produces unusually long messages, which are unstable. Thus, at least for polymerase II, the termination of transcription obeys a sequence signal in the template. This signal is reflected in the terminus of its RNA in the form of a special hairpin structure which apparently facilitates post-transcriptional cleavage. The requirement of an endonucleolytic cleavage process on a small nuclear ribonucleoprotein "snRNP" has been clearly demonstrated with histone genes.[238,239] Presumably, the sequence AAUAAA and perhaps another frequently occurring sequence CApyUG[56] near the poly A addition site are being recognized by a small nuclear RNA, U4, in the "snRNP". The situation is parallel to the splicing process for mRNA maturation, in which another class of "snRNP" is involved in aligning the intron-donor-acceptor sites with its U1 RNA,[240] demonstrable by specific antibody inactivation.[241] In yeast, the coding sequence for U1-like RNA constitutes a portion of the 3' acceptor site of the intron of the same gene.[119]

In eukaryotes, sequences further downstream from a gene also convey termination signals.

Thus, for β-major globin gene a stretch of DNA 1 kb 3′ to the coding sequence also signals termination beyond which no transcription is observed to proceed.[67,242] The precise nature and generality of this termination signal in eukaryotes is yet to be determined. Gene fusion techniques have been applied to examine the action of terminator sequences.[243]

Another sequence, TGTAAATA, is conserved in the 3′ untranslated region of several genes, including the gene for metallothionein,[107] as is a T rich sequence about 60 nucleotides downstream of the termination codon. As has been noted,[215] the repetitive poly (dG-dT) sequences are present in and around many genes, particularly 3′ of the polyadenylation site. Signals for polyadenylation of matured mRNA in most genes share the AATAAA sequence.

IV. INTERACTIONS OF TRANSCRIPTIONAL REGULATORY ELEMENTS

A. Cellular Transcription Factors

A major effort in the understanding of the transcription process and its regulation is devoted to defining conditions in vitro that mimic those in vivo. Promoter sequences deduced from experiments in vitro that have been mapped for several cellular and viral genes contribute only a subset of those sequence signals required for efficient transcription in vivo, most of which lack regulatory specificity. This is probably due to incompleteness of the artificial assay conditions in vitro. Already it has been shown that circular DNAs, native configurations for some genomes, provide a greater coincidence in sequence specificity for initiation both in vivo and in vitro, than in their linear forms.[244] More importantly, a series of factors has been identified with fidelity in transcription.

Several factors for RNA polymerase I transcription in vitro have been identified[245] and have been shown to be species specific. Human, mouse, and protozoan ribosomal DNAs are accurately transcribed in vitro only if cellular components from a homologous source are provided. Four major components have been characterized chromatographically: fraction A enhances transcription, B suppresses random initiation, C contains most of the polymerase, and D is indispensable for accurate initiation (see Reference 73).

For RNA polymerase II, at least 5 factors are essential and sufficient for accurate initiation at Ad2 major late promoters in a relatively crude cell extract.[246,247] One, TFIIC, purified to homogeneity, inhibits random transcription in crude systems but is not required in more purified systems. The presence of another, TFIIA, maximizes the transcription level, presumably due to its carrying an RNase inhibitor. The others are factors which bind to RNA polymerase (TFIIB), ATPase (TFIIE), or the template (TFIID). The hydrolysis of ATP by ATPase is presumably important in the initiation step. However, the requirement of these factors apparently applies only to certain late promoters.

Independent studies show that with HeLa cell S100 extracts, two components can be identified as necessary for efficient and accurate transcription by pol II in vitro. One of these factors is strikingly similar to actin and acts at the pre-initiation step by recognizing a specific promoter sequence, the TATA box.[248] By differentiating the requirements for XTP, procedures have now been developed to form and isolate transcriptional initiation as well as elongation complexes.[249] Cellular/host factors specific for SV40 and *Drosophila* transcription in vivo have also been described.[250-252]

Some cellular factors inhibit rather than enhance viral gene transcription in vitro, suggesting that regulation in cell-free systems can be demonstrated. The early promoter of SV40 T antigen and adenovirus DNA binding protein[253] are blocked preferentially by such factors as are other early adenovirus (EII and EIII) but not major late or EIV promoters. Thus, for polymerase II control at the transcription level, several cellular components are required.

Several factors have been identified to be essential for the accurate initiation and termination of transcription by polymerase III in vitro.[254,255] Factor TFIIIA is uniquely required for 5S genes. It binds to 5S promoters independent of other proteins and may be involved in initiation. A stable preinitiation complex with 50 bp of intragenic sequence of the 5S gene and other factors (TFIIIB, TFIIIC) has been observed. A suggestion was made earlier that factor A is also involved in a feedback inhibition/autoregulatory mechanism when its supply is low. Factor C, on the other hand, probably plays a key role in the transcription of other viral genes but forms stable transcription complexes. Certain viral gene products are known to activate cloned genes such as β-globin.[256]

B. Tissue and Temporal Specificity of Gene Expression

A genome codes for more information than is needed at a given time and site. Thus, it is not surprising to find that different genes are regulated either according to a predetermined genetic program or otherwise in a concerted response to external stimulants, e.g., inducers or repressors.

An essential feature of prokaryotic gene regulation lies in timely interaction between the genome and sequence-specific DNA binding proteins. As discussed earlier, initiation and termination involve σ and ρ factor recognition of DNA sequence signals which have been well established. In eukaryotes regulation through protein-DNA interaction is complicated by the structure of their genomes, in which DNA, RNA, and proteins are arranged through subunits such as nucleosomes to form chromatin. Furthermore, while prokaryotic genes may be arranged as operons, many eukaryotic genes may be organized into clusters. If gene clusters are indeed the basic genetic functional units,[259] mechanisms regulating the coordinated expression must exist. The organization of chromatin is dynamic; its DNA, RNA, and protein juxtaposition change according to developmental stages or tissue types.

Transcriptional regulation in differentiation may be exerted at various levels and may involve exogenous substances, nutrients and stimulants or endogenous gene products, hormones,[258,259] and control factors. Methylation and demethylation may be examples of such controls.[260,261] Sequential expression of distantly linked, functionally related genes such as embryonic, fetal, and adult globin and chorion genes, as well as systematic switching and rearrangement of genes for variable and constant regions of immunoglobulins, has been well documented.

Viral genomes such as that of adenovirus,[103] offer a more simple system to understand for temporal and tissue-specific regulation. Expression of viral genes first involves the choice of promoters. Early promoters are either active without the need for viral proteins or responsive to positive or negative regulators encoded by the virus. These gene products in turn regulate late viral gene expression in a cascade of regulatory events. The presence of a given promoter ensures the choice of 5′ capping termini of transcripts, as several transcripts may share the same initiation point, but vary in their mature sequence. The mature sequence is determined by the specific polyadenylation site(s) selected, which decides the relative abundance of each of the mRNA species in this family. During the processing steps, preferential splicing defines the final product, mRNA, available to the cell. Clearly a series of subtle regulation steps, presumably through specific protein-DNA interactions, is needed to yield the complex but precise transcriptional endpoints.

Indeed animal viruses are transmitted only in selective tissues and often require specific interaction between their own transacting activator and cellular control elements. The human T cell leukemia virus genome carries an extra gene, X, whose product has been shown to act in a positive feedback manner, enhancing transcription of its own LTR and of other transcriptional units in T cells.[262,263] Alternatively, LTR function may be suppressed by specific proteins in nonlymphoid cells.[143] Genetic development may well involve a progressive reduction of the number of expressed genes.

In a eukaryotic cell, transcriptionally active chromatin is more sensitive toward DNase I than inactive chromatin. The hypersensitive site has been located at the regulatory region preceding the coding sequence of several structural genes.[264-270] The sensitivity is conferred by chromosomal protein(s) that is associated with DNA instead of the sequence of DNA itself. Tissue-specific patterns of DNase I sensitivity have been demonstrated for genes coding for chicken β-globin, chicken lysozyme, *Drosophila* heat shock proteins, histones, and rat preproinsulin.

The region of an active gene that is hypersensitive to DNase I varies; some involve enhancer elements,[271] others involve the promoter.[272] Mouse mammary tumor viral enhancer is transiently sensitive to nuclease when it is responding to glucocorticoid induction.[273]

The expression of retroviral genes has also been correlated with conformation in host chromatin.[274] An unexpressed ev-1 locus of chicken endogenous retrovirus is no more sensitive to DNase I than bulk DNA, whereas when it becomes transcriptional it is preferentially sensitive to DNase I digestion. However, the site of hypersensitivity for 3v-3 gene lies within the LTR sequences on both the 5′ and 3′ termini of the proviral coding regions. These regions share a 15 to 20 bp sequence homology with tandemly repeated regions in SV40 responsible for initiation in vivo of early transcription.

Specific factors have recently been identified which would allow reconstitution of a DNase I hypersensitive site in the chicken β-globin[272] and *Drosophila* heat shock genes.[275] This progress promises further understanding of chromatin structural dynamics and gene regulation.

V. SUMMARY AND PROSPECTUS

The concept of a gene as a contiguous region of DNA co-linear with its protein product has been extensively extended. The coding region of a gene may be interrupted; genes of related function may be clustered and tandemly repeated, but separated by spacer sequences; duplicated sequences of a gene may lose their expression and exist as regions of nonexpressed DNA displaying significant homology to a corresponding functional gene; DNA may be peripatetic, with gene function depending on the ability to move about within a genome. Therefore, diversity is seen not only in the organization of a gene but also in the gene's regulatory elements.

As more DNA sequences are determined, many regulatory sequences will not fit the consensus, even allowing for certain probabilities of substitution. These observations should not pose a dilemma, as most of the regulatory consensus sequences for transcription were deduced from mutational analysis, particularly by deletions, for which the loss of transcriptional initiation or efficiency was measured within a boundary. Flanking sequences, helix handedness, and higher order folding of the chromatin may also play a role in regulation; indeed, this has already been shown for many genes. Further understanding of the phylogeny of regulatory sequences depends upon additional knowledge about the interactions between these sequences and proteins.

The question of how regulatory sequences evolve does not yet have an answer. It seems that the main driving force in the diversification of multicellular organisms results with the organization and distribution of control signals. The regulation of structural information may evolve through rearrangement or amplification of these signals.

Even as one examines the relatively simple monotonic system of *E. coli*, one cannot escape being impressed by the fact that there already are multiple levels of feedback and the interacting homeostatic control at work. Each gene is under precise temporal scrutiny of its control sequences, be it a promoter, attenuator, or other pausing and termination signals. Special recognition by RNA polymerase and its ancillary proteins of unique DNA segments in the genome is the basic rule of transcriptional regulation.

Other prokaryotes, the life cycle of which involves alternations between vegetative and sporulated stages, rely on multiple differentiating regulatory sequences or rearrangement of the same. In other cases, the same promoter sequence serves to provide a dual orientation. The mode of rearranged regulatory signals is particularly prominent in the single cell eukaryotes; yeast mating type cassette is one better characterized example. Higher eukaryotes deviate from simpler to one polymerase-promoter regulation to encompass multiple forms of polymerases each with its own regulatory sequences for different sets of genes. Eukaryotic genes are not organized into operons, they may exist as clusters, and many of them are complicated by exon/intron structures. Thus, additional sequence signals must be created to accommodate post-transcriptional processing, which includes splicing and transport from the nucleus to the cytoplasm, yielding mature, functional mRNA.

Clearly during development and growth, sets of genes are activated or inactivated as evidenced by dynamic structural changes in the chromatin and/or chemical modifications of DNA. These require additional sets of regulatory effectors.

There appears to be no clear and simplistic rule in the promoter sequences known, which is absolutely essential to the process of transcription. Most DNA sequences interact with regulatory proteins showing dyad symmetry; the promoter sequences, however, are asymmetric. This feature may well provide directionality of transcription for which a separate and unique path of evolution must be followed.

ACKNOWLEDGMENTS

The author wishes to acknowledge the able assistance of Betty Chinn Smith and Mary Cismowski in processing this manuscript and the support of NSF Grant PCM 8104369.

REFERENCES

1. **Davanloo, P., Rosenberg, A. H., Dunn, J. J., and Studier, F. W.,** Cloning and expression of the gene for bacteriophage T7 RNA polymerase, *Proc. Natl. Acad. Sci. U.S.A.,* 81, 2035, 1984.
2. **Chamberlin, M.,** RNA polymerase — 1976, in *RNA Polymerase,* Losick, R. and Chamberlin, M., Eds., Cold Spring Harbor Laboratory, Cold Spring Harbor, N.Y., 1976, 159.
3. **Gilman, M. Z. and Chamberlin, M. J.,** Developmental and genetic regulation of *Bacillus subtilis* genes transcribed by sigma^{28}-RNA polymerase, *Cell,* 35, 285, 1983.
4. **Zuber, P. and Losick, R.,** Use of a lacZ fusion to study the role of the spo0 genes of *Bacillus subtilis* in developmental regulation, *Cell,* 35, 275, 1983.
5. **Ingles, C. J., Himmelfarb, H. J., Shales, M., Greenleaf, A. L., and Friesen, J. D.,** Identification, molecular cloning, and mutagenesis of *Saccharomyces cerevisiae* RNA polymerase genes, *Proc. Natl. Acad. Sci. U.S.A.,* 81, 2157, 1984.
6. **Spencer, D. F., Schnare, M. N., and Gray, M. W.,** Pronounced structural similarities between the small subunit ribosomal RNA genes of wheat mitochondria and *Escherichia coli, Proc. Natl. Acad. Sci. U.S.A.,* 81, 493, 1984.
7. **Stern, D. B. and Palmer, J. D.,** Extensive and widespread homologies between mitochondrial DNA and chloroplast DNA in plants, *Proc. Natl. Acad. Sci. U.S.A.,* 81, 1946, 1984.
8. **Bardwell, J. C. A. and Craig, E. A.,** Major heat shock gene of *Drosophila* and the *Escherichia coli* heat-inducible DNA K gene are homologous, *Proc. Natl. Acad. Sci. U.S.A.,* 81, 848, 1984.
9. **Landick, R., Vaughn, V., Lau, E. T., Van Bogelen, R. A., Erickson, J. W., and Neidhardt, F. C.,** Nucleotide sequence of the heat shock regulatory gene of *E. coli* suggests its protein product may be a transcription factor, *Cell,* 38, 175, 1984.
10. **Normark, S., Bergstrom, S., Edlind, T., Grundstrom, T., Bengtake, J., Lindberg, F. P., and Olsson, O.,** Overlapping genes, *Ann. Rev. Genet.,* 17, 499, 1983.
11. **Sanger, F., Air, G. M., Barrell, B. G., Brown, N. L., Coulson, A. R., Fiddes, J. C., Hutchison, C. A., III, Slocombe, P. M., and Smith, M.,** Nucleotide sequence of bacteriophage phiX174 DNA, *Nature (London),* 265, 687, 1977.
12. **Shaw, D. C., Walker, J. E., Northrup, F. D., Barrell, B. G., Godson, G. N., and Fiddes, J. C.,** Gene K, a new overlapping gene in bacteriophage G4, *Nature (London),* 272, 510, 1978.
13. **Attardi, G.,** Organization and expression of the mammalian mitochondrial genome: a lesson in economy, *Trends Biochem. Sci.,* 6, 100, 1981.

14. **Chang, D. D. and Clayton, D. A.,** Precise identification of individual promoters for transcription of each strand of human mitochondrial DNA, *Cell,* 36, 635, 1984.

15. **Bogenhagen, D. F., Applegate, E. F., and Yoza, B. K.,** Identification of a promoter for transcription of the heavy strand of human MT DNA: in vitro transcription and deletion mutagenesis, *Cell,* 36, 1105, 1984.

16. **Ohno, S.,** Segmental homology and internal repetitiousness identified in putative nucleic acid polymerase and human hepatitis B surface antigen of human hepatitis B virus, *Proc. Natl. Acad. Sci. U.S.A.,* 81, 3781, 1984.

17. **Brunel, F., Thi, V. H., Pilaete, M.-F., and Davison, J.,** Transcription regulatory elements in the late region of bacteriophage T5 DNA, *Nucleic Acids Res.,* 11, 7649, 1983.

18. **Baty, D., Barrera-Saldana, H. A., Everett, R. D., Vigeron, M., and Chambon, P.,** Mutational dissection of the 21 bp repeat region of the SV40 early promoter reveals that it contains overlapping elements of the early-early and late-late promoters, *Nucleic Acids Res.,* 12, 915, 1984.

19. **Cunin, R., Eckhardt, T., Piette, J., Boyen, A., Pierard, A., and Glansdorff, N.,** Molecular basis for modulated regulation of gene expression in the arginine regulon of *Escherichia coli* K-12, *Nucleic Acids Res.,* 11, 5007, 1983.

20. **Yanofsky, C.,** Comparison of regulatory and structural regions of genes of tryptophan metabolism, *Mol. Biol. Evol.,* 1, 143, 1984.

21. **Ames, B. N., Tsang, T. H., Buck, M., and Christman, M. F.,** The leader mRNA of the histidine attenuator region resembles tRNA[his]: possible general regulatory implications, *Proc. Natl. Acad. Sci. U.S.A.,* 80, 5240, 1983.

22. **Eiserbeis, S. J. and Parker, J.,** The nucleotide sequence of the promoter region of hisS, the structural gene for histidyl-tRNA synthetase, *Gene,* 18, 107, 1982.

23. **Travers, A.,** Protein contacts for promoter location in eukaryotes, *Nature (London),* 303, 755, 1983.

24. **Attardi, G., Huang, P. C., and Kabat, S.,** Recognition of ribosomal RNA sites in DNA. II. Analysis of the *E. coli* system, *Proc. Natl. Acad. Sci. U.S.A.,* 53, 1490, 1965a.

25. **Attardi, G., Huang, P. C., and Kabat, S.,** Recognition of ribosomal RNA sites in DNA. The HeLa cell system, *Proc. Natl. Acad. Sci. U.S.A.,* 54, 185, 1965b.

26. **Dutta, S. K., Williams, N. P., and Mukhopadhyay, D. K.,** Ribosomal RNA genes of *Neurospora crassa:* multiple copies and specificities, *Mol. Gen. Genet.,* 189, 207, 1983.

27. **Jacq, C., Miller, J. R., and Brownlee, G. G.,** A pseudogene structure in 5S DNA of *Xenopus zaevis, Cell,* 12, 109, 1977.

28. **Nishioka, Y., Leder, A., and Leder, P.,** Unusual α-globin-like gene that has clearly lost both globin intervening sequences, *Proc. Natl. Acad. Sci. U.S.A.,* 77, 2806, 1980.

29. **Hollis, G. F., Hieter, P. A., McBride, O. W., Swan, D., and Leder, P.,** Processed genes: a dispersed human immunoglobulin gene bearing evidence of RNA-type processing, *Nature (London),* 296, 321, 1982.

30. **Proudfoot, N. J. and Maniatis, T.,** The structure of a human α-globin pseudogene and its relationship to α-globin gene duplication, *Cell,* 21, 537, 1980.

31. **Lacy, E. and Maniatis, T.,** The nucleotide sequence of a rabbit beta-globin pseudogene, *Cell,* 21, 545, 1980.

32. **Wong, W. M., Abrahamson, J. L., and Nazar, R. N.,** Are DNA spacers relics of gene amplification events? *Proc. Natl. Acad. Sci. U.S.A.,* 81, 1768, 1984.

33. **Maurer, B. J., Barker, P. E., Masters, J. N., Ruddle, F. H., and Attardi, G.,** Human dihydrofolate reductase gene is located in chromosome 5 and is unlinked to the related pseudogenes, *Proc. Natl. Acad. Sci. U.S.A.,* 81, 1484, 1984.

34. **Childs, G., Maxson, R., Cohn, R. H., and Kedes, L.,** Orphons: dispersed genetic elements derived from tandem repetitive genes of eukaryotes, *Cell,* 23, 651, 1981.

35. **Huang, P. C.,** DNA, RNA and protein interactions, *Prog. Biophys. Mol. Biol.,* 23, 103, 1971.

36. **Bujard, H., Niemann, A., Breunig, K., Roisch, U., Dresel, A., von Gabain, A., Gentz, R., Stuber, D., and Weiher, H.,** in *Promoters,* Rodriquez, R. L. and Chamberlin, M. J., Eds., Praeger, New York, 1982, 121.

37. **Kajitani, M. and Ishihama, A.,** Determination of the promoter strength in the mixed transcription system. II. Promoters of ribosomal RNA, ribosomal protein S1 and recA protein operons from *Escherichia coli, Nucleic Acids Res.,* 11, 3873, 1983.

38. **Pribnow, D.,** Nucleotide sequence of an RNA polymerase binding site at an early T7 promoter, *Proc. Natl. Acad. Sci. U.S.A.,* 72, 784, 1975.

39. **Lewin, B.,** *Genes,* John Wiley & Sons, New York, 1983.

40. **Roberts, T. M.,** A lac promoter system for the over expression of prokaryotic and eukaryotic genes in *E. coli,* in *Promoters,* Rodriquez, R. L. and Chamberlin, M. J., Eds., Praeger, New York, 1982, 452.

41. **Travers, A. A., Lamond, A. I., Mace, H. A. F., and Berman, M. L.,** RNA polymerase interactions with upstream region of the *E. coli* tryT promoter, *Cell,* 35, 265, 1983.

42. **Shih, M.-C. and Gussin, G. N.,** Differential effects of mutations on discrete steps in transcription initiation at the lambda P$_{RE}$ promoter, *Cell,* 34, 941, 1983.

43. **Stohrer, G., Osband, J. A., and Alvarado-Urbino, G.,** Site-specific modification of the lactose operator with acetylaminofluorene, *Nucleic Acids Res.,* 11, 5093, 1983.

44. **deHaseth, P. L., Goldman, R. A., Cech, C. L., and Caruthers, M. H.,** Chemical synthesis and biochemical reactivity of bacteriophage lambda P$_R$ promoter, *Nucleic Acids Res.,* 11, 773, 1983.

45. **Levy, S., Sures, I., and Kedes, L. H.,** Sequence at the 5′ end of *S. purpuratus* H2B histone mRNA and its location within histone DNA, *Nature (London),* 279, 737, 1979.

46. **Konkel, D. A., Tilghman, S. M., and Leder, P.,** The sequences of the chromosomal mouse beta-globin major gene: homologies in capping, splicing and poly(A) sites, *Cell,* 15, 1125, 1978.

47. **Wright, S., Rosenthal, A., Flavell, R., and Grosveld, F.,** DNA sequences required for regulated expression of beta-globin genes in murine erythroleukemia cells, *Cell,* 38, 265, 1984.

48. **Cordell, B., Bell, G., Tischer, E., DeNoto, F., Ulrich, A., Pictet, R., Rutter, W. J., and Goodman, H. M.,** Isolation and characterization of a cloned rat insulin gene, *Cell,* 18, 533, 1979.

49. **Gannon, F., O'Hare, K., Perrin, F., LePennec, J. P., Benoist, C., Cochet, M., Breathnach, R., Royal, A., Garapin, A., Gami, G., and Chambon, P.,** Organization and sequences at the 5′ end of a cloned complete ovalbumin gene, *Nature (London),* 278, 428, 1979.

50. **Lai, E. C., Woo, L. C., Dugaiczyk, A., and O'Malley, B. W.,** The ovalbumin gene: alleles created by mutations in the intervening sequences of the natural gene, *Cell,* 16, 201, 1979a.

51. **Lai, E. C., Stein, J. P., Catteral, F. C., Woo, L. C., Mace, M. L., Means, A. R., and O'Malley, B. W.,** Molecular structure and flanking nucleotide sequences of the natural chicken ovomucoid gene, *Cell,* 18, 829, 1979b.

52. **Tonegawa, S., Maxam, A. M., Tizard, R., Bernard, O., and Gilbert, N.,** Sequence of a mouse germ line gene for a variable region of an immunoglobulin light chain, *Proc. Natl. Acad. Sci. U.S.A.,* 75, 1485, 1978.

53. **Parslow, T. G., Blair, D. L., Murphy, W. J., and Granner, D. K.,** Structure of the 5′ ends of immunoglobulin genes, *Proc. Natl. Acad. Sci. U.S.A.,* 81, 2650, 1984.

54. **Kozak, M.,** Compilation and analysis of sequences upstream from the translational start site in eukaryotic mRNA, *Nucleic Acids Res.,* 12, 857, 1984.

55. **Cochet, M., Cannon, F., Hen, R., Maroteaux, L., Perrin, F., and Chambon, P.,** Organization and sequence studies of the 17-Pieces chicken conalbumin gene, *Nature (London),* 282, 567, 1979.

56. **Benoist, C., O'Hare, K., Breithnach, R., and Chambon, P.,** The ovalbumin gene sequences of putative control regions, *Nucleic Acids Res.,* 8, 127, 1980.

57. **Wasylyk, B., Buchwalder, A., Bassone, A., Kedinger, R., and Chambon, P.,** Promoter sequences of eukaryotic protein-coding genes, *Science,* 209, 1406, 1978.

58. **Baker, C. C., Merisse, J., Coutois, G., Galibert, F., and Ziff, E.,** Messenger RNA for the adenovirus-2 DNA binding protein, *Cell,* 18, 569, 1979.

59. **Grosschedl, R., Wastlyk, B., Chambon, P., and Birnstiel, M. L.,** Point mutation in the TATA box curtails expression of sea urchin H2A histone gene in vivo, *Nature (London),* 294, 178, 1981.

60. **McKnight, S. L.,** Functional relationships between transcriptional control signals of the thymidine kinase gene of herpes simplex virus, *Cell,* 31, 355, 1982.

61. **McKnight, S. L. and Kingsbury, R.,** Transcriptional control signals of a eukaryotic protein-coding gene, *Science,* 217, 316, 1982.

62. **Geiser, M., Mattaj, I. W., Wilks, A. F., Seldran, M., and Jost, J.-P.,** Structure and sequence of the promoter area and of a 5′ upstream demethylation site of the estrogen-regulated chicken vitellogenin II gene, *J. Biol. Chem.,* 258, 9024, 1983.

63. **Falkner, F. G. and Zachau, H. G.,** Correct transcription of an immunoglobulin K gene requires an upstream fragment containing conserved sequence elements, *Nature (London),* 310, 71, 1984.

64. **Zinn, K., DiMaio, D., and Maniatis, T.,** Identification of two distinct regulatory regions adjacent to the human beta-interferon gene, *Cell,* 34, 865, 1983.

65. **DeLange, T., Liu, Y. C., Van der Ploeg, L. H. T., Borst, P., Tromp, M. C., and Van Boom, J. H.,** Tandem repetition of the 5′ mini-exon of variant transcription? *Cell,* 34, 891, 1983.

66. **Allan, M., Lanyon, G., and Paul, J.,** Multiple origins of transcription in the 4.5 kb upstream of the epsilon-globin gene, *Cell,* 35, 187, 1983.

67. **Hofer, E., Hofer-Warbinek, R., and Darnell, J. E., Jr.,** Globin RNA transcription: a possible termination site and demonstration of transcriptional control correlated with altered chromatin structure, *Cell,* 29, 887, 1982.

68. **Langridge, P. and Feix, G.,** A zein gene of maize is transcribed from two widely separated promoter regions, *Cell,* 34, 1015, 1983.

69. **Bensimhon, M., Gabarro-Arpa, J., Ehrlich, R., and Reiss, C.,** Physical characteristics in eucaryotic promoter, *Nucleic Acids Res.,* 11, 4521, 1983.

70. **Concino, M. F., Lee, R. F., Merryweather, J. P., and Weinmann, R.,** The adenovirus major late promoter TATA box and initiation site are both necessary for transcription in vitro, *Nucleic Acids Res.,* 12, 7423, 1984.

71. **Charnay, P., Treisman, R., Mellon, P., Chao, M., Axel, R., and Maniatis, T.,** Differences in human alpha- and beta-globin gene expression in mouse erythroleukemia cells: the role of intragenic sequences, *Cell,* 38, 251, 1984.

72. **Merrill, G. F., Hauschka, S. D., and McKnight, S. L.,** tk Enzyme expression in differentiating muscle cells is regulated through an internal segment of the cellular tk gene, *Mol. Cell. Biol.,* 4, 1777, 1984.

73. **Sommerville, J.,** RNA polymerase I promoters and transcription factors, *Nature (London),* 310, 189, 1984.

74. **Sollner-Webb, B., Wilkinson, J. A. K., Roan, J., and Reeder, R. H.,** Nested control regions promote *Xenopus* ribosomal RNA synthesis by RNA polymerase I, *Cell,* 35, 199, 1983.

75. **Financsek, I., Mizumoto, K., and Maramatsu, M.,** Nucleotide sequence of the transcription initiation region of a rat ribosomal RNA gene, *Gene,* 18, 115, 1982.

76. **Busby, S. J. and Reeder, R. H.,** Spacer sequences regulate transcription of ribosomal gene plasmids injected into *Xenopus* embryos, *Cell,* 34, 989, 1983.

77. **Morgan, G. T., Reeder, R. H., and Bakken, A. H.,** Transcription in cloned spacers of *Xenopus laevis* ribosomal DNA, *Proc. Natl. Acad. Sci. U.S.A.,* 80, 6490, 1983.

78. **Reeder, R. H.,** Enhancers and ribosomal gene operons, *Cell,* 38, 349, 1984.

79. **Bakken, A., Morgan, G., Sollner-Webb, B., Roan, J., Busby, S., and Reeder, R.,** Mapping of transcription initiation and termination signals on *Xenopus laevis* ribosomal DNA, *Proc. Natl. Acad. Sci. U.S.A.,* 79, 56, 1982.

80. **Kohorn, B. D. and Rae, P. M.,** A component of *Drosophila* RNA polymerase I promoter lies within rRNA transcription unit, *Nature (London),* 304, 179, 1983.

81. **Kolata, G.,** Control element found within structural gene, *Science,* 220, 294, 1983a.

82. **Kressman, A., Hofstetter, H., Di Capua, E., Grosschedl, R., and Birnstiel, M. L.,** A tRNA gene of *Xenopus laevis* contains at least two sites promoting transcription, *Nucleic Acids Res.,* 7, 1749, 1979.

83. **Newman, A. J., Ogden, R. C., and Abelson, J.,** tRNA gene transcription in yeast: effects of specified base substitutions in the intragenic promoter, *Cell,* 35, 117, 1983.

84. **Folk, W. R. and Hofstetter, H.,** A detailed mutational analysis of the eucaryotic tRNA$_1^{met}$ gene promoter, *Cell,* 33, 585, 1983.

85. **Bogenhagen, D. F., Sakonju, S., and Brown, D. D.,** A control region in the center of the 5S RNA gene directs specific initiation of transcription. II. The 3′ border of the region, *Cell,* 19, 27, 1980.

86. **Bogenhagen, D., Wormington, W. M., and Brown, D. D.,** The 3′ border of the region, *Cell,* 28, 413, 1982.

87. **Fowlkes, D. M. and Shenk, T.,** Transcription control regions of the adenovirus VAI RNA gene, *Cell,* 22, 405, 1980.

88. **Guilfoyle, R. and Weinmann, R.,** Control region for adenovirus VA RNA transcription, *Proc. Natl. Acad. Sci. U.S.A.,* 78, 3378, 1981.

89. **Bhat, R. A. and Thimmappaya, B.,** Adenovirus mutants with DNA sequence perturbations in the intragenic promoter of VAI RNA gene allow the enhanced transcription of VAII RNA gene in HeLa cells, *Nucleic Acids Res.,* 12, 7377, 1984.

90. **Galli, G., Hofstetter, H., and Birnstiel, M. L.,** Two conserved sequence blocks within eukaryotic tRNA genes are major promoter elements, *Nature (London),* 294, 626, 1981.

91. **Hipskind, R. A. and Clarkson, S. G.,** 5′-flanking sequences that inhibit in vitro transcription of a *Xenopus laevis* tRNA gene, *Cell,* 34, 881, 1983.

92. **Irani, M. H., Orosz, L., and Adhya, S.,** A control element within a structural gene: the gal operon of *Escherichia coli, Cell,* 32, 783, 1983.

93. **Gilbert, W.,** Why genes in pieces, *Nature (London),* 271, 501, 1978.

94. **Wellauer, P. K. and Dawid, I. B.,** The structural organization of ribosomal DNA in *D. melanogaster, Cell,* 10, 193, 1977.

95. **Jeffreys, A. J. and Flavell, R. A.,** The rabbit beta-globin gene contains a long insert in the coding sequence, *Cell,* 12, 1097, 1977.

96. **Tilghman, S. M., Tiemeier, D. C., Seidman, J. G., Peterlin, B. M., Sullivan, M., Maizel, J. V., and Leder, P.,** Intervening sequences of DNA identified in the structural portion of a mouse globin gene, *Proc. Natl. Acad. Sci. U.S.A.,* 75, 725, 1978.

97. **Breathnach, R., Mandel, J. L., and Chambon, P.,** Ovalbumin gene is split in chicken DNA, *Nature (London),* 270, 5652, 1977.

98. **Royal, A., Garapin, A., Gami, B., Perrin, F., Mandel, J. L., LeMear, M., Bregegegre, F., Gannon, F., LePennec, F., Chambon, P., and Koubilsky, P.,** The ovalbumin gene region: common features in the organization of three genes expressed in chicken oviduct under hormonal control, *Nature (London),* 279, 125, 1979.

99. **Nguyen-Hum, M. C., Stratmann, M., Groner, B., Wurtz, T., Land, H., Giesecke, K., Sippel, A. F., and Schutz, G.,** Chicken lysozyme gene contains several intervening sequences, *Proc. Natl. Acad. Sci. U.S.A.,* 76, 76, 1979.

100. **Catterall, J. F., Stein, J. P., Lai, E. C., Woo, S. L. C., Dugaiczyk, A., Mace, M. L., Means, A. R., and O'Malley, B. W.,** The chicken ovomucoid gene contains at least six intervening sequences, *Nature (London),* 278, 323, 1979.

101. **Goodman, H. M., Olson, M. V., and Hall, B. D.,** Nucleotide sequence of mutant eucaryotic gene: the yeast tyrosine inserting ochre suppressor sup40, *Proc. Natl. Acad. Sci. U.S.A.,* 74, 5453, 1977.

102. **Valenzuela, P., Venegas, A., Weinberg, F., Bishop, R., and Rutter, W. J.,** Structure of yeast phenylalanin t-RNA gene: an intervening DNA segment within the region coding for tRNA, *Proc. Natl. Acad. Sci. U.S.A.,* 75, 190, 1978.

103. **Chow, L. T., Gelinas, R. E., Broker, J. R., and Roberts, R. J.,** An amazing sequence arrangement at the 5' ends of adenovirus-2 mRNA, *Cell,* 12, 1, 1977.

104. **Aloni, Y., Dhar, R., Laub, O., Morowitz, M., and Khoury, G.,** Novel mechanism for RNA maturation, *Proc. Natl. Acad. Sci. U.S.A.,* 74, 3686, 1977.

105. **Quax, W., Egberts, W. V., Hendricks, W., Quax-Jeuken, Y., and Bloemendal, H.,** The structure of the vimentin gene, *Cell,* 35, 215, 1983.

106. **Nathans, J. and Hogness, D. S.,** Isolation, sequence analysis, and intron-exon arrangement of the gene encoding bovine rhodopsin, *Cell,* 34, 807, 1983.

107. **Searle, P. F., Davison, B. L., Stuart, G. W., Wilkie, T. M., Norseth, G., and Palmiter, R. D.,** Regulation, linkage, and sequence of mouse metallothionein I and II genes, *Mol. Cell. Biol.,* 4, 1221, 1984.

108. **Griffith, B. B., Walters, R. A., Enger, M. D., Hilderbrand, C. E., and Griffith, J. K.,** cDNA cloning and nucleotide sequence comparison of Chinese hamster metallothionein I and II mRNAs, *Nucleic Acids Res.,* 11, 901, 1983.

109. **Porter, A. G., Barber, C., Garey, N. H., Hallwell, R. A., Threefall, G., and Emtage, J. S.,** Complete nucleotide sequence of an influenzae virus haemagglutinin gene from cloned DNA, *Nature (London),* 278, 471, 1979.

110. **Schaffner, W., Kunz, G., Daetwyler, H., Telford, J., Smith, H. O., and Birnstiel, M. L.,** Genes and spacers of cloned sea urchin histone DNA analyzed by sequencing, *Cell,* 14, 655, 1978.

111. **Crick, F.,** Split genes and RNA splicing, *Science,* 204, 264, 1979.

112. **Chambon, P.,** Split genes, *Sci. Am.,* May, 60, 1981.

113. **Deno, H. and Sugiura, M.,** Chloroplast tRNA^{gly} gene contains a long intron in the D stem: nucleotide sequences of tobacco chloroplast genes for tRNA^{gly} (UCC) and tRNA^{arg} (UCU), *Proc. Natl. Acad. Sci. U.S.A.,* 81, 405, 1984.

114. **Montminy, M. R., Goodman, R. H., Horovitch, S. J., and Habener, J. F.,** Primary structure of the gene encoding rat preprosomatostatin, *Proc. Natl. Acad. Sci. U.S.A.,* 81, 3337, 1984.

115. **Blake, C. C. F.,** Exons encode protein functional units, *Nature (London),* 277, 598, 1979.

116. **Go, M.,** Correlation of DNA exonic regions with protein structural units in haemoglobin, *Nature (London),* 291, 90, 1981.

117. **Doolittle, W. F. and Sapienza, C.,** Selfish genes, the phenotype paradigm and genome evolution, *Nature (London),* 284, 601, 1980.

118. **Fritsch, E. F., Lawn, R. M., and Maniatis, T.,** Molecular cloning and characterization of human beta-globin gene cluster, *Cell,* 19, 959, 1979.

119. **Langford, C. J., Klinz, F.-J., Donath, C., and Gallwitz, D.,** Point mutations identify the conserved, intron-contained TACTAAC box as essential splicing signal sequence in yeast, *Cell,* 36, 645, 1984.

120. **Milner, R. J., Bloom, F. E., Lai, C., Lerner, R. A., and Sutcliffe, J. G.,** Brain-specific genes have identifier sequences in their introns, *Proc. Natl. Acad. Sci. U.S.A.,* 81, 713, 1984.

121. **Sutcliffe, J. G., Milner, R. J., Gottesfeld, J. M., and Lerner, R. A.,** Identifier sequences are transcribed specifically in brain, *Nature (London),* 308, 237, 1984.

122. **Cech, T. R.,** RNA splicing: three themes with variations, *Cell,* 34, 713, 1983.

123. **Miller, J. H. and Rezinkoff, W. S.,** *The Operon,* Cold Spring Harbor Laboratory, Cold Spring Harbor, N.Y., 1980.

124. **Britten, R. J. and Kohne, D. E.,** Repeated sequences in DNA, *Science,* 161, 529, 1968.

125. **Davidson, E. H., Klein, W. H., and Britten, R. J.,** Sequence organization in animal DNA and speculation on hn RNA as a coordinate regulatory transcript, *Dev. Biol.,* 55, 69, 1977.

126. **Singer, M. F.,** Highly repeated sequence in mammalian genomes, *Int. Rev. Cytol.,* 76, 67, 1982.

127. **Houck, C. M., Rinehart, F. P., and Schmid, C. W.,** An ubiquitous family of repeated DNA sequences in the human genome, *J. Mol. Biol.,* 132, 289, 1979.

128. **Schmid, C. W. and Jelinek, W. R.,** The Alu family of dispersed repetitive sequences, *Science,* 216, 1065, 1982.

129. **Sharp, P. A.,** Conversion of RNA to DNA in mammals: Alu-like elements and pseudogenes, *Nature (London)*, 301, 471, 1983.

130. **Haynes, S. R. and Jelinek, W. R.,** Low molecular weight RNAs transcribed in vitro by RNA polymerase III from Alu-type dispersed repeats in Chinese hamster DNA are also found in vivo, *Proc. Natl. Acad. Sci. U.S.A.,* 78, 6130, 1981.

131. **Padgett, R. A., Wahl, G. M., and Stark, G. R.,** Properties of dispersed highly repeated DNA within and near the hamster CAD gene, *Mol. Cell. Biol.,* 2, 302, 1982.

132. **Duncan, C. H., Jagadeeswaran, P., Wang, R. R. C., and Weissman, S. M.,** Structural analysis of templates and RNA polymerase III transcripts of Alu family sequences interspersed among the human beta-like globin genes, *Gene,* 13, 185, 1981.

132a. **Shen, C.-J. and Maniatis, T.,** Nucleotide sequence, DNA modification and in vitro transcription of Alu family repeats in the human alpha-like globin gene, in *Genetic Engineering Techniques: Recent Advances,* Huang, P. C., Kuo, T. T., and Wu, R., Eds., Academic Press, New York, 1982, 129.

133. **Young, P. R., Scott, R. W., Hamer, D. H., and Tilghman, S. M.,** Construction and expression in vivo of an internally deleted mouse alpha-fetoprotein gene: presence of a transcribed Alu-like repeat within the first intervening sequence, *Nucleic Acids Res.,* 10, 3099, 1982.

134. **Ito, R., Sato, K., Helmer, T., Jay, G., and Agarwal, K.,** Structural analysis of the gene encoding human gastrin: the large intron contains an Alu sequence, *Proc. Natl. Acad. Sci. U.S.A.,* 81, 4662, 1984.

135. **Bell, G. I., Pictet, R., and Rutter, W. J.,** Analysis of the regions flanking the human insulin gene and sequence of an alu gene family member, *Nucleic Acids Res.,* 8, 4091, 1980.

136. **Temin, H.,** Function of the retrovirus long terminal repeat, *Cell,* 28, 3, 1982.

137. **Khoury, G. and Gruss, P.,** Enhancer elements, *Cell,* 33, 313, 1983.

138. **Majors, J. and Varmus, H. E.,** A small region of the mouse mammary tumor virus long terminal repeat confers glucocorticoid hormone regulation on a linked heterologous gene, *Proc. Natl. Acad. Sci. U.S.A.,* 80, 5866, 1983.

139. **Hayward, W. S., Neel, B. G., and Astrin, S. M.,** Activation of a cellular onc gene by promoter insertion in ALV-induced lymphoid leukosis, *Nature (London),* 290, 475, 1981.

140. **Chen, H. R. and Barker, W. C.,** Nucleotide sequences of the retroviral long terminal repeats and their adjacent regions, *Nucleic Acids Res.,* 12, 1767, 1984.

141. **Seiki, M., Hattori, S., Hirayama, Y., and Yoshida, M.,** Human adult T-cell leukemia virus: complete nucleotide sequence of the provirus genome integrated in leukemia cell DNA, *Proc. Natl. Acad. Sci. U.S.A.,* 80, 3618, 1983.

142. **Shimotohno, K., Golde, D. W., Miwa, M., Sugimura, T., and Chen, I. S. Y.,** Nucleotide sequence analysis of the long terminal repeat of human T-cell leukemia virus type II, *Proc. Natl. Acad. Sci. U.S.A.,* 81, 1079, 1984.

143. **Chen, I. S., McLaughlin, J., and Golde, D. W.,** Long terminal repeats of human T-cell leukemia virus II genome determine target cell specificity, *Nature (London),* 309, 276, 1984.

144. **Ono, M., Cole, M. D., White, A. T., and Huang, R. C. C.,** Sequence organization of cloned intracisternal A particle genes, *Cell,* 21, 465, 1980.

145. **Cole, M. D., Ono, M., and Huang, R. C.,** Terminally redundant sequences in cellular intracisternal A-particle genes, *J. Virol.,* 38, 680, 1981.

146. **Cole, M. D., Ono, M. and Huang, R. C.,** Intracisternal A-particle genes: structure of adjacent genes and mapping of the boundaries of the transcriptional unit, *J. Virol.,* 42, 123, 1982.

147. **Huang, R. C.,** Some approaches for analyzing transcriptional processes in prokaryotic and eukaryotic systems, in *Genetic Engineering Techniques: Recent Developments,* Huang, P. C., Kuo, T. T., and Wu, R., Eds., Academic Press, New York, 1982, 93.

148. **Davidson, E. H., Jacobs, H. T., and Britten, R. J.,** Very short repeats and coordinate induction of genes, *Nature (London),* 301, 468, 1983.

149. **Cheung, S., Arndt, K., and Lu, P.,** Correlation of lac operator DNA imino proton exchange kinetics with its function, *Proc. Natl. Acad. Sci. U.S.A.,* 81, 3665, 1984.

150. **Benoist, C. and Chambon, P.,** In vivo sequence requirements of the SV40 early promoter region, *Nature (London),* 290, 304, 1981.

151. **Gruss, P., Dhar, R., and Khoury, G.,** Simian virus 40 tandem repeated sequences as an element of the early promoter, *Proc. Natl. Acad. Sci. U.S.A.,* 78, 943, 1981.

152. **Sassone-Corsi, P., Dougherty, J. P., Wasylyk, B., and Chambon, P.,** Stimulation of in vitro transcription from heterologous promoters by the simian virus 40 enhancer, *Proc. Natl. Acad. Sci. U.S.A.,* 81, 308, 1984.

153. **Fromm, M. and Berg, P.,** Deletion mapping of DNA regions required for SV40 early region promoter function in vivo, *J. Mol. Appl. Genet.,* 1, 457, 1982.

154. **Moreau, P., Hen, R., Wasylyk, B., Gaub, M. P., and Chambon, P.,** The SV40 72 base repair repeat has a striking effect on gene expression both in SV40 and other chimeric recombinants, *Nucleic Acids Res.,* 9, 6047, 1981.

155. **Gluzman, Y. and Shenk, T., Eds.,** *Enhancers and Eukaryotic Gene Expression,* Cold Spring Harbor Laboratory, Cold Spring Harbor, N.Y., 1983.

156. **Levinson, B., Khoury, G., Vande Woude, G., and Gruss, P.,** Activation of SV40 genome by 72-base pair tandem repeats of Moloney sarcoma virus, *Nature (London),* 295, 568, 1982.

157. **Imperiale, M. J., Feldman, L. T., and Nevins, J. R.,** Activation of gene expression by adenovirus and herpes virus regulatory genes acting in trans and by a cis-acting adenovirus enhancer element, *Cell,* 35, 127, 1983.

158. **Hartzell, S. W., Byrne, B. J., and Subramanian, K. N.,** Mapping of the late promoter of simian virus 40, *Proc. Natl. Acad. Sci. U.S.A.,* 81, 23, 1984.

159. **Mather, E. and Perry, R.,** Transcriptional regulation of immunoglobulin V genes, *Nucleic Acids Res.,* 9, 6855, 1981.

160. **Humphries, R. K., Ley, T., Turner, P., Moulton, A. D., and Nienhuis, A. W.,** Differences in human alpha-, beta- and delta-globin gene expression in monkey kidney cells, *Cell,* 30, 173, 1982.

161. **Rosenthal, N., Kress, M., Gruss, P., and Khoury, G.,** BK viral enhancer element and a human cellular homolog, *Science,* 222, 749, 1983.

162. **Alt, F. W., Rosenberg, N., Casanova, R. J., Thomas, E., and Baltimore, D.,** Immunoglobulin heavy chain class switching and inducible expression in an Abelson murine leukaemia virus transformed cell line, *Nature (London),* 296, 325, 1982.

163. **Mercola, M., Wang, X.-F., Olsen, J., and Calame, K.,** Transcriptional enhancer elements in the mouse immunoglobulin heavy chain locus, *Science,* 221, 663, 1983.

164. **Guarante, L.,** Yeast promoters: positive and negative elements, *Cell,* 36, 799, 1984.

165. **Guarante, L., Lalonde, B., Gifford, P., and Alani, E.,** Distinctly regulated tandem upstream activation sites mediate catabolite repression of the CYC1 gene of *S. cerevisiae, Cell,* 36, 503, 1984.

166. **Hayday, A. C., Gillies, S. D., Saito, H., Wood, C., Wiman, K., Hayward, W. S., and Tonegawa, S.,** A human immunoglobulin gene-associated enhancer element: a role for transcriptional enhancement in C-myc translocation, *Nature (London),* 307, 334, 1984.

167. **Payne, G. S., Bishop, J. M., and Varmus, H. E.,** Multiple arrangements of viral DNA and an activated host oncogene in bursal lymphomas, *Nature (London),* 295, 209, 1982.

168. **Marcu, K. B., Harris, L. J., Stanton, L. W., Erilkson, J., Watt, R., and Croce, C. M.,** Transcriptionally active c-myc oncogene is contained within NIARD, a DNA sequence associated with chromosome translocations in B-cell neoplasia, *Proc. Natl. Acad. Sci. U.S.A.,* 80, 519, 1983.

169. **Battey, J., Moulding, C., Taub, R., Murphy, W., Stewart, T., Potter, H., Lenoir, G., and Leder, P.,** The human c-myc oncogene: structural consequences of translocation into the IgH locus in Burkitt lymphoma, *Cell,* 34, 779, 1983.

170. **Adams, J. M., Gerondakis, S., Webb, E., Mitchell, J., Bernard, O., and Cory, S.,** Transcriptionally active DNA region that rearranges frequently in murine lymphoid tumors, *Proc. Natl. Acad. Sci. U.S.A.,* 79, 6966, 1982.

171. **Rowley, J. D.,** Identification of the constant chromosome regions involved in human hematologic malignant disease, *Science,* 216, 749, 1982.

172. **Shimizu, A. and Honjo, T.,** Immunoglobulin class switching, *Cell,* 36, 801, 1984.

173. **Banerji, J., Olson, L., and Schaffner, W.,** A lymphocyte-specific cellular enhancer is located downstream of the joining region in immunoglobulin heavy chain genes, *Cell,* 33, 729, 1983.

174. **Scholer, H. R. and Gruss, P.,** Specific interaction between enhancer- containing molecules and cellular components, *Cell,* 36, 403, 1984.

175. **Parker, M.,** Enhancer elements activated by steroid hormones, *Nature (London),* 304, 687, 1983.

176. **Gillies, S. D., Morrison, S. L., Oi, V. T., and Tonegawa, S.,** A tissue-specific transcription enhancer element is located in the major intron of a rearranged immunoglobulin heavy chain gene, *Cell,* 33, 717, 1983.

177. **Gillies, S. D., Folsom, V., and Tonegawa, S.,** Cell type-specific enhancer element associated with a mouse MHC gene, E_B, *Nature (London),* 310, 594, 1984.

178. **Tuan, D. and London, I. M.,** Mapping of DNase I-hypersensitive sites in the upstream of DNA of human embryonic epsilon-globin gene in K562 leukemia cells, *Proc. Natl. Acad. Sci. U.S.A.,* 81, 2718, 1984.

179. **Berg, P. E. and Anderson, W. F.,** Correlation of gene expression and transformation frequency with the presence of an enhancing sequence in the transforming DNA, *Mol. Cell. Biol.,* 4, 368, 1984.

180. **Benjamin, R. C., Fitzmaurice, W. P., Huang, P. C., and Scocca, J. J.,** Nucleotide sequence and properties of the cohesive chromosomal termini from bacteriophage HP1c1 of Haemophilus influenzae Rd., *Gene,* 31, 207, 1984.

181. **Hamer, D. and Khoury, G.,** Enhancers and control elements, in *Enhancers and Eukaryotic Gene Expression,* Gluzman, Y. and Shenk, T., Eds., Cold Spring Harbor Laboratory, Cold Spring Harbor, N.Y., 1983.

182. **Weber, F., deVilliers, J., and Schaffner, W.,** An SV40 "enhancer trap" incorporates exogenous enhancers or generates enhancers from its own sequences, *Cell,* 36, 983, 1984.

183. **Wasylyk, B., Wasylyk, C., Augereau, P., and Chambon, P.,** The SV40 72 bp repeat preferentially potentiates transcription starting from proximal natural or substitute promoter elements, *Cell,* 32, 503, 1983.

184. **Varshavsky, A. J., Sundin, O. H., and Bohn, M. J.,** SV40 viral minichromosome: preferential exposure of the origin of replication as probed by restriction endonucleases, *Nucleic Acids Res.,* 5, 3469, 1978.

185. **Saragosti, S., Cereghini, S., and Yaniv, M.,** Fine structure of the regulatory region of simian virus 40 minichromosomes revealed by DNase I digestion, *J. Mol. Biol.,* 160, 133, 1982.

186. **Klug, A., Rhodes, D., Smith, J., Finch, J. T., and Thomas, J. O.,** A low resolution structure for the histone core of the nucleosome, *Nature (London),* 287, 509, 1980.

187. **Wu, H.-M. and Crothers, D. M.,** The locus of sequence-directed and protein-induced DNA bending, *Nature (London),* 308, 509, 1984.

188. **Frederick, C. A., Grable, J., Melia, M., Samudzi, C., Jen-Jacobson, L., Wang, B.-C., Greene, P., Boyer, H. W., and Rosenberg, J. M.,** Kinked DNA in crystalline complex with Eco RI endonuclease, *Nature (London),* 309, 327, 1984.

189. **Pabo, C. and Sauer, R. T.,** Protein-DNA recognition, *Ann. Rev. Biochem.,* 53, 293, 1984.

190. **Smith, G. R.,** DNA supercoiling: another level for regulating gene expression, *Cell,* 24, 599, 1981.

191. **Hanna, M. M. and Meares, C. F.,** Topography of transcription: path of the leading end of nascent RNA through the *Escherichia coli* transcription complex, *Proc. Natl. Acad. Sci. U.S.A.,* 80, 4238, 1983.

192. **Fisher, M.,** DNA supercoiling and gene expression, *Nature (London),* 307, 686, 1984.

193. **Wang, J. C.,** Unwinding at the promoter and the modulation of transcription by DNA supercoiling, in *Promoters,* Rodriquez, R. L. and Chamberlin, M. J., Eds., Praeger, New York, 1982, 229.

194. **Eisenberg, H.,** From A to Z: new twists to an old helix, *Trends Biochem. Sci.,* 40, 1984.

195. **Chen, W., Yang, H., Zubay, A., and Gellert, M.,** The effects of supercoiling on promoter function in prokaryotes, in *Promoters,* Rodriquez, R. L. and Chamberlin, M. J., Eds., Praeger, New York, 1982, 242.

196. **Lescure, B. and Arcangioli, B.,** Yeast RNA polymerase II initiates transcription in vitro at TATA sequences proximal to potential non-B forms of the DNA template, *EMBO J.,* 3, 4837, 1984.

197. **Peck, L. J., Nordheim, A., Rich, A., and Wang, J. C.,** Flipping of cloned d(pCpG)ₙ·d(pCpG)ₙ DNA sequences from right- to left-handed helical structure by salt, Co(III), or negative supercoiling, *Proc. Natl. Acad. Sci. U.S.A.,* 79, 4560, 1982.

198. **Butzow, J. J., Shin, Y. A., and Eichhorn, G. L.,** Effect of template conversion from the B to the Z conformation on RNA polymerase activity, *Biochemistry,* 23, 4837, 1984.

199. **Wang, A. H.-J., Quigley, G. J., Kolpak, F. J., Crawford, J. L., van Boom, J. H., van der Marel, G., and Rich, A.,** Molecular structure of a left-handed double helicle DNA fragment at atomic resolution, *Nature (London),* 282, 680, 1979.

200. **Drew, H., Takano, T., Tanaka, S., Itakura, K., and Dickerson, R. E.,** High-salt d(CpGpCpG), a left-handed Z' DNA double helix, *Nature (London),* 286, 567, 1980.

201. **Kolata, G.,** Z-DNA moves toward "real biology", *Science,* 222, 495, 1983b.

202. **Richmond, T. J., Finch, J. T., Rushton, B., Rhodes, D., and Klug, A.,** Structure of the nucleosome core particle at 7A resolution, *Nature (London),* 311, 532, 1984.

203. **Pilet, J. and Leng, M.,** Comparison of poly(dG-dC).poly(dG-dC) conformations in oriented films and in solution, *Proc. Natl. Acad. Sci. U.S.A.,* 79, 26, 1982.

204. **Nordheim, R., Tesser, P., Azorin, F., Kwon, Y. H., Moller, A., and Rich, A.,** Isolation of *Drosophila* proteins that bind selectively to left-handed Z-DNA, *Proc. Natl. Acad. Sci. U.S.A.,* 79, 7729, 1982.

205. **Haniford, D. B. and Pulleybank, D. E.,** Facile transition of poly(d(TG).d(CA)) into a left-handed helix in physiological conditions, *Nature (London),* 302, 632, 1983.

206. **Singleton, C. K., Klysik, J., Stirdivant, S. M., and Wells, R. D.,** Left-handed Z-DNA is induced by supercoiling in physiological ionic conditions, *Nature (London),* 299, 312, 1982.

207. **Pohl, F. M. and Jovin, T. M.,** Salt-induced co-operative conformational change of a synthetic DNA: equilibrium and kinetic studies with poly(dG-dC), *J. Mol. Biol.,* 67, 375, 1972.

208. **Behe, M. and Felsenfeld, G.,** Effects of methylation on a synthetic polynucleotide: the B-Z transition in poly(dG-m⁵dC).poly(dG-m⁵dC), *Proc. Natl. Acad. Sci. U.S.A.,* 78, 1619, 1981.

209. **Vardimon, L. and Rich, A.,** In Z-DNA the sequence of GCGC is neither methylated by Hha I methyltransferase nor cleaved by Hha I restriction endonuclease, *Proc. Natl. Acad. Sci. U.S.A.,* 81, 3268, 1984.

210. **Nordheim, A. and Rich, A.,** The sequence (dC-dA)ₙ·(dG-dT)ₙ forms left-handed Z-DNA in negatively supercoiled plasmids, *Proc. Natl. Acad. Sci. U.S.A.,* 80, 1821, 1983.

211. **Lu, P., Cheung, S., and Arndt, K.,** Possible molecular detent in the DNA structure at regulatory sequences, *J. Biomol. Struct. Dyn.,* 1, 509, 1983.

212. **Nordheim, A., Pardue, M. L., Lafer, E. M., Moller, A., Stollar, B. D., and Rich, A.,** Antibodies to left-handed Z-DNA bind to interband regions of *Drosophila* polytene chromosome, *Nature (London),* 294, 417, 1981.

213. **Hamada, H. and Kakunaga, T.,** Potential Z-DNA forming sequences are highly dispersed in the human genome, *Nature (London),* 298, 396, 1982.

214. **Walmsley, R. M., Szostk, J. W., and Petes, T. D.,** Is there left-handed DNA at the ends of yeast chromosomes? *Nature (London),* 302, 84, 1983.

215. **Rogers, J.,** CACA sequences — the ends and the means? *Nature (London),* 305, 101, 1983.

216. **Slightom, J. L., Blechl, A. E., and Smithies, O.,** Human fetal ᴳgamma- and ᴬgamma-globin genes: complete nucleotide sequences suggest that DNA can be exchanged between these duplicated genes, *Cell,* 21, 627, 1980.

217. **Shen, C.-J. and Maniatis, T.,** Nucleotide sequence, DNA modification and in vitro transcription of Alu family repeats in the human alpha-like globin gene, in *Genetic Engineering Techniques: Recent Advances,* Huang, P. C., Kuo, T. T., and Wu, R., Eds., Academic Press, New York, 1982, 129.

218. **Proudfoot, N. J. and Maniatis, T.,** The structure of a human α-globin pseudogene and its relationship to α-globin gene duplication, *Cell,* 21, 537, 1980.

219. **Saffer, J. D. and Lerman, M. I.,** Unusual class of Alu sequences containing a potential Z-DNA segment, *Mol. Cell. Biol.,* 3, 960, 1983.

220. **Hamada, H., Petrino, M. G., and Kakunaga, T.,** A novel repeated element with Z-DNA-forming potential is widely found in evolutionary diverse eukaryotic genomes, *Proc. Natl. Acad. Sci. U.S.A.,* 79, 6465, 1982.

221. **Hochtl, J. and Zachau, H. G.,** A novel type of aberrant recombination in immunoglobulin genes and its implications for V-J joining mechanism, *Nature (London),* 302, 260, 1983.

222. **Steinmetz, M., Moore, K. W., Frelinger, J. G., Sher, B. T., Shen, F.-W., Boyse, E. A., and Hood, L.,** A pseudogene homologous to mouse transplantation antigens: transplantation antigens are encoded by eight exons that correlate with protein domains, *Cell,* 25, 683, 1981.

223. **Miesfeld, R., Krystal, M., and Arnheim, N.,** A member of a new repeated sequence family which is conserved throughout eucaryotic evolution is found between the human delta and beta globin genes, *Nucleic Acids Res.,* 9, 5931, 1981.

224. **Lafer, E. M., Moller, A., Nordheim, A., Stollar, B. D., and Rich, A.,** Antibodies specific for left-handed Z-DNA, *Proc. Natl. Acad. Sci. U.S.A.,* 78, 3546, 1981.

225. **Pohl, F. M., Thomas, R., and DiCapua, E.,** Antibodies to Z-DNA interact with form V DNA, *Nature (London),* 300, 545, 1982.

226. **Moller, A., Gabriels, J. E., Lafer, E. M., Nordheim, A., Rich, A., and Stollar, B. D.,** Monoclonal antibodies recognize different parts of Z-DNA, *J. Biol. Chem.,* 257, 12081, 1982.

227. **Lipps, H. J., Nordheim, A., Lafer, E. M., Ammermann, D., Stollar, B. D., and Rich, A.,** Antibodies against Z-DNA react with the macronucleus but not the micronucleus of the hypotrichous ciliate stylonychia mytilus, *Cell,* 32, 435, 1983.

228. **Veigas-Pequignot, E., Derbin, C., Malfoy, B., Taillandier, E., Leng, M., and Dutrillaux, B.,** Z-DNA immunoreactivity in fixed metaphase chromosomes of primates, *Proc. Natl. Acad. Sci. U.S.A.,* 80, 5890, 1983.

229. **Engel, D. and Richardson, J. P.,** Conformational alterations of transcription termination protein rho induced by ATP and by RNA, *Nucleic Acids Res.,* 12, 7389, 1984.

230. **Pinkham, J. L. and Platt, T.,** The nucleotide sequence of the rho gene of *E. coli* K-12, *Nucleic Acids Res.,* 11, 3531, 1983.

231. **von Hippel, P. H., Bear, D. G., Morgan, N. D., and McSwigger, J. A.,** Protein nucleic acid interactions in transcriptions or molecular analysis, *Ann. Rev. Biochem.,* 53, 389, 1984.

232. **Holmes, W. M., Platt, T., and Rosenberg, M.,** Termination of transcription in *E. coli, Cell,* 32, 1029, 1983.

233. **Brennan, S. M. and Geiduschek, E. P.,** Regions specifying transcriptional termination and pausing in the bacteriophage SP01 terminal repeat, *Nucleic Acids Res.,* 11, 4157, 1983.

234. **Yanofsky, C.,** Comparison of regulatory and structural regions of genes of tryptophan metabolism, *Mol. Biol. Evol.,* 1, 143, 1984.

235. **Cozzarelli, N. R., Gerrard, S. P., Schlissel, M., Brown, D. D., and Bogenhagen, D. F.,** Purified RNA polymerase III accurately and efficiently terminates transcription of 5S RNA genes, *Cell,* 34, 829, 1983.

236. **Maderious, A. and Chen-Kiang, S.,** Pausing and premature termination of human RNA polymerase II during transcription of adenovirus in vivo and in vitro, *Proc. Natl. Acad. Sci. U.S.A.,* 81, 5931, 1984.

237. **Proudfoot, N. J.,** The end of the message and beyond, *Nature (London),* 307, 412, 1984.

238. **Galli, G., Hofstetter, H., Stunnenberg, H. G., and Birnstiel, M. L.,** Biochemical complementation with RNA in the *Xenopus* oocyte: a small RNA is required for the generation of 3' histone mRNA termini, *Cell,* 34, 823, 1983.

239. **Birchmeier, C., Folk, W., and Birnstiel, M. L.,** The terminal RNA stem-loop structure and 80 bp of spacer DNA are required for the formation of 3' termini of sea urchin H2A mRNA, *Cell,* 35, 433, 1983.

240. **Lerner, M. R. and Steitz, J.,** Snurps and scyrps, *Cell,* 25, 298, 1981.

241. **Padgett, R. A., Hardy, S. F., and Sharp, P. A.,** Splicing of adenovirus RNA in a cell-free transcription system, *Proc. Natl. Acad. Sci. U.S.A.,* 80, 5230, 1983.

242. **Salditt-Georgieff, M. and Darnell, J. E., Jr.,** A precise termination site in the mouse B^major^-globin transcription unit, *Proc. Natl. Acad. Sci. U.S.A.,* 80, 4694, 1983.

243. **Rosenberg, M., Chepelinsky, A. B., and McKenney, K.,** Studying promoters and terminators by gene fusion, *Science,* 222, 734, 1983.

244. **Hen, R., Sassone-Corsi, P., Corden, J., Gaub, M. P., and Chambon, P.,** Sequences upstream from the TATA box are required in vivo and in vitro for efficient transcription from the adenovirus serotype 2 major late promoter, *Proc. Natl. Acad. Sci. U.S.A.,* 79, 7132, 1982.

245. **Weil, P. A., Luse, D., Segall, J., and Roeder, R. G.,** Selective and accurate initiation at the adenovirus 2 major late promoter in a soluble system dependent on purified RNA polymerase II and DNA, *Cell,* 18, 469, 1979.

246. **Heintz, N. and Roeder, R. G.,** Transcription of eukaryotic genes in soluble cell-free systems, in *Genetic Engineering,* Vol. 4, Setlow, J. K. and Hollaender, A., Eds., Plenum Press, New York, 1982, 57.

247. **Dignam, J. D., Lebovitz, R., and Roeder, R. G.,** Accurate transcription initiation by RNA polymerase II in a soluble system derived from isolated mammalian nuclei, *Nucleic Acids Res.,* 11, 1475, 1983.

248. **Egly, J. M., Miyamotot, N. G., Moncollin, V., and Chambon, P.,** Is actin a transcription initiation factor for RNA polymerase B? *EMBO J.,* 3, 2363, 1984.

249. **Tolunay, H. E., Yang, L., Anderson, W. F., and Safer, B.,** Isolation of an active transcription initiation complex from HeLa cell-free extract, *Proc. Natl. Acad. Sci. U.S.A.,* 81, 5916, 1984.

250. **Dynan, W. S. and Tjian, R.,** Isolation of transcription factors that discriminate between different promoters recognized by RNA polymerase II, *Cell,* 32, 669, 1983a.

251. **Dynan, W. S. and Tjian, R.,** The promoter-specific transcription factor Sp 1 binds to upstream sequences in the SV40 early promoter, *Cell,* 35, 79, 1983b.

252. **Parker, C. S. and Topol, J.,** A drosophila RNA polymerase II transcription factor contains a promoter-region-specific DNA-binding activity, *Cell,* 36, 357, 1984.

253. **Myers, R. M., Rio, D. C., Robbins, A. K., and Tjian, R.,** SV40 gene expression is modulated by the cooperative binding of T antigen to DNA, *Cell,* 25, 373, 1981.

254. **Sakonju, S. and Brown, D. D.,** Contact points between a positive transcription factor and the *Xenopus* 5S RNA gene, *Cell,* 31, 395, 1982.

255. **Shastry, B. S. and Ng, S.-Y.,** Multiple factors involved in the transcription of class III genes in *Xenopus laevis, J. Biol. Chem.,* 257, 12979, 1982.

256. **Green, M. R., Treisman, R., and Maniatis, T.,** Transcriptional activation of cloned human beta-globin genes by viral immediate-early gene products, *Cell,* 35, 137, 1983.

257. **Bodmer, W. F.,** Evolution of proteins, in *Evolution from Molecules to Men,* Bendall, D. S., Ed., Cambridge University Press, London, 1983.

258. **Cata, A. C. B.,** How do steroid hormones function to induce the transcription of specific genes? *Biosci. Rep.,* 3, 101, 1983.

259. **Chandler, V. L., Maler, B. A., and Yamamoto, K. R.,** DNA sequences bound specifically by gluco-corticoid receptor in vitro render a heterologous promoter hormone responsive in vivo, *Cell,* 33, 489, 1983.

260. **Chisholm, R. L.,** Methylation and developmental regulation of gene expression, *Trends Biochem. Sci.,* 7, 421, 1982.

261. **Langner, K.-D., Vardimon, L., Renz, D., and Doerfler, W.,** DNA methylation of three 5′CCGG3′ sites in the promoter and 5′ region inactivate the E2a gene of adenovirus type 2, *Proc. Natl. Acad. Sci. U.S.A.,* 81, 2950, 1984.

262. **Haseltine, W. A., Sodroski, J., Patarca, R., Briggs, D., Perkins, D., and Wong-Staal, F.,** Structure of 3′ terminal region of type II human T lymphotropic virus: evidence for new coding region, *Science,* 225, 419, 1984.

263. **Sodroski, J. G., Rosen, C. A., and Haseltine, W. A.,** Trans-acting transcriptional activation of the long terminal repeat of human T lymphotropic viruses in infected cells, *Science,* 225, 381, 1984.

264. **Wu, C.,** The 5′ ends of *Drosophila* heat shock genes in chromatin are hypersensitive to DNase I, *Nature (London),* 286, 854, 1980.

265. **Wu, C. and Gilbert, W.,** Tissue-specific exposure of chromatin structure at the 5′ terminus of the rat preproinsulin II gene, *Proc. Natl. Acad. Sci. U.S.A.,* 78, 1577, 1981.

266. **Weintraub, H., Larsen, A., and Groudine, M.,** Alpha-globin gene switching during the development of chicken embryos: expression and chromosome structure, *Cell,* 24, 333, 1981.

267. **Samal, B. and Worcel, A.,** Chromatin structure of the histone genes of *D. melanogaster, Cell,* 23, 401, 1981.

268. **Schon, E., Evans, T., Welsh, J., and Efstradiadis, A.,** Conformation of promoter DNA: fine mapping of S₁ hypersensitive sites, *Cell,* 35, 837, 1983.

269. **Mace, H. A. F., Pelham, H. R. B., and Travers, A. A.,** Association of an S₁ nuclease-sensitive structure with short direct repeats 5′ of *Drosophila* heat shock genes, *Nature (London),* 304, 1983.

270. **Fritton, H. P., Igo-Kemenes, T., Nowock, J., Strech-Jurk, U., Thiesen, M., and Sippel, A. E.,** Alternative sets of DNase I-hypersensitive sites characterize the various functional states of the chicken lysozyme gene, *Nature (London),* 311, 163, 1984.

271. **Jongstra, J., Reudelhuber, T. L., Oudet, P., Benoist, C., Chae, C.-B., Jeltsch, J.-M., Mathis, D. J., and Chambon, P.,** Induction of altered chromatin structures by simian virus 40 enhancer and promoter elements, *Nature (London),* 307, 708, 1984.

272. **Emerson, B. M. and Felsenfeld, G.,** Specific factor conferring nuclease hypersensitivity at the 5' end of the chicken adult beta-globin gene, *Proc. Natl. Acad. Sci. U.S.A.,* 81, 95, 1984.

273. **Zaret, K. S. and Yamamoto, K. R.,** Reversible and persistent changes in chromatin structure accompany activation of a glucocorticoid-dependent enhancer element, *Cell,* 38, 29, 1984.

274. **Groudine, M., Eisenman, R., and Weintraub, H.,** Chromatin structure of endogenous retroviral genes and activation by an inhibitor of DNA methylation, *Nature (London),* 292, 311, 1981.

275. **Wu, C.,** Activating protein factor binds in vitro to upstream control sequences in heat shock gene chromatin, *Nature (London),* 311, 81, 1984.

INDEX

A

Agrobacterium tumefaciens, small RNAs in, 91
Alignment matrix, 9, 10
Alpha chain, 177—179
Alu sequences, 27, 199—200
α-Amanitin, 28
Amphibians, organization of histone genes in, 142
Animals, ntDNA diversity in, 121
APL, advantages of, 3
Artificial intelligence, 4
Aspergillus
    active ATPase subunit 9 gene in, 109
    intron variation in, 112, 114
    mtDNA variation in, 113
    nuclear variation in, 129
Aspergillus nidulans, size of mitochondrial genomes
    in, 110

B

B1 and B2 sequences, 27, 33
Bacillus, rRNA processing of, 50
Bacteria, rRNA processing of, 49—50
Base composition, calculation of, 3
BASIC, disadvantages of, 3
Basidiomycetes, mitochondrial recombination in,
    113
Beta chain, 172, 179
Beta genes, 177
Bohr effect, 185
Bolt Beranek and Newman, Inc. (BBN), 4, 6, 7

C

Cat, adult hemoglobin of, 171
Chicken
    codon usage in histone genes of, 158, 160
    histone genes of, 152
        organization, 143, 144
        sequences, 155
        silent site divergence, 149, 153, 156
Chlamydomonas
    mitochondrial recombination in, 113
    mtDNA in, 108
Chloroplast DNA, organization of, 68—69
Chloroplast ribosomal RNA transcription units, 69
Chromosomes, role of histones in, see also Histone
        genes, 140
CIS-acting regulatory sequences
    control through DNA topology, 203—206
    enhancers-activators, 201—203
    promoter specificity, 193—201
    terminal and downstream signals for transcription,
        206—208
Cleavage sites

secondary structure folding around, 62—63
    sequences at, 61
Cnemidophorus, mtDNA analysis of, 127
Coding region, pattern of nucleotide changes in,
        163—164
Coding sequences, introns as, 198—199
Compatibility methods, 13
Computers, in estimation of DNA nucleotide se-
        quence divergence, 1—15
    acquisition of nucleotide sequence data, 2—4
    assessment of nucleotide sequence homology, 9—
        12
    interference of phylogenetic relationships, 12—15
    nucleic acid sequence databases, 4—9
Concerted evolution, of globin genes, 176
Control regions, identification of, 3
Cooperativity, 185
"Copia-like elements", 20
Crossovers
    in globin genome, 180
    unequal, 32, 38
Crown-gall tumors, 96—98
ct-lethals, formation of, 28
Cucurbitaceae, mtDNA of, 112

D

Databases, nucleic acid sequence, 4—9
Deletions
    in globin genome, 180
    mobile elements and, 31
Delta genes, 177
Dictyostelium discoideum, organization of rDNA in,
        56
Displacement loop (D-loop) region, 109, 126, 128
Domain, of protein, 173, 176
DNA
    extrachromosomal circular, 23
    topology, 203
        base modification, 205
        binding to proteins, 205—206
        negative supercoiling, 204
    mtDNA, see Mitochondrial DNA
    rDNA, 56, 58
    Z-DNA, 203—206
DNA sequencing, rapid techniques for, see also
        Mobile dispersed genetic elements; Sequence
        data; Sequence homology, 2
Drift, 182
Drosophila
    heat shock gene of, 191
    mitochondrial genome of, 111
    mtDNA evolution in, 128
    organization of histone genes in, 142
    variation in mtDNA in, 127
Drosophila melanogaster
    hybrid dysgenesis systems in, 37